D1694804

Biological Field Emission Scanning Electron Microscopy

Current and Future Titles in the Royal Microscopical Society—John Wiley Series

Published

Principles and Practice of Variable Pressure/Environmental Scanning Electron Microscopy (VP-ESEM) – Debbie Stokes

Aberration-Corrected Analytical Electron Microscopy – Edited by Rik Brydson

Diagnostic Electron Microscopy – A Practical Guide to Interpretation and Technique – Edited by John W. Stirling, Alan Curry and Brian Eyden

Low Voltage Electron Microscopy – Principles and Applications – Edited by David C. Bell and Natasha Erdman

Standard and Super-Resolution Bioimaging Data Analysis: A Primer – Edited by Ann Wheeler and Ricardo Henriques

Electron Beam-Specimen Interactions and Applications in Microscopy – Budhika Mendis

Biological Field Emission Scanning Electron Microscopy, 2 Volume Set – Edited by Roland A. Fleck and Bruno M. Humbel

Forthcoming

Understanding Practical Light Microscopy – Jeremy Sanderson

Correlative Microscopy in the Biomedical Sciences – Edited by Paul Verkade and Lucy Collinson

The Preparation of Geomaterials for Microscopical Study: A Laboratory Manual – Owen Green and Jonathan Wells

Electron Energy Loss Spectroscopy – Edited by Rik Brydson and Ian MacLaren

Biological Field Emission Scanning Electron Microscopy

Volume I

Edited by

Roland A. Fleck

Centre for Ultrastructural Imaging
King's College London
United Kingdom

Bruno M. Humbel

Imaging Section
Okinawa Institute of Science and Technology
Japan

WILEY

This edition first published 2019

Registered Offices
John Wiley & Sons, Inc., 111 River Street, Hoboken, NJ 07030, USA
John Wiley & Sons Ltd, The Atrium, Southern Gate, Chichester, West Sussex, PO19 8SQ, UK

Editorial Office
John Wiley & Sons Ltd, The Atrium, Southern Gate, Chichester, West Sussex, PO19 8SQ, UK

For details of our global editorial offices, customer services, and more information about Wiley products visit us at www.wiley.com.

Wiley also publishes its books in a variety of electronic formats and by print-on-demand. Some content that appears in standard print versions of this book may not be available in other formats.

Library of Congress Cataloging-in-Publication Data

Names: Fleck, Roland A., 1970- editor. | Humbel, Bruno M., editor.
Title: Biological field emission scanning electron microscopy / edited by
Roland A. Fleck, Bruno M. Humbel.
Description: First edition. | Hoboken, NJ : Wiley, 2019. | Series: Royal
Microscopical Society - John Wiley series | Includes bibliographical
references and index. |
Identifiers: LCCN 2018021231 (print) | LCCN 2018021980 (ebook) | ISBN
9781118663240 (Adobe PDF) | ISBN 9781118663264 (ePub) | ISBN 9781118654064
(hardback)
Subjects: | MESH: Microscopy, Electron, Scanning–methods | Microscopy,
Electron, Scanning–instrumentation
Classification: LCC QH212.S3 (ebook) | LCC QH212.S3 (print) | NLM QH 212.S3 |
DDC 502.8/25–dc23
LC record available at https://lccn.loc.gov/2018021231

Cover Design: Wiley
Cover Image: Courtesy of Roland Fleck with coloring by Willy Blanchard, Electron Microscopy Facility, University of Lausanne, Lausanne Switzerland

Set in 10/12pt SabonLTStd by SPi Global, Chennai, India
Printed and bound in Singapore by Markono Print Media Pte Ltd

10 9 8 7 6 5 4 3 2 1

Contents to Volume I

Contents to Volume II

About the Editors

ROLAND A. FLECK, PhD, FRCPath, FRMS, is a Professor in Ultrastructural Imaging and Director of the Centre for Ultrastructural Imaging at King's College London. Having specialised in basic research into cellular injury at low temperatures and during cryo-preservation regimes he has developed specialist knowledge of freeze fracture/freeze etch preparation of tissues and wider cryo-microscopic techniques. As director of the Centre for Ultrastructural Imaging he supports advanced three dimensional studies of cells and tissues by both conventional room temperature and cryo electron microscopy. He is a visiting Professor of the Faculty of Health and Medical Sciences, University of Copenhagen and Professor of the UNESCO Chair in Cryobiology, National Academy of Sciences of Ukraine, Institute for Problems of Cryobiology, Kharkiv, Ukraine.

BRUNO M. HUMBEL, Dr. sc. nat. ETH, is head of the Imaging Section at the Okinawa Institute of Science and Technology, Onna son, Okinawa, Japan. He is awarded a research professorship at Juntendo University, Tokyo, Japan. He got his PhD at the Federal Institute of Technology, ETH, Zurich, Switzerland, with Prof. Hans Moor and Dr. Martin Müller, both pioneers in cryo-electron microscopy (high-pressure freezing, freeze-fracturing, freeze-substitution and low-temperature embedding, cryo-SEM, cryo-sectioning). His research focuses on sample preparation for optimal, life-like imaging of biological objects in the electron microscope. The main interests are preparation methods based on cryo-fixation applied in Cell Biology. From here, hybrid follow-up methods like freeze-substitution or freeze-fracturing are used. He is also involved in immunolabelling technology, e.g., ultra-small gold particles and has been working on techniques for correlative microscopy and volume microscopy for a couple of years. He teaches cryo-techniques and immunolabelling and correlative microscopy in international workshops and has professional affiliations with Zhejiang University, Hangzhou, People's Republic of China as a distinguished professor and co-director of the Center of Cryo-Electron Microscopy and with the Federal University of Minas Gerais, Belo Horizonte, Brazil, as a FAPEMIG visiting professor at the Centro de Microscopia da UFMG.

List of Contributors

Isabel Angert, Carl Zeiss Microscopy GmbH, Oberkochen, Germany

Christian Böker, Carl Zeiss Microscopy GmbH, Oberkochen, Germany

Damien De Bellis, Electron Microscopy Facility, University of Lausanne, Lausanne, Switzerland

Willy Blanchard, Electron Microscopy Facility, University of Lausanne, Lausanne, Switzerland

Faysal Boughorbel, Thermo Fisher Scientific, Eindhoven, The Netherlands

Anwen Bullen, UCL Ear Institute, University College London, United Kingdom

Jean Daraspe, Electron Microscopy Facility, University of Lausanne, Lausanne, Switzerland

Stefan Diller, Scientific Photography, Wuerzburg, Germany

Goran Dražić, National Institute of Chemistry, Laboratory for Materials Chemistry, Ljubljana, Slovenia

Damjana Drobne, Department of Biology, Biotechnical Faculty, University of Ljubljana, Slovenia

Martin Edelman, Carl Zeiss Microscopy GmbH, Oberkochen, Germany

Dennis Eggert, Heinrich Pette Institute, Leibniz Institute for Experimental Virology, Hamburg, Germany

Andreja Erman, Institute of Cell Biology, Faculty of Medicine, University of Ljubljana, Slovenia

Dean Flanders, The Friedrich Miescher Institute for Biomedical Research, Basel, Switzerland

Roland A. Fleck, Centre for Ultrastructural Imaging, King's College London, UK

Andrew Forge, UCL Ear Institute,University College London, United Kingdom

Yoshiyuki Fukuda, Department of Structural Biology, Max Planck Institute of Biochemistry, Martinsried, Germany and Department of Cell Biology and Anatomy, Graduate School of Medicine, University of Tokyo, Tokyo, Japan

Christel Genoud, Friedrich Miescher Institute for Biomedical Research, Basel, Switzerland

Lucille A. Giannuzzi, L.A. Giannuzzi & Associates LLC, Fort Myers, FL, USA and EXpressLO LLC, Lehigh Acres, FL, USA

Martin W. Goldberg, Science Laboratories, School of Biological and Biomedical Sciences, Durham University, Durham, United Kingdom

Urs Gomez, Imagic Bildverarbeitung AG, Glattbrugg, Switzerland

Heinz Gross, ETH ScopeM, Swiss Federal Institute of Technology, ETH-Hönggerberg, Zurich, Switzerland

Paul Gunning, Smith & Nephew Advanced Wound Management, Hull, UK

Richard H.R. Hahnloser, Institute of Neuroinformatics, University of Zurich and ETH Zurich, Neuroscience Center, Zurich, Switzerland

Bärbel Hauröeder, Zentrales Institut des Sanitätsdienstes der Bundeswehr, Koblenz, Germany

Mike F. Hayles, Cryo-FIB-SEM Technology, Eindhoven, The Netherlands

Johan Hazekamp, Unilever R&D Vlaardingen, Vlaardingen, The Netherlands

Stephan Hiller, Carl Zeiss Microscopy GmbH, Oberkochen, Germany

Heinrich Hohenberg, Heinrich Pette Institute, Leibniz Institute for Experimental Virology, Hamburg, Germany

Bruno M. Humbel, Electron Microscopy Facility, University of Lausanne, Switzerland and Imaging Section, Okinawa Institute of Science and Technology, Onna-son, Okinawa, Japan

Takeshi Ishikawa, Science and Medical Systems Business Group, Hitachi High-Technologies Corporation, Minato-ku, Tokyo, Japan

Jaroslav Jiruše, TESCAN, Brno, Czech Republic

David C. Joy, 232 Science and Energy Research, Facility and Department of Materials Science and Engineering, The University of Tennessee, Knoxville, TN, USA

Caroline Kizilyaprak, Electron Microscopy Facility, University of Lausanne, Lausanne, Switzerland

Vratislav Košťál, TESCAN, Brno, Czech Republic

Mami Konomi, Science and Medical Systems Business Group, Hitachi High-Technologies Corporation, Minato-ku, Tokyo, Japan

Emine Korkmaz, Thermo Fisher Scientific, Eindhoven, The Netherlands

Andrew Leis, Bio21 Molecular Science and Biotechnology Institute, The University of Melbourne, 30 Flemington Road, Parkville, Victoria 3010, Australia

Bohumila Lencová, TESCAN, Brno, Czech Republic

Ben Lich, Thermo Fisher Scientific, Eindhoven, The Netherlands

Falk Lucas, ETH ScopeM, Swiss Federal Institute of Technology, ETH-Hönggerberg, Zurich, Switzerland

Maria Marosvölgyi, arivis AG, Business Unit arivis Vision, Rostock, Germany

Oliver Meckes, Eye of Science, Reutlingen, Germany

Arno Merkle, Carl Zeiss Microscopy GmbH, Oberkochen, Germany

Robert Morrison, Quorum Technologies, Ltd, Laughton, UK

Sara Novak, Department of Biology, Biotechnical Faculty, University of Ljubljana, Slovenia

Kazumichi Ogura, JEOL Ltd, Akishima, Tokyo, Japan

Masako Osumi, Japan Woman's University, Tokyo, Japan and NPO Integrated Imaging Research Support, Tokyo, Japan

Nicole Ottawa, Eye of Science, Reutlingen, Germany

Pavel Potocek, Thermo Fisher Scientific, Eindhoven, The Netherlands

Rudolf Reimer, Heinrich Pette Institute, Leibniz Institute for Experimental Virology, Hamburg, Germany

Guenter P. Resch, Nexperion e.U. – Solutions for Electron Microscopy, Wien, Austria

Alexander Rigort, Thermo Fisher Scientific, FEI Deutschland GmbH, Germany

Marjolein van Ruijven, Unilever R&D Vlaardingen, Vlaardingen, The Netherlands

Mitsugu Sato, Research and Development Division, Hitachi High-Technologies Corporation, Hitachinaka, Ibaraki-ben, Japan

Patrick Schwarb, Imagic Bildverarbeitung AG, Glattbrugg, Switzerland

Heinz Schwarz, Max Planck Institute for Developmental Biology, Tübingen, Germany

Eyal Shimoni, Department of Chemical Research Support, Weizmann Institute of Science, Rehovot, Israel

Jeremy Skepper, Cambridge Advanced Imaging Centre, University of Cambridge, Cambridge, United Kingdom

Sebastian Tacke, ETH ScopeM, Swiss Federal Institute of Technology, ETH-Hönggerberg, Zurich, Switzerland

Ryuichiro Tamochi, Science and Medical Systems Business Group, Hitachi High-Technologies Corporation, Minato-ku, Tokyo, Japan

Ruth Taylor, UCL Ear Institute, University College London, United Kingdom

Thomas Templier, Institute of Neuroinformatics, University of Zurich and ETH Zurich, Neuroscience Center, Zurich, Switzerland

Erin M. Tranfield, Institute Gulbenkian de Ciência,Oeiras, Portugal

Paul Walther, Central Facility for ElectronMicroscopy, Ulm University, Ulm, Germany

Alice Warley, Centre for Ultrastructural Imaging, King's College London, United Kindom and Visiting Professor, Department of Histology and Cell Biology Faculty of Medicine, University of Granada, Granada, Spain

Roger Wepf, ETH ScopeM, Swiss Federal Institute of Technology, ETH-Hönggerberg, Zurich, Switzerland and UQ, CMM University of Queensland, Brisbane, Australia

Martyn Winn, STFC Daresbury Laboratory, Warrington, United Kingdom

Jeremy Woodward, SBRU, University of Cape Town, South Africa

Jeremy D. Woodward, Department of Integrative Biomedical Sciences in the division of Medical Biochemistry & Structural Biology and Structural Biology Research Unit, University of Cape Town, Cape Town, South Africa

Andrew Yarwood, JEOL (UK) Ltd, Welwyn Garden City, Hertfordshire, UK

Dirk Zeitler, Carl Zeiss Microscopy GmbH, Oberkochen, Germany

Foreword

The task of microscopy in biology is to provide the structural information for correlation of structure and function in complex biological systems. Specimen preparation and imaging techniques should be directed towards the preservation and imaging of the smallest possible significant details in order to fully exploit this unique, integrating feature of biological microscopy, complementing the more linear biochemical procedures. High spatial and temporal resolution is required to describe an aqueous dynamic biological system closely related to the living state. A quantitative description of biological structures down to molecular dimensions may remain a dream, but one must nevertheless attempt to realize it.

With the advent of the field emission electron source paired by important progress in related electron optics, scanning electron microscopy has reached a level of structural and analytical resolution comparable to transmission electron microscopy. Cryoimmobilization techniques, in addition, can rapidly arrest the living processes.

The main problems during the preparation of biological material for electron microscopy arise from the necessity to transform the living sample into a solid state in which it can resist the physical impact of the electron microscope. Basically, preparative procedures are identical for TEM and SEM. In SEM the much higher dose and the necessity to localize the signal at the specimen surface demand additional solutions.

Biological electron microscopy laboratories are often established and financed to service biological research projects, with the biological question being academically the only relevant part. The efforts, however, to investigate new preparative ideas and to set up, modify and/or improve existing approaches are high. Sound methodological research directed to the solution of relevant biological questions helps to educate greater understanding of the physical and chemical changes a sample may encounter during processing and imaging. Thus, advancing electron microscopy into a source of primary information not only for systems of reduced complexity, such as vitrified layers of macromolecules, will enable microscopists to discuss with bioscientists directly linkages between structure and function and for micrographs to be more often selected for the data they contain rather than for mere illustrative purposes.

Biologists normally choose the electron micrographs according to their expectation and thanks to the introduction of more stable and user friendly instruments many biologists routinely operate advanced instrumentation, however, high resolution (highly significant) information is not obtained only because a field emission SEM (FEGSEM) is used. All the steps involved in specimen preparation and imaging must be carefully understood in order to appreciate the possible level of primary information.

Sound basic research has been possible only in a few laboratories and the present books (Volume I and II) aim to collate knowledge of sample preparation procedures and electron optics to promote best practice in the use of the FEGSEM to answer basic biological questions.

Martin Müller

1

Scanning Electron Microscopy: Theory, History and Development of the Field Emission Scanning Electron Microscope

David C. Joy

232 Science and Engineering Research Facility, Department of Materials Science and Engineering, University of Tennessee, Knoxville, TN, USA

1.1 THE SCANNING ELECTRON MICROSCOPE

Since its initial development (Everhart and Thornley, 1958) the scanning electron microscope (SEM) has earned a reputation for being the most widely used, high performance, imaging technology that is available for applications ranging from imaging, fabrication, patterning, and chemical analysis, and for materials of all types and applications. It is estimated that 150 000 or so such instruments are now currently in use worldwide, varying in performance and complexity from simple desk-top systems to state-of-the-art field emission gun systems that can now cost in excess of $5 million.

The basic principle of the scanning electron microscope is simple. An incident electron beam is brought to a focus that typically varies in size from a fraction of a centimeter in diameter down to a spot that can be smaller by a factor of many thousands of times, and with an energy varying from 100 eV or less to a maximum of 30 keV or more. This beam spot is typically then scanned (Figure 1.1) in a linear "raster" pattern across the region of interest, although other patterns – such as a radial beam – are sometimes employed for special purposes. Typically the final deposited pattern will contain of the order of 1000 × 1000 or more individual imaging points.

The incident beam electrons can interact with the sample atoms through either elastic or inelastic scattering. Elastic scattering is where the incident electrons are deflected with no loss of energy. Inelastic scattering involves a loss of energy, often by ionizing the sample atoms. The incident electrons will scatter (both elastically and inelastically) many times in

Biological Field Emission Scanning Electron Microscopy, First Edition.
Edited by Roland A. Fleck and Bruno M. Humbel.

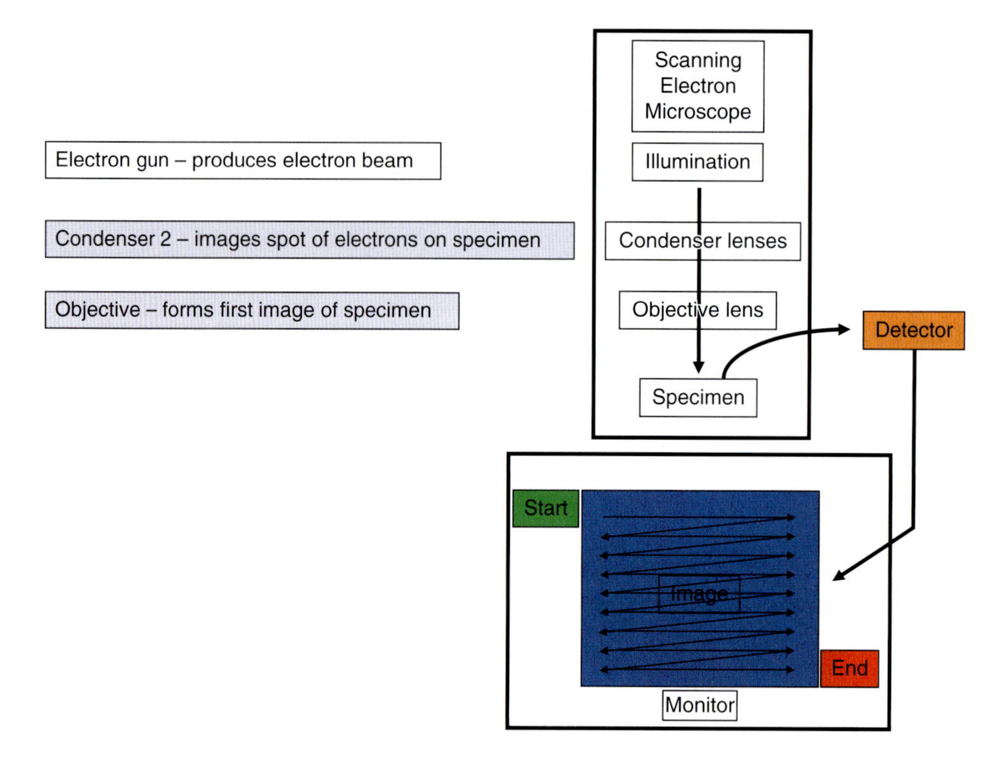

Figure 1.1 The SEM scan raster.

a region of the sample known as the interaction volume. The size of the interaction volume will depend on the incident energy and the nature of the sample, but can be of the order of a micrometer in diameter. A number of different types of signal generated by the beam–sample interaction can be detected. The intensity of the signal detected can be plotted as a function of probe position to form an image. Two important signals are secondary electrons (SEs) and back-scattered electrons (BSEs). Secondary electrons are electrons from the sample atoms that are released through ionization. They are relatively low in energy $<\sim 25$ eV and tend to only escape from the top few tens of nanometers of the surface. They provide strong topographical imaging of surfaces. Back-scattered electrons are incident electrons that have been multiply scattered and emerge again from the surface. The strength of the scattering that can return the electrons to the surface depends strongly on atomic number, Z, and so BSE imaging gives compositional contrast. Another common signal detected is X-rays from the decay of the ionized atoms. The energy of the X-ray photon emitted is characteristic of the element ionized, and so energy-dispersive X-ray (EDX) spectroscopy allows mapping of element species.

Most modern SEMs will likely have, and make use of, several types of detector so as to optimally detect, capture, collect, and display other analytical and imaging modes as desired.

In operation the electron source must be carefully set up and optimized so as to generate the smallest spot size for the electrons while still ensuring that the beam current reaching the specimen is adequately stable for periods of many hours without the need for any further operator interactions. The overall measure of imaging performance for the electron source is determined by its brightness β, which is defined as

$$\beta = 4I/(d^2.\pi^2.\alpha^2) \text{ amps/cm}^2\text{/steradian}$$

where d is the diameter of the spot size of the beam at the target, I is the incident beam current, and α is the solid angle subtended by the illumination at the specimen.

For an electron beam source of some specified energy the beam brightness is said to be "conserved", which means that varying the beam current – as, for example, by varying the beam spot size or the convergence angle of the beam – will always result in compensating changes in the other parameters in the system so that the magnitude of β in the equation remains constant. As a result, the intensity of the incident beam current I varies as $d^2\alpha^2$ and if either the beam spot size d or the beam convergence angle α are reduced, then the beam current will decrease, which may ultimately result in the beam becoming lost in the background noise of the instrument. The imaging performance of an SEM is very important and therefore is always very dependent on optimum alignment.

1.2 THE THERMIONIC GUN

For the first 25 years or so of the SEM era the only available sources of the energetic electrons required for microscopy were the so-called thermionic ("hot beam") emitters mentioned above. Even today so-called "table-top" SEM instruments remain in widespread use because of their low cost, good resolution, and operating convenience.

In operation the required electron beam current is generated by heating a tungsten wire filament. This so-called "thermionic emitter" is usually fabricated from high quality tungsten wire that has been bent into a "V" shape and is maintained at a temperature of about 2700 K by means of a separate power supply that heats the tip region. The "V" shape noted above is maintained at some negative voltage typically from about ~1 keV – 30 keV with reference to ground potential. The corresponding incident beam currents typically can vary from 10^{-6} down to 10^{-12} amps or so.

To optimize the yield of the emitted beam current that is generated a "grid cap", or a "Wehnelt" cylinder – with a circular aperture centered on the tip of the emitted beam current– is employed. The cap is maintained in position by a potential source that is set to about 50 volts or greater, so that the emitted beam from the source can be brought to a focused crossover at a point chosen some distance beyond the column grid cap. The generated electron beam can then be accelerated down the column and on to the specimen. For a given beam energy the intensity of the imaging incident beam will be restricted by, and will be highly dependent upon, the emission performance of the gun, so advanced electron microscopes in particular always require carefully optimized beam sources and hardware.

In typical current SEMs equipped with such a thermionic gun the smallest usable beam spot size will be of the order of a few nanometers, and can provide a beam current of between 10 and 1000 picoamps. To achieve an acceptable signal-to-noise ratio in the chosen area of the image range typically requires exposure times of between 30 and 100 seconds depending on the performance of the gun. Higher performance gun sources, discussed later, can reduce the exposure time required by several orders of magnitude, but the ultimate resolution of an SEM with such a thermal emitter is limited both by the need to maintain an adequate incident beam current and by the inherent energy spread of the emitting source. Despite these limitations, thermionic emitters are still in widespread use as they are well suited for imaging at magnifications below 50 k×, although they can only offer relatively poor imaging resolution because the low brightness of the source sets a minimum limit to the useful spot size and the high temperature of the emitting tip tends to broaden the energy spread of the electron beam.

Some further enhancement in imaging performance can by achieved by employing "pointed filaments". As their name implies, in these devices the emitter tip region of the "V" shaped filament is sharpened so as to further increase the field present at the top of the tip. This then results both in an improvement of the electron yield and in a reduction in the apparent source size of the emitter. However, the improvement in performance so achieved is not much better than modest, and the lifetime of the emitter is reduced by the modifications that must be made to the tip. Other sources of even higher performance are therefore still required.

1.3 THE LANTHANUM HEXABORIDE ("LaB_6") SOURCE

This was first described by Lafferty (Lafferty et al. 1951) and later on was further developed and optimized by Broers (Broers et al. 1960) at IBM in the late 1960s. A LaB_6 source can provide significantly better performance than the conventional tungsten emitter described earlier because the LaB_6 has a much lower work function (temperature). The resultant performance enhancement is of importance because for the typical "hairpin" beam sources discussed above each 10% reduction in the work function of the source will increase the emission current density *J* by a factor of about 1.5 times. As a result of this a LaB_6 emitter operating at 1500 K can generate a significantly higher brightness image than that generated by a conventional tungsten thermionic source operating at 2700 K. In addition, the sharply pointed tip geometry of the LaB_6 emitter results in an effective source size, which is lower than that of a conventional tungsten thermionic emitter and so further enhances the image resolution.

The reduction in the operating temperature that can be employed also serves to enhance the lifetime of the LaB_6 source itself. However, it must be noted that LaB_6 is itself extremely reactive and rapidly forms compounds with all materials other than carbon and rhenium. As a result, once an LaB_6 emitter achieves, and is able to maintain, good stability and brightness it must never be allowed to completely cool down again, and nor should it be exposed to the atmosphere. In summary, LaB_6 emitters are competitive, high brightness, sources with a good signal production, but they must be kept running and under vacuum continuously to maintain and provide the desired stability and performance.

1.4 OTHER ENHANCED "HIGHER BRIGHTNESS" SOURCES

The thermionic electron sources discussed above are simple to build, adequately reliable in use, low in cost, and can offer more than adequate performance for low and medium resolution imaging. However, they cannot provide the significantly higher beam currents, nor the smaller beam source sizes, that are required when imaging specimens in the nanometer scale range. An additional problem is that the thermionic emitter itself must operate at a high temperature, and this in turn generates chromatic aberrations that give rise to a loss of imaging resolution, particularly at low beam energies. Because such uses are increasingly important, there is an increasing need for electron sources that can offer not only higher brightness performance but also smaller source sizes and a reduction in the energy width of the beam.

The development of efficient, high brightness, electron sources began with Thompson's discovery of the electron (Thompson et al, 1895), which demonstrated that charged particles could be emitted from the interior of a conducting material provided that sufficiently large electric fields were applied. Some twenty years later the phenomena of field emission was re-visited by Fowler and Nordheim in 1928 (Fowler and Nordheim, 1928) who proposed the existence of the process now known as "quantum tunneling". While strictly speaking their work only applies to field emission generation from bulk crystalline solids, it does provide a convenient way of understanding what processes are involved.

In 1937 Erwin Muller (Muller, 1937) developed the first practical applications of this technology with his "field ion microscope". This device consisted of a sealed chamber containing a low pressure of hydrogen gas. At one end of the chamber was a stiff tungsten wire, terminating in a sharp point and maintained at a low temperature by the cooled hydrogen gas. At the other end of the chamber was a fluorescent screen. When a sufficiently large potential was provided to the tip, this resulted in the generation of very high electric fields ($\sim 10^7$ volts/cm) in the region around the tip, which was held at a negative potential with respect to ground and so attracted positive ions towards itself. The negative ions that were being generated at the same time were accelerated away from the tip region by the drift field and allowed to travel towards a viewing screen. The drift region itself was field free and so the ions traveled in diverging straight lines to produce bright spots on the screen, each of which could be traced back to the particular atom on the tip from which it had originally come. Subsequent work after the Second World War with this simple device ultimately led to the production of the first ever images of individual atoms (Muller and Bahadur, 1950).

All of the requirements noted above can now be satisfied by employing a field emission gun ("FEG") source. The first attempts to use such an emission source for imaging in a scanning microscope were in fact made by Cosslet and Haines (Cosslet and Haines, 1954) but their efforts were not really successful because they were unable to achieve a sufficiently high vacuum to guarantee stable emissions. Fortunately the rapid improvements in ultra-high vacuum technology that occurred in the 1950s helped to eliminate these problems and led to the publication of a book by Gomer (1961), which discussed the practical advantages of a field emission source for electron microscopy. It was recognized that the most important step in achieving the goal of imaging at near-atomic levels required developing an electron source that offered both the highest possible brightness and the highest resolution at typical SEM energies, that is, 10 to 30 keV. From Gomers' work it was already evident that an optimized field emission source would be the best choice provided that a variety of practical problems could be solved.

The important breakthrough was made in 1968 by Professor Albert Crew and his group at the University of Chicago (Crew, Isaacson, and Johnson, 1968). Their instrument used a field emission cathode in the form of a tungsten rod with a very sharp tip (<100 nm diameter) at one end. When the cathode is held at a negative potential relative to the anode then the electric field at the tip is so strong (typically of the order of 10^{10} volts/cm) that the potential barrier effectively becomes very narrow as well as being reduced in height. As a result electrons can then tunnel directly through the energy barrier and leave the cathode without requiring any additional thermal energy. The anodes act as a pair of electrostatic lenses and form a real image of the emitter tip at a distance of a few centimeters below the second anode. This arrangement, although simple, can produce a beam probe just a few nm in diameter or smaller, and with a beam current of 10^{-12} A or higher.

1.5 THE TWENTY-FIRST CENTURY SEM

The current versions of "FEG SEM" instruments are usually housed in an ultra-high vacuum environment and can operate at energies from 1 keV or less and up to 30 keV or more. Immersion optics that are able to identify materials and to optimize the resolution and high resolution performance, can be obtained for both iSE and BSE modes of operation when the operating conditions are properly optimized. For example, the Crewe type "electron beam" source and its descendants can now routinely provide nanometer scale imaging resolution when used under optimized conditions. Although achieving still higher imaging resolution is possible, it becomes increasingly more challenging because no electron–optical system is ever going to be perfect. In reality, every sample that is made inherently contains a finite number of errors and deviations from perfection, and these serve to limit the effective resolution level that can be achieved. For example, electron beams can be focused into a sub-angstrom diameter spot, but the resultant depth of field of the image so formed may then become too restricted to yield anything of practical value.

1.6 THE FUTURE FOR ION BEAM IMAGING – ABOVE AND BEYOND

The twenty-first century solution to improving the resolution and imaging capability of the SEM is to make the change from using an electron beam and to use ion beams instead. This is advantageous because the wavelength of a hydrogen beam at a given energy is a factor of about 750× shorter than that of an electron beam of the same energy and so the physical size of the beam will no longer be the limiting factor for imaging resolution. It will ultimately then be possible for the operator to image and process a much wider range of materials and in real time. Current ion beam instruments usually make use of beam materials such as hydrogen, helium, neon, or gallium beams, but other chemistries for special applications are also possible and could be the basis of some very interesting studies.

REFERENCES

Broers, A.N. (1968) High brightness electron gun using field emission cathode, in *Proceedings of the 26th A. Mtg EMSA*, p. 294.

Cosslett, V.E. and Haine, M.E. (1956) The tungsten point cathode as an electron source, in *Proceedings of Conference on Electron Microscopy*, London, 1954, Royal Microscopical Society, London, 639 pp.

Crewe, A.V., Isaacson, M., and Johnson, D (1959) A simple scanning electron microscope. *Rev. Sci. Instrum.*, 41, 241.

Everhart, T. and Thornley, T. (1960) Wide-band detector for micro-microampere low-energy electron currents. *J. Sci. Instr.*, 37, 246.

Fowler, R.H. and Nordheim, L.W. (1928) Electron emission in intense elecric fields. *Proc. Royal Society*, A119.

Gomer, R. (1961) *Field Emission and Field Ionization*, Harvard University Press.

Lafferty, J.M. (1951) Boride cathodes. *J. App. Phys.*, 22, 299.

Muller, E.W. (1937) Elektronenmikroskopische Beobachtungen von Feldkathoden. *Zeitschrift*, 106, 541–550.

Muller, E. and Bahadur, K. (1956) Field Iinization of gases at a metal surface and the resolution of the field ion microscope. *Phys. Rev.*, 102, 624.

Thompson, J.J. (1897) Cathode rays. *Phil. Mag.*, 44, 293.

2

Akashi Seisakusho Ltd – SEM Development 1972–1986

Michael F. Hayles
Cryo-FIB/FEG-SEM Technologist, Eindhoven, The Netherlands

2.1 INTRODUCTION

The company Akashi Seisakusho Ltd was founded in 1916 in Tokyo by Dr Kazue Akashi as a manufacturer of precision scientific instruments. Cooperation with various Japanese universities meant research and development was undertaken with close relationships with leading scientists. Over the next four decades the company's research and development department was responsible for high quality precision seismology, vibration measuring and materials testing equipment.

2.2 TEM DEVELOPMENT

At the beginning of the 1950s the Electro-technical Laboratory, Ministry of Communication, headed by Dr Shigeo Suzuki started design work on the first transmission electron microscope (TEM) for Akashi. It was released to the public in 1953 as the SUM-80, a vertical 3 lens system with 5 nm resolution and 80 kV acceleration voltage. During the 1950s, Akashi produced a series of horizontal columned TEMs following the trend of the time, such as the Philips EM100 and the AEI EM4. From development of a single pole piece lens in 1960 Dr Suzuki invented the S-zone lens (second zone lens) (Suzuki, Akashi and Tochigi, 1968). This consisted of a highly excited condenser-objective lens with the specimen located on the image side of the lens center. Dr Ernst Ruska also worked, about the same time, on a single field condenser-objective lens where the specimen was placed in the center (Riecke and Ruska, 1966), a future basis for the DS-130. The S-zone lens of Dr Suzuki was

Biological Field Emission Scanning Electron Microscopy, First Edition.
Edited by Roland A. Fleck and Bruno M. Humbel.

incorporated into the vertical S-500 transmission electron microscope of 1970. Horizontal column instruments were temporarily shelved in favor of vertical types, probably due to the inherent problem of gravity effect on the accelerating beam path.

2.3 SEM DEVELOPMENT, WITH TEM REPERCUSSIONS

Research into SEM instruments started in parallel with TEM during the 1970s and took over as the prior manufactured instrumentation for the decade. Akashi produced SEMs from 'Table-Top' to 'Bench-Top' in the MSM, Alpha and ISI systems, ISI being International Scientific Instruments, USA, an American company owned by Akashi and the first company to sell Akashi SEMs outside of Japan.

In 1979 Akashi produced the DS-130 SEM, a complete break from the previous instruments with a CPU controlled two-stage design housed in a TEM-like column and resolution of <3 nm, previously not realized in a SEM with a tungsten filament. With renewed research in TEM the horizontal column came back to life again in 1980 as the LEM-2000, an 80 kV, first of its kind, light/electron correlated microscope ideal for medical and pathological research. This was the first light/electron microscope (LEM) that could accept microtome serial sectioned specimens on a 7 mm grid that would be mapped in the light optics and automatically transferred into the electron optics with motor drive to the same recorded coordinates (Jones *et al.*, 1982). It had a unique facility to record many positions over the same section or serial sections to make a patchwork type large image stitched automatically by image recognition. The DS-130 and LEM-2000 instruments were the brainchild of talented engineers from International Precision Incorporated, a Japanese society commissioned by Dr Akashi to bring their inventive ideas to fruition.

2.4 TEM AGAIN, BUT SEM LIVES ON

By 1982 the vertical TEM systems progressed to the 002A, a CPU-controlled fourstage condenser lens system combined with a strongly excited condenser-objective lens, which resolved the behavior of atoms in crystals at 0.2 nm resolution. TEM once more became the company's priority with the help of International Precision and in 1984 the 002B followed with improvements aimed at easier usage while retaining performance specifications. This was Akashi's final TEM and by 1986 all development on TEM had stopped. Akashi Electron Optics division also stopped marketing through International Scientific Instruments by the end of 1986. Advanced Beam Technology (ABT) Ltd, Japan, was formed and took over the manufacturing and marketing of the previous Akashi Electron Optics instrumentation at the start of 1987. To replace the marketing network originally formed as ISI, ABT use a world-wide subsidiary instrument company of Toshiba called TOPCON. Over the next few years TOPCON became a corporation expanding its business by taking on board other product lines. By 1991 ABT transferred all SEM manufacturing and marketing to TOPCON. ABT Ltd ceased trading.

This review of Akashi SEMs goes into detail as far as 1986, from then on the name Akashi Electron Optics is lost and some mention of ABT and TOPCON instruments is made to finalize the story of a producer of unique, ultra-high resolution SEM/TEM instruments, some of which, seen in hindsight, were 'way before their time'.

2.5 MSM (MINI SCANNING MICROSCOPE) TABLE-TOP SERIES

SEM development started in the early 1970s and the MSM-2 (Mini-SEM) was released in 1972 (Figure 2.1). This was the first table-top scanning electron microscope and had a pre-aligned three-stage electron lens system (Figure 2.2a) with an *X*, *Y* alignable gun carrying a Wehnelt in cartridge form for easy replacement. Accelerating voltage was a fixed 15 kV, chosen because of the compactness of the lens system. The standard 1 inch sample chamber was with an *X*, *Y* micrometer and tilt movements. The vacuum system was a ½ inch diffusion pump and small rotary pump with direct simple valve mechanisms operated by rotating knobs. The control console with monitor was in 'box' form and sat on the table next to the column.

The MSM series went on to six models from 1972 to 1976 and could be upgraded to larger stages for larger samples. During this time experimental systems were produced such as an MRS (mini-rapid-scan) model with an air-cooled pumping system and a TV-MSM using television scanning speed only. A WDX model MSM-102W was introduced around the time of the MSM-6 using single crystal vertical spectrometers, technology learnt from the TEM S-500.

2.5.1 Alpha Bench-Top Series

The Alpha series was started as a version of the TV-MSM but mounted on a bench top with the control console. All these models had a Polaroid camera that swung down over the viewing monitor to record the image. The table-top electron microscope had had its day by 1975, selling more than 2000 systems worldwide and gaining awards in America as well as Japan. The success of the Mini-SEM series fueled the Akashi research to develop better, more sophisticated SEMs.

2.5.2 SMS Bench-Mounted Series

Super-Mini-SEM SMS-1 was released in 1974. This was the first model with bench-mounted column and chamber with a partly suspended vacuum system. The electronics were now

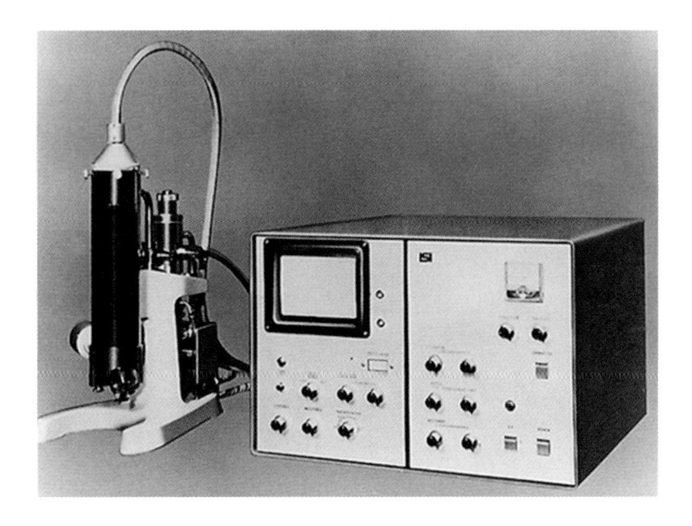

Figure 2.1 MSM-2, the first table-top model.

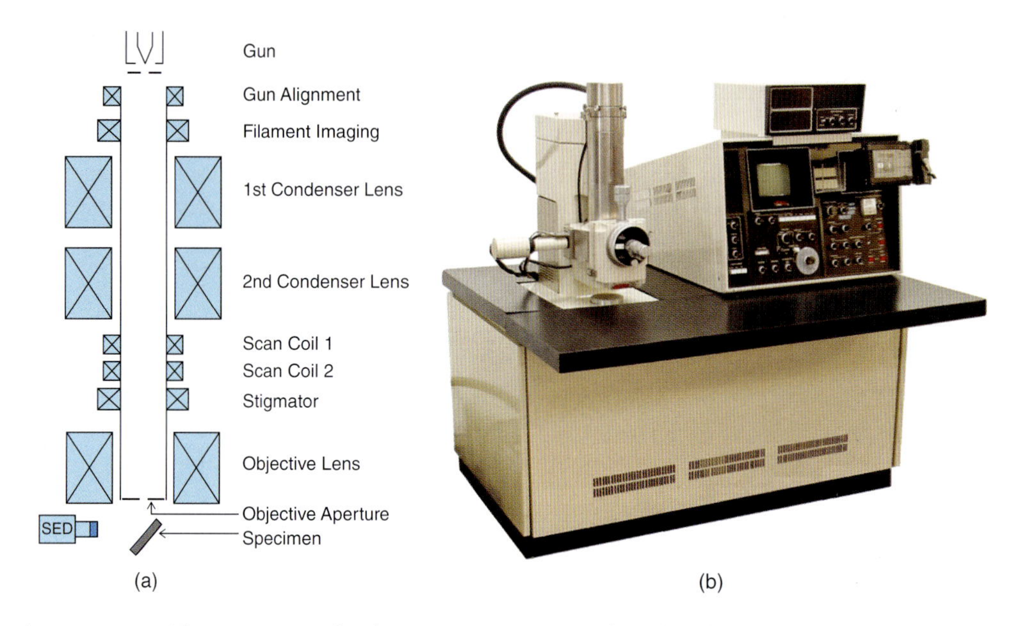

Figure 2.2 (a) The conventional column construction attributed to the MSM, SMS, Alpha and SX series. (b) The Super 3A with a dedicated bench-type console, later typical of the ISI models.

housed in a reclining modular frame as part of the bench-top and gave the choice of multiple accelerating voltage operation. Models SMS-1, 2, 3 and 3A (Figure 2.2b) were developed for the next 2 years. The Super 3A was the last of the SMS vacuum design, although the bench console continued with the ISI range. With resolution at 7.5 nm and facilitating 2, 5, 10, 15 and 25 kV, it became a 'workhorse' for materials research when combined with the then newly available energy dispersive X-ray (EDX) systems. Recording images was now using Polaroid, roll film or 35 mm from a separate record cathode ray tube (CRT). The Alpha series, also bench mounted, continued as a lower cost 15 kV scanning electron microscope, emulating its predecessor the TV-MSM.

2.5.3 ISI Series

Akashi, since the early 1970s, had been marketing the Mini-SEMs in America and Europe under the name of International Scientific Instruments (ISI). By 1977 demands were made from the emerging IC manufacturing industry on scanning electron microscope manufacturers to produce more sophisticated systems with larger stages. Akashi rose to the challenge and brought out models such as the ISI-40, 60 and 100, which had more operating kV choice per model and stage sizes to suit. The vacuum system was now suspended in the bench mounting to act as counterbalance for antivibration purposes. The ISI-40 could be equipped with up to three optional single-channel crystal spectrometers and was commercially labelled the ISI-40W. The ISI-100 could accept a 4 inch wafer (standard at the time) with total coverage. Further upgrades to the ISI-60 and 100, were models ISI-60A and 100B, which followed to accommodate the IC and general research industries. The 100B had an auto-beam system that controlled the column and gun alignment between a spot or kV change. User functionality was therefore simplified as alignments would remain corrected and deliver the best illumination and distortion-free conditions irrespective of accelerating voltage operation. This was the top Akashi SEM and resolution for this instrument stood at 6 nm, the best for conventional SEMs of the time.

2.5.4 DS130 (W) Tungsten

In 1979 Akashi announced a major breakthrough in scanning electron microscopy that cut the then known secondary electron image resolution barrier (6 nm) by half. The DS-130 SEM (Figure 2.3a) was a computer controlled SEM, with dual stages and dual detectors, a 1 to 40 kV high brightness tungsten filament Wehnelt, five lenses and an ultra-high vacuum system (Figure 2.4c). The in-lens top stage (Suzuki, Akashi and Tochigi, 1968) was capable of <3 nm with through-the-lens SE detection and a large sample lower stage capable of <5 nm with conventional SE/BSE detection. The top stage could house several detectors, SED, BSED, STEM and EDX (Figure 2.3b), with the ability to detect and display individually, or at the same time, on the large dual monitors via a split-screen mode. Image processing was built in and included gamma, derivative, inversion, waveform, Y modulation and auto contrast and brightness as standard. An unusual feature was embedded in the DS-130 in that it could perform electron channeling patterns (ECPs) at rocking angles from 5° to 30° on areas as small as 3 μm² on prepared surfaces. The changing pattern, while moving across the sample, could be displayed on one of the monitors, which gave another dimension to analyzing crystalline samples.

The top stage worked with a back-focal plane aperture house just below the condenser lenses, unlike conventional pin-hole type objective lens apertures found in most SEMs, as well as in the bottom stage of the DS-130 (Koike *et al.*, 1979). In practice the back-focal plane aperture was projected and aligned over a common point on the sample for continuity during alignment for all operating conditions recorded through the CPU, making usage for what may seem a complex instrument relatively easy. Because of the combination of the condenser system with either objective, a long camera length could be made by operating the top-stage optics with the bottom-stage pin-hole aperture removed, while observing a specimen on the bottom stage. This created better depth of focus for larger objects at short working distances and a slightly better resolution than claimed for the bottom stage at

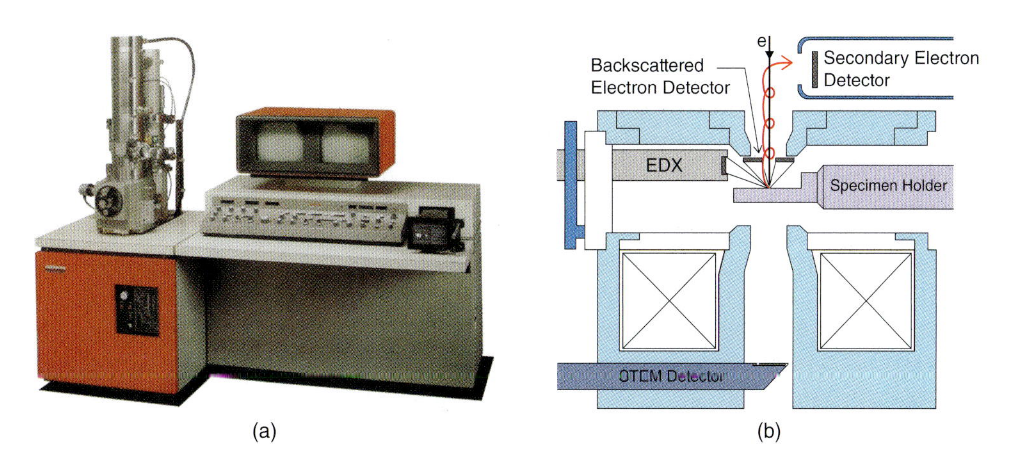

Figure 2.3 (a) The first production DS-130 (tungsten filament version). (b) Multiple detectors shown in position relative to the top-stage immersion objective lens.

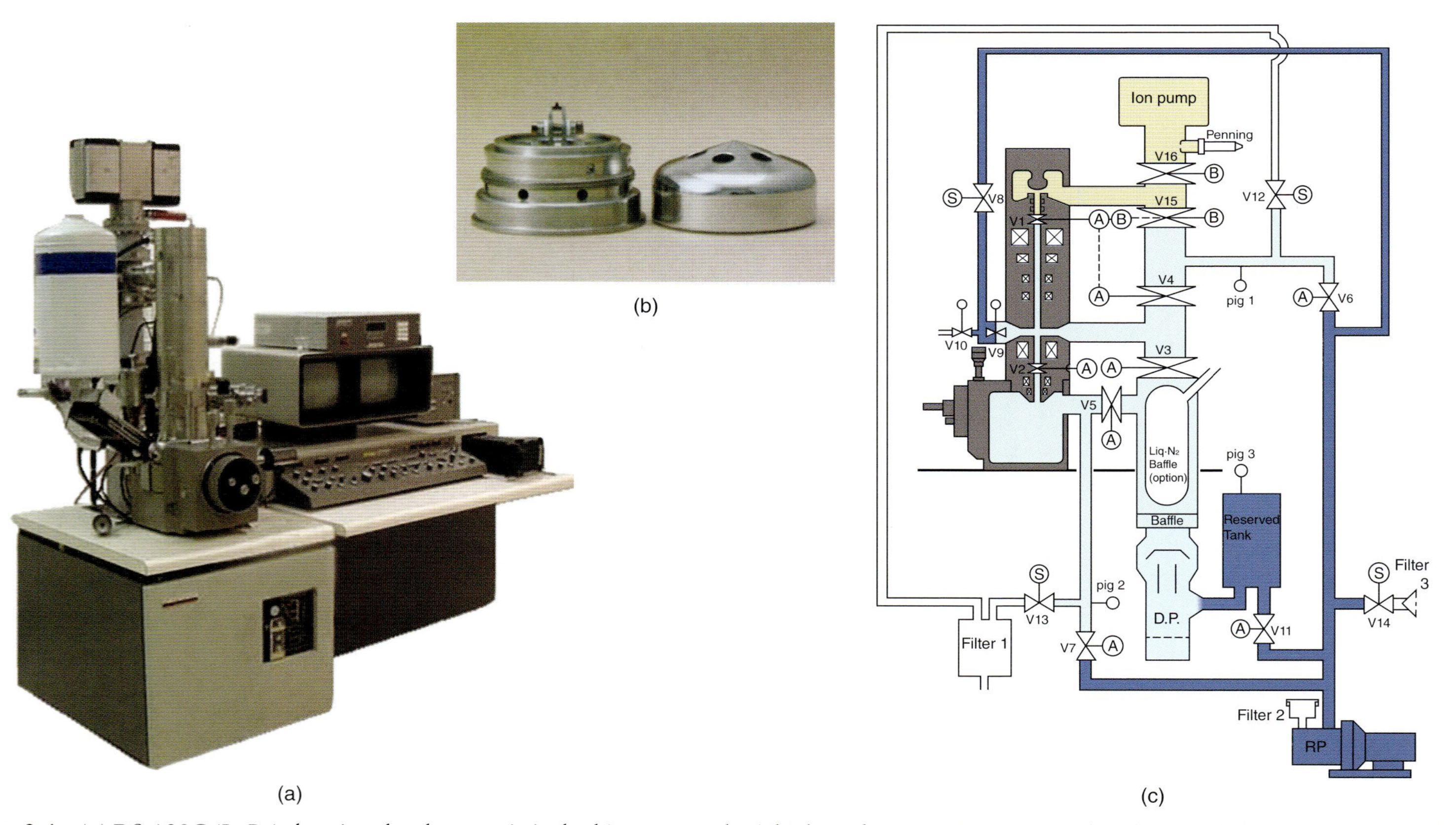

Figure 2.4 (a) DS-130C (LaB_6) showing the characteristic dual ion pump. (b) A high performance LaB_6 Conical Wehnelt with Denka M1 tip installed. (c) DS-130C vacuum schematic showing the dual ion pump and adjoining pipework to the gun area distinguishing the region of higher vacuum operated by the ion pump. Valve 16 operated for LaB_6 only, whereas Valve 15 operated for W or LaB_6. There was a source dependent differential aperture above Valve 'V1'. Valve 'V1' was operated when vacuum was achieved at the gun area, depending on the programming for W or LaB_6. Valves labeled 'S' are gas-admittance valves for bringing the system to atmosphere.

high accelerating voltages. This was undoubtedly a preliminary step to the later Sigma/SS systems design.

An energy dispersive X-ray analyzer (EDX) system could also be fitted to either stage. The Mylar-window end of the EDX system (without collimator) sat just outside the high strength lens field of the top stage, therefore avoiding much of the backscattered electrons initially trapped by the lens field. The background free X-ray spectrum could be discreet even at a thousand counts, making elemental analysis possible without using background subtraction. An EDX system fitted to the bottom stage worked in the conventional way for analysis of larger samples.

2.5.4.1 DS-130C LaB_6 (Lanthanum Hexaboride)

Akashi quickly provided a DS-130C model (Figure 2.4a) with LaB_6 tipped Wehnelt and STEM (scanning transmission electron microscope) detector for the in-lens top stage, although this was an option on the original tungsten DS-130. Resolution for the top stage was now <2 nm with the in-lens detector and <1 nm in the STEM mode. STEM, known these days as TSEM (a rearrangement of words), was exceptional in performance for, besides being able to work with standard thickness sectioned material on 3.05 mm grids, the specimen could be tilted between -5° and $+95^{\circ}$. Imaging in STEM mode was with high contrast due to the choice of low accelerating voltages, unlike the TEM systems of the time. Imaging of some biological structures could only be seen at less than 40 kV and the DS-130 STEM has been known to have competed with TEM instruments in this area very successfully. The additional vacuum and electronic equipment to accommodate an LaB_6 module included a double ion pump with isolating valves from the rest of the vacuum system. Once the vacuum of 5×10^{-6} mbar was reached by the liquid nitrogen trapped diffusion pump, the gun area was isolated and the ion pump took over, bringing the vacuum at the Wehnelt into the 10^{-7} mbar range. With recommended cleaning of the Wehnelt once per year, the full Vogel mount Denka M1 LaB_6 filament could last with performance more than 3 years and has been recorded at a maximum of 7 years in daily use. The unique pre-align design of the four piece conical Wehnelt (Figure 2.4b) meant long lifetime operation with very little lanthanum oxide build-up, which was unusual for LaB_6 in SEMs at the time. The condenser lens system was a dual gap first lens and single gap second lens. This could work as a three lens system under computer controlled combinations of two condenser lenses with either the top or bottom objective lenses.

The DS-130C had all attributes of the original DS-130 but with further improvements over the original DS-130 with respect to the top-stage high strength objective lens. The DS-130C top stage had two sample positions with dedicated lens operation: Mode 1 was similar to the original DS-130 covering the complete accelerating voltage range and Mode 2 optimizing a second zonal position, which could be used for a range of low accelerating voltages from 500 V to 3 kV. Both featured lower Cs and Cc values (Figure 2.5). Loading samples in the top stage was by a rod mechanism through and air-lock, similar to a TEM. Different rods were available for the modes of operation but also for other applications such as STEM and EDX. A major improvement to the bottom stage was to move from a 45° angled objective lens to one of conical shape at a 70° angle. This immediately improved IC wafer tilting for high angle observation.

This instrument was without doubt the best performing LaB_6 SEM of its day and for many years after and was the instrument of choice in top research laboratories around the world during the 1980s.

DS-130	Focal length *f* (mm)	Spherical aberration Cs (mm)	Chromatic aberration Cc (mm)
Original and DS-130S (Mode 1)	9.3	5.6	7.0
Type C (Mode 1)	5.3	3.2	4.0
Type C (Mode 2)	2.6	1.5	1.9

Figure 2.5 Objective lens aberration characteristics for the DS-130, DS-130S and DS-130C.

2.5.4.2 DS-130S

The DS-130S replaced the original tungsten model with a few improvements of the DS-130C. These included the bottom-stage conical lens and an option for LaB_6 with high performance and long-life Wehnelt cartridges.

2.5.4.3 DS-130F Field Emission Gun (FEG)

The research into this instrument started in the 1980s while the DS-130 LaB_6 was in use but only came into service by the end of Akashi's ownership of the Electron Optics group. It was taken over by ABT and immediately marketed in the IC industry, which by the mid-1980s demanded field emission instrumentation. The competition was from Cold-Field Emission (Cold-FE) systems, which although unreliable to some extent and needing daily conditioning were gathering market share in this industry. During the design phase Akashi was looking for field emission expertise and found it in a group of researchers at Oregon Graduate Centre, a private university for postgraduates who were developing a new Hot-FE source. By 1984 they had produced a working Schottky Emitter (Tuggle and Watson, 1984). Oregon researchers along with Akashi developed the field emission gun (FEG) for the DS-130F. With slightly more energy spread than a cold-field emitter, the Schottky system was very stable and did not need daily routine attention. Vacuum was less demanding being 2 decades higher at 10^{-8} mbar compared to 10^{-10} mbar for cold-FE and only 1 decade lower than used for LaB_6. The DS-130 was the ideal platform that already functioned throughout via a CPU; therefore building control of the FE gun, lenses and vacuum system was not from scratch.

The DS-130F came with high resolution combined with a large sample handling capability. Top-stage resolution was 1 nm while the bottom stage delivered 2 nm, even with a high tilt angle, making use of the 70° conical lens. Sample size was a standard 150 mm diameter (6 inch wafer), but with the optional bottom chamber/motor, driven stage sizes up to 200 mm diameter (8 inch wafer) could be fully observed.

Although the DS-130F was a logical step for Akashi and ABT to take, Schottky FE (thermal FE) was slow to take off with other SEM manufacturers, making the DS-130F meet cold-FE competition head-on. The reason for the slow advancement of what seemed to be a solution to cold-FE was cost. The thermal FE and its supporting electronics were expensive compared to cold-FE and not all manufacturers at the time thought it a wise investment. The delay allowed cold-FE to prove its superiority and make significant headway into the IC industry market, which in turn made it difficult to successfully sell Schottky systems for some years. In 1991 TOPCON took over and marketed the DS-130F with new colors and model number.

2.6 SIGMA (SS) SERIES

To replace the aging ISI-40, 60A and 100B, three new SEMs were developed during the early 1980s based on the DS-130 top-stage lens design and were called Vari-Zone Lens SEMs (Figure 2.6a). These emerged in the early 1980s as Sigma systems in Japan and soon after as SS systems in America and Europe. The SS-40, 60 and 130 (Figure 2.6b) had similar stage design as the lower stage of the DS-130 for large samples and conventional work, but could accommodate an in-lens/through-the-lens sample module that could then be placed inside the lens for ultra-high resolution. This brought versatility to a lower cost system that could be used in a variety of research domains. The SS-40 and 60 had accelerating voltages to 30 kV whereas the SS-130 had 40 kV, all with the option of tungsten (W) or LaB_6. Like the predecessor, the 100B, the SS-130 had the Akashi auto-beam control. All three systems announced the arrival of the motor driven stage, which was a standard feature. Resolution was measured for the SS-40 at 4 nm with W and 3 nm LaB_6, the SS-60 at 3.5 nm with W and 2.5 nm $LabB_6$ and the SS-130 at 3 nm with W and 2 nm LaB_6.

2.6.1 Vari-Zone Lens Modes of Operation

2.6.1.1 Conventional Pin-Hole Design (Mode 2)

Scanning electron microscopes until the Akashi DS-130 and SS instruments were of a standard pin-hole lens design. This meant that the objective lens housed an aperture in the final pole piece just above the specimen (Figure 2.7a). The lens field would be concentrated within the pole of the lens and therefore have little or no influence on the trajectory of secondary

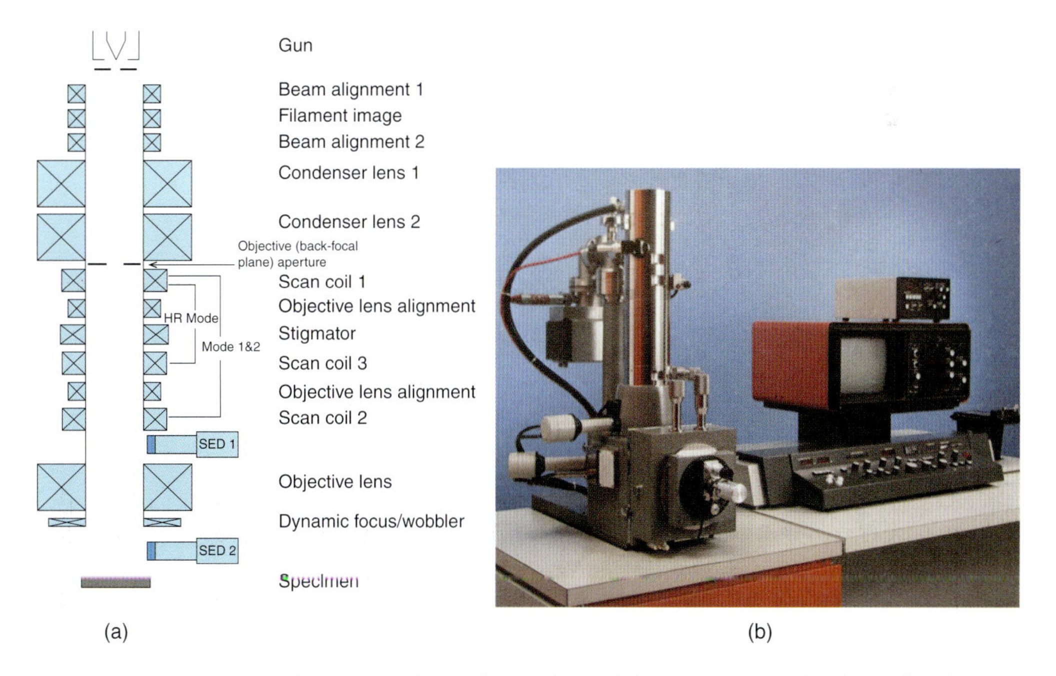

Figure 2.6 (a) Vari-zone lens (SS) column design derived from DS-130 technology. (b) The most accomplished vari-zone SS-130 LaB_6.

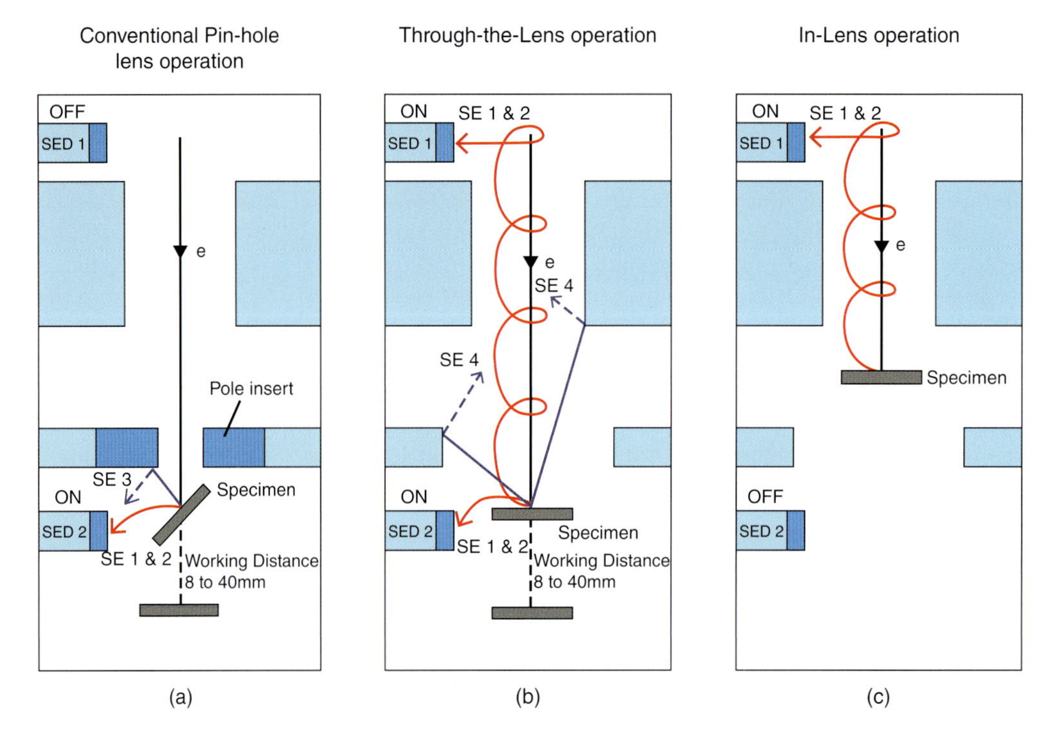

Figure 2.7 Three modes of operation, where (b) and (c) defied SEM convention. (a) Conventional pin-hole lens operation. (b) Through-the-lens operation. (c) In-lens operation.

or backscattered electrons produced at the specimen. In the SS system with the top SE detector switched to OFF, biasing of the bottom SE detector had a greater effect on secondary electrons, drawing in SE-1s immediately from the beam impact area on the surface and SE-2s produced by backscattered electrons leaving the surface within a few nanometers of the impact area. Backscatter electrons that leave the specimen surface and react with the surrounding structures such as the final pole piece create SE-3s, which could then enter the SE detector adding a strong topographical element to the image.

2.6.1.2 Through-the-Lens Design (Mode 1)

It was recognized that with high strength lens design one could create an open pole (Figure 2.7b), bringing a number of advantages. Because of the lower aberration characteristics for this through-the-lens design image, resolution is far higher compared to the pin-hole lens. Unlike the pin-hole lens the magnetic field extends out below the open lens and traps SE-1s and 2s leaving the specimen at a short working distance. SEs are forced to spiral back up the column to the top SE detector. In this case no SE-3s can get to the detector and only few long trajectory SE-4s may enter by incidence. At the same time the bottom SE detector collects few SE-1s and 2s at short working distance, but as the specimen is lowered more SE-1s and 2s enter the bottom SE detector. Switching modes of operation from the console with this configuration one could mix or separate signals and create an almost pure SE component with the top detector and separated surface topography with the bottom detector of the same area being investigated.

2.6.1.3 *In-Lens Design (HR Mode)*

One step further, by switching the bottom SE detector to OFF and placing the specimen within the lens produces the highest resolving SE images, having collected almost entirely SE-1s and 2s that have spiraled back up the column (Figure 2.7c).

2.6.1.4 *Return of the Alpha Series*

Conventional Alpha systems, SX-25, SX-30, SX-40 and SR-50 were produced over the next 5 years. Variations of these models could be found with spectrometers labeled as Alpha-30W and 30AWA. Instruments in this line were generally taken up by small industries for failure and materials analysis.

2.6.2 Environmental SEM

2.6.2.1 *WET-SEM*

Environmental research had by 1983 gathered momentum and researchers were asking for a scanning electron microscope where one could visualize material in its native state instead of it collapsing or drying out under vacuum. The WS-250 was a WET-SEM system produced to fill the demand for environmental research and was an improvement over its predecessor the WS-300. It worked on controlled differential pressure between the specimen chamber and the column, allowing the sample to be in almost atmosphere while the electron column remained at working vacuum. Imaging at high pressure between 0.06 mbar and 2.6 mbar in the specimen chamber meant that hydrated and vacuum sensitive samples could now be seen without adverse effect from the vacuum. An uncoated sample viewed under high pressure charge-free conditions was a revelation of the time when non-conductive samples were normally coated with a conductive metal such as gold. Imaging was created by use of the Robinson Backscattered Electron Detector (invented by Dr Vivien Robinson of EPT-SEMRA Pty Ltd, Sydney, Australia) and contracted to ISI. By collecting only backscattered electrons at low voltages for imaging such samples, detailed real surface topography could now be analyzed. Unlike later dedicated ESEM type systems the WET-SEM was a simple solution that could be accommodated as an upgrade for a variety of Akashi models if necessary. This entailed a differential pumping aperture placed between the chamber and the remainder of the vacuum system and the chamber pumped via the Robinson detector at a higher pressure. This system was first known as ECM (environmental control module) but later, because of conflict with the word 'environmental', was changed to CFAS (charge free anti-contamination system). These systems were the first attachments to SEMs anywhere to be produced and marketed. Several hundred systems were sold mainly in Japan between 1978 and the mid-1980s. Because of the increasing market share other SEM manufacturers followed the trend with their own attachments or SEM models, generally calling them 'Low-Vacuum', or even taking the name of 'WET-SEM'.

2.6.3 Integrated Circuit (IC) Orientated SEM

2.6.3.1 IC-130

The IC-130 LaB_6 was similar in form and replaced the SS-130 mainly as a wafer inspection instrument.

2.6.3.2 WB-6

The WB-6 (wide bore) was a computer controlled SEM with a vari-zone high strength lens system and dual secondary detectors (through-the-lens), similar to the SS series. The difference with this microscope is that the end lens had a wider opening to accept larger specimens and shorter focal distance effective to collect most of the secondary electron yield from the very surface of the specimen. It could incorporate the optional electron retarding field coil developed (Menzel and Buchanan, 1985) to filter very low voltage electrons to the upper secondary electron detector for study of insulators associated with IC research. Resolution was 4 nm with a tungsten filament and 3 nm with LaB_6. There were two accelerating voltage ranges, one optimized for low voltage range work and started at 500 V to 3 kV in 100 V steps and one for more conventional work with 3 kV to 30 kV in 1 kV steps. Computer control allowed modes of operation including: high resolution, standard, low magnification, large current, free and external control, where functional conditions could be optimized to suit any external mode, such as EDX. Although this instrument was promoted into the IC market, because of low voltage facilities it also proved very capable as a standard vari-zone lens SEM.

2.6.3.3 CL-6 and CL-8

The CL computer controlled system had a special 30°/60° conical objective lens for low accelerating voltage with high resolution imaging on whole wafers. Two tilt angles of 30° and 60° were possible with this instrument at 1 kV to give resolutions of 12 nm to 20 nm. The short working distance possible coupled with low voltage capability made it ideal for wafer inspection. A choice could be made between two chamber sizes, 6 inch or 8 inch, to suit the size of wafer being investigated. Both chambers had the option of a wafer loader airlock. The stage of the CL-8 had motorized drive to X and Y with a high continuous speed of 2500 micrometers/second down to 2 micrometers/second pulse speeds with 0.25 micrometers/pulse. The other movements of Z (height), R (rotation) and tilt were manually operated.

2.6.4 Dedicated Integrated Circuit (IC) SEM

2.6.4.1 MEA-3000

The MEA-3000 was a dedicated IC wafer inspection machine especially for 'line-width measurement', whereas all other previous IC systems could be used as general SEMs as well. What placed the MEA-3000 apart was the fact that it only had a low accelerating voltage, 500 V to 3 kV in 100 V steps. The LaB_6 emitter could deliver more than 1×10^{-6}

ampere with a beam diameter of 15 nm. The column was built with four lenses with a low aberration objective, all controlled by the Akashi auto-beam system. Sample coverage was 150 mm × 150 mm, stepping motors being used for *X*, *Y* and R (rotation). Stage driving modes included cross, line, mapping and manual mode by button control or joystick. Important rotation alignment correction could be made between the stage and the IC being investigated. The scan and display system was digital with freeze-frame, line-width measuring on-screen and real-time image processing with storage of 512 × 512 pixels × 8 bit, which, although dated now, was a definite asset in 1986. A computer was used for the line-width measurement control and data processing. Data included wafer identification, measurement positions and measured values. This was necessary due to the fact that wafers enter the system via a cassette auto feed load-lock. The cassette loading system took 25 wafers at one time and different types of cassette for the size of wafers. The load and unload function was fully automatic.

2.6.4.2 EBT-1000

The EBT-100 was a stroboscopic electron beam tester that started out as a WB-6. It incorporated a combination of an electron retarding field coil (Menzel and Buchanan, 1985), which in operation could filter low voltage electrons to the upper secondary electron detector and a stroboscopic pulse generator to feed current into LSI/VLSI wafer circuitry without mechanical contact. This option converted the instrument into a beam tester tool able to image live circuit operation on a wafer. Resolution was 100 nm at 1 kV and, similar to the MEA-3000, only used a low accelerating voltage range. Sample wafers of up to 127 mm diameter could be accommodated.

2.6.4.3 SR-50, the Last Alpha Systems

This model consisted of a column with a conical high strength lens system allowing imaging at short focal length (1 mm) and specimen tilting at short working distance (25° tilt at 1.5 mm WD). A maximum sample size of 150 mm diameter could be accommodated on the 5 speed motor driven TXYZ stage. The SR-50 had a resolution of 5 nm with a tungsten filament and 3.5 nm with LaB_6. The accelerating voltage was in 12 steps from 1 kV to 30 kV, with beam current ranging from 10^{-7} to 10^{-12} amperes. Many features developed in other instruments over time of Akashi SEMs seem to have accumulated in this instrument. Unfortunately, these advances came too late as this was Akashi's last small instrument. The SR-50A became an ABT/TOPCON product along with the DS-130F.

2.6.4.4 ISI-ABT-Topcon

Although ABT controlled all that was left of Akashi–ISI SEM product lines, their main concern was the IC orientated systems as the IC market was still thriving in 1986, especially with the advent of personal computers. It is not openly known what 'Specials' were made following the MEA-3000 and the EBT-1000, but it is known that the WB-6 and CL systems continued, as seen in (Figure 2.8). When TOPCON took over the instrumentation and marketing from ABT in 1991 the dedicated IC wafer inspection SEMs that were sold

Table -Top		Desk -Top		HR SEM	Vari-Zone	Analysis	IC Systems	Approx
MSM	SMS	Alpha	SMS/ISI	DS-130	Sigma/SS	WDX	Specials	Year
MSM-2								1972
MRS-2								
MSM-4	SMS-1							1974
MSM-6	SMS-2					MSM-102W		
TV-MSM								
MSM-7	SMS-3							1976
MSM-4C	SMS-3A							
MSM-7C			ISI-60					
			ISI-100					
		ALPHA-9	ISI-40			ISI-40W		
		ALPHA-10	ISI-60A					
			ISI-100B					1979
			ISI-30					
						WS-300		
				DS-130				1980
		SX-25/30			SS-40-60-130	ALPHA-30W		1981/2
		SX-30E		DS-130C/S	WB-6/IC-130	WS-250	MEA-3000	1982/3
		SX-40			CL-6	ALPHA-30AWA	EBT-1000	
		SR-50		DS-130F	CL-8			1986
		SR-50A		DS-130F	WB-6, CL-6/8		?	
					ABT / TOPCON			

Figure 2.8 Table of Akashi SEMs (with approximate year of introduction).

by ABT from 1986 came to an end, and only a limited number of SEM models were now made available. Instruments like the DS-130F lost its momentum, unlike the LaB_6 years, and other manufacturers were now making up ground, especially with Cold-FE and later more sophisticated Schottky FEG-SEM instrumentation. In some parts of the world there still remains some confusion as to the naming of the SEM models from the years 1986 to 1991; some instruments are found labeled 'ISI-ABT' and others 'TOPCON-ABT'. After 1991 all are labeled only with the 'TOPCON' name. TOPCON slowly phased out the Akashi styled SEMs in favor of more dedicated systems with commercial OS computer control in line with competitor corporate trends.

2.6.5 Measurement Notation and Bar

The measurement notation and defined bar varied over the different series of Akashi SEMs. The MSM/SMS and ISI instruments displayed a simple left facing 0.1 micrometer bar, which changed in size, with short bars to the right each representing ×10, here seen in Figure 2.10b (10 micrometers). For the SX and SS/IC/WB (Vari-Zone series) the same bar system is used but right facing as seen in (Figure 2.10c, d and e). The DS-130 original 1979 model had a varying length bar below crude text (U = micrometer, N = nanometer), represented in (Figure 2.9a, b, d, e and f and Figure 2.10a). This changed to a static bar above more refined notation for models DS-130C and S, seen in (Figure 2.9c). The SR-50(A) had a static solid looking bar with varying size notation above, seen in (Figure 2.10f).

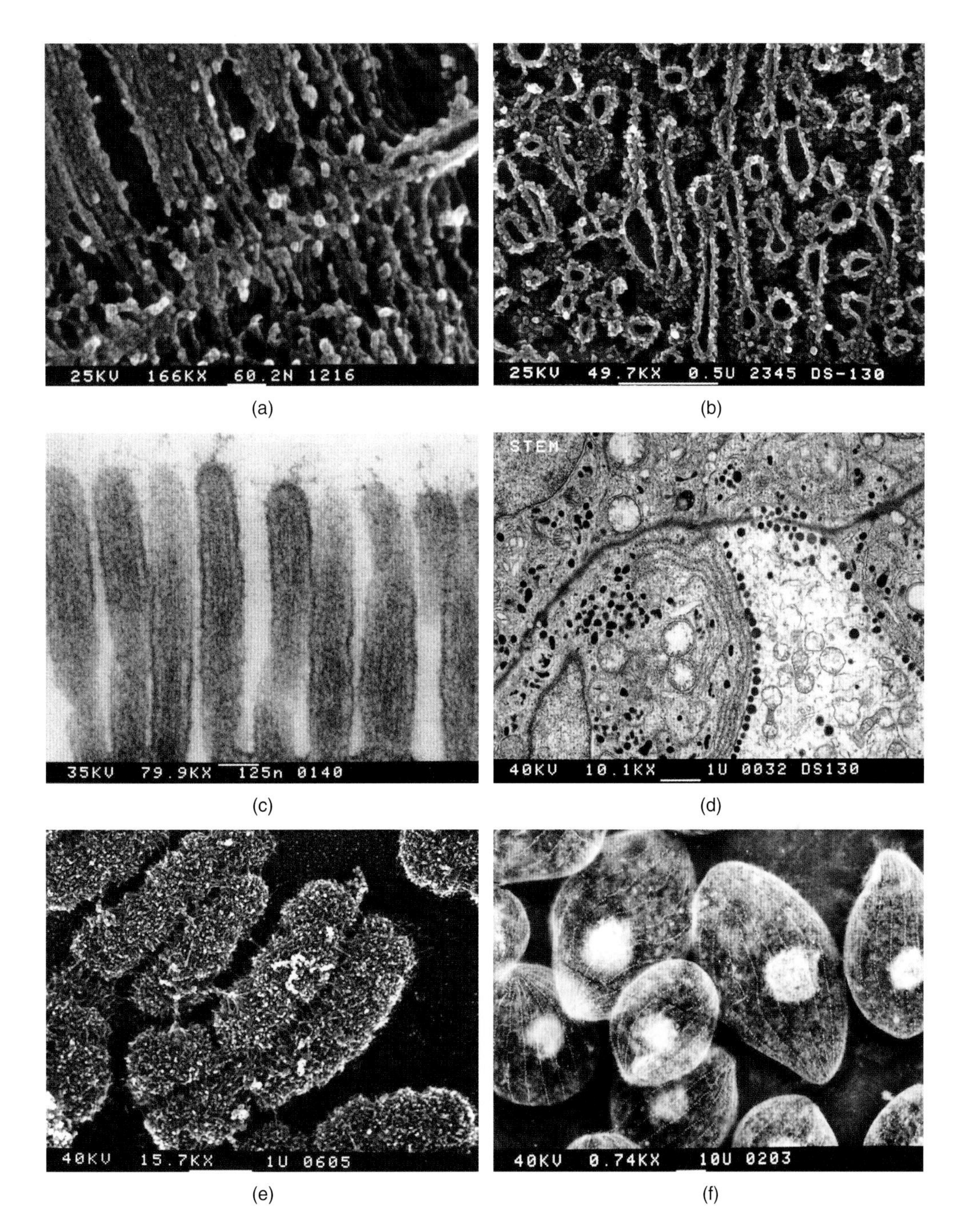

Figure 2.9 (a) Heart muscle, Pt coated, SE, DS-130 LaB_6. (b) Pancreas of rat, Pt coated, SE, DS-130(W). (c) Microvilli showing actin filaments and enterocyte digestive enzymes at the plus end, STEM, DS-130 LaB_6. (d) Human pancreas, STEM, DS-130(W). (e) Human chromosomes, uncoated, SE, DS-130(W). (f) Tetrahymena Pyriformis, silver stained, BSE, DS-130(W).

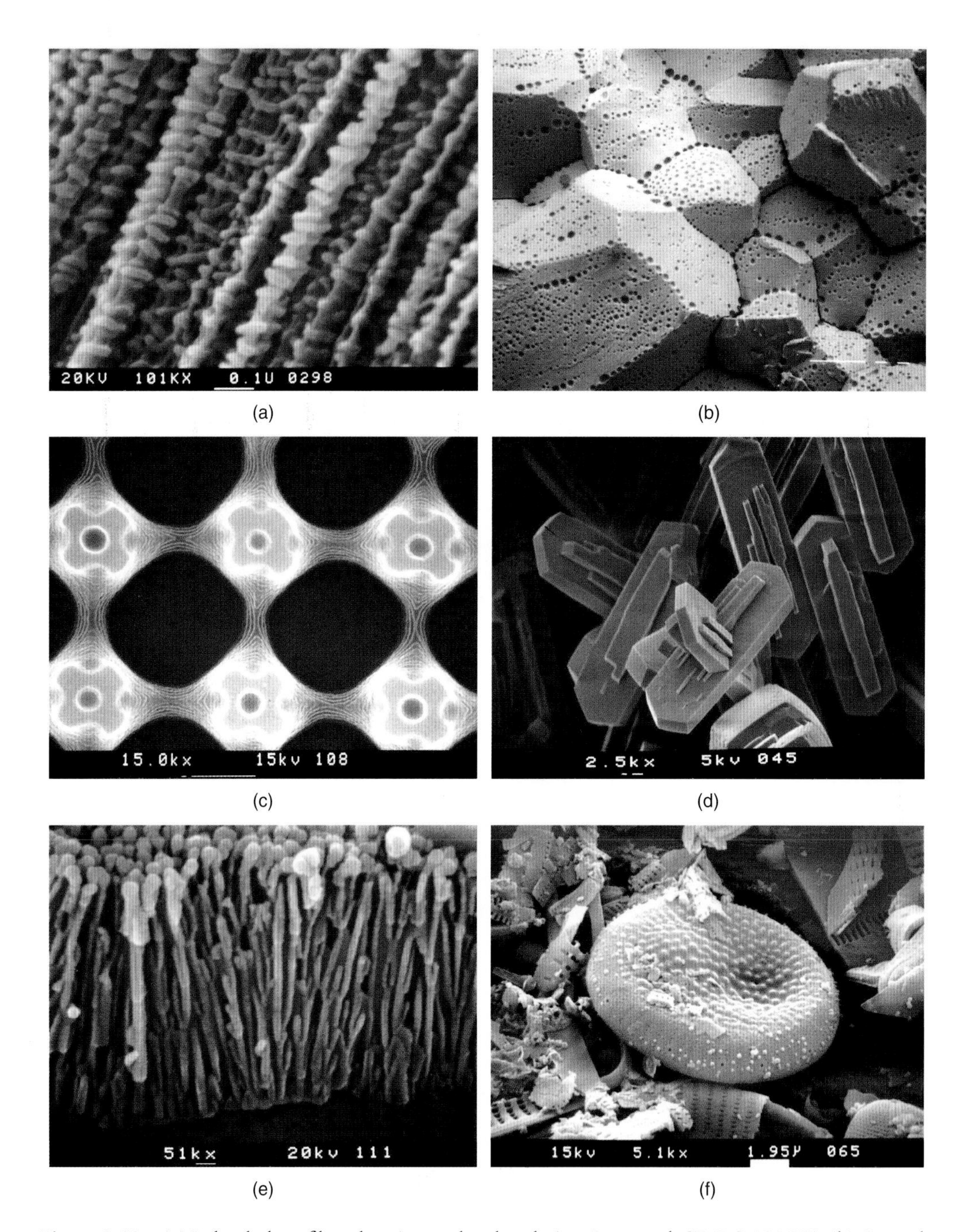

Figure 2.10 (a) Polyethylene fiber showing molecular chains, Pt coated, SE, DS-130(W). (b) Crystals formed on a tungsten filament, BSE, ISI-40. (c) IC wafer showing 'standing-wave pattern' after etch process, uncoated, SE, SS-60. (d) Polysynthetic crystal aromatic, uncoated, SE, SX-30. (e) Anodized aluminum layer, uncoated, SE, SS-40. (f) Foraminifera, Pt coated, SE, SR-50.

ACKNOWLEDGEMENTS

Figures 2.9a, 2.9b are from the Akashi DS-130 brochure of 1982 and are attributed to Dr. T. Inoue, Anatomy Laboratory 2, Faculty of Medicine, Tottori University, Japan. Neither the author, nor Wiley, have been able to trace the rights-holder to these images during the writing of this contribution, and would therefore welcome contact from any persons who can provide information relating to them.

REFERENCES

Jones, S., Chapman, S.K., Crocker, P.R., Carson, G. and Levison, D.A. (1982) Combined light and electron microscope in routine histopathology. *Journal of Clinical Pathology*, 35, 425–429.

Koike, H., Kyogoku, H., Yanaka, T., Watanabe, M. and Uchida, H. (1979) Development of dual stage scanning electron microscope ISI DS-130, in *Proceedings of the 37th EMSA*, p. 612.

Menzel, E. and Buchanan, R. (1985) Some recent developments in low voltage E-beam testing of ICs. *Journal of Microscopy*, 140 (3), 331–349.

Riecke, W.D. and Ruska, E. (1966) A 100 kV transmission electron microscope with single-field condenser-objective, in *Proceedings of the 6th International Congress for Electron Microscopy*, Kyoto, Japan, vol. 1, pp. 19–20.

Suzuki, S., Akashi, K. and Tochigi, H. (1968) Objective lens properties of very strong excitation, in *Proceedings of the 26th EMSA*, p. 320.

Tuggle, D.W. and Watson, S.G. (1984) A low voltage field emission column with a Schottky emitter. *Proceedings of EMSA*, 42, 454–457.

3

Development of FE-SEM Technologies for Life Science Fields

Mitsugu Sato[1], Mami Konomi[2], Ryuichiro Tamochi[2] and Takeshi Ishikawa[2]

[1]*Research and Development Division, Hitachi High-Technologies Corporation, Hitachinaka-shi, Ibaraki-ken, Japan*

[2]*Science and Medical Systems Business Group, Hitachi High-Technologies Corporation, Tokyo, Minato-ku, Japan*

3.1 INTRODUCTION

Since the first commercial product development by Cambridge Instruments in 1965, the SEM (scanning electron microscope) has been improving in terms of resolution, image contrast and operability, etc., to show remarkable progress in today's instruments. A high brightness electron source is essential for attaining high resolution SEM. In 1968, Prof. A.V. Crewe of the University of Chicago developed the FE (field emission) electron source (Crewe *et al.* 1968) which provides 1000 times higher brightness than a conventional thermionic electron source with tungsten filament and a correspondent substantial leap in attainable SEM resolution. The FE electron source was commercialized in 1972 as Hitachi HFS-2 FE-SEM with the introduction of FE technology from Prof. Crewe (Komoda and Saito 1972). Around the same time, Coates and Welter developed CWIKSCAN/100 FE-SEM (see the advertisement of *Analytic Chemistry*, 1974, 44 (12), 60A). Since the early 1970s both the reliability and operability have been greatly improved by technical developments to stabilize electron emission and automatically control the voltage to extract electrons. Furthermore, SEM performance has been drastically increased by shortening the focal length of the objective lens, improving sensitivity of signal detection and through the use of signal discrimination technology. Concomitant with this, applications have also shown remarkable progress in response to these performance improvements, as attested to by the various chapters in this book discussing the wide range of applications to which the FE-SEM is now applied in biological and life science research.

Biological Field Emission Scanning Electron Microscopy, First Edition.
Edited by Roland A. Fleck and Bruno M. Humbel.

In this chapter, the advancement of Hitachi FE-SEM technology and its related applications in the biological field will be described. To help understanding the technical background of these developments, the principle of SEM and general approaches for achieving higher resolution will be reviewed first.

3.2 PRINCIPLE OF SEM AND MECHANISM OF RESOLUTION

3.2.1 Principle of SEM

Figure 3.1 shows the structure of SEM and the principle of image formation (Sato 2014). In SEM, electrons are extracted from an electron source and accelerated to form an electron beam (primary electron beam). Primary electrons are focused on to the sample by the condenser and objective lenses and the imaged area is irradiated by the electron beam in a sequential scanning pattern, controlled using scanning coils. The amount of signal (as electrons) emitted from the sample is displayed as brightness to form an SEM image. Signal electrons forming the SEM image consist mainly of a low energy secondary electron (SE-1) and a high energy backscattered electron (BSE) emitted directly from the sample upon irradiation of a primary electron beam, but additional interactions include a secondary electron (SE-2) excited at the sample surface by BSE escaping out of the sample and the secondary electron (SE-3) generated by BSE colliding with the chamber and/or magnetic pole of the objective lens. Inside the SEM column, there are also a secondary electron (SE-4) and a backscattered electron generated by the primary electron irradiating an objective lens aperture plate, but these electrons are hardly detected due to the structure of the electron optical column and have little influence on the image formation. Figure 3.2 shows the principle of the electron lens used for both the condenser lens and the objective lens. The electron lens consists of a magnet coil and yoke. The magnetic field caused by the magnetic poles (N-pole and S-pole) acts as a convex lens to focus primary electrons. Primary electrons focused by

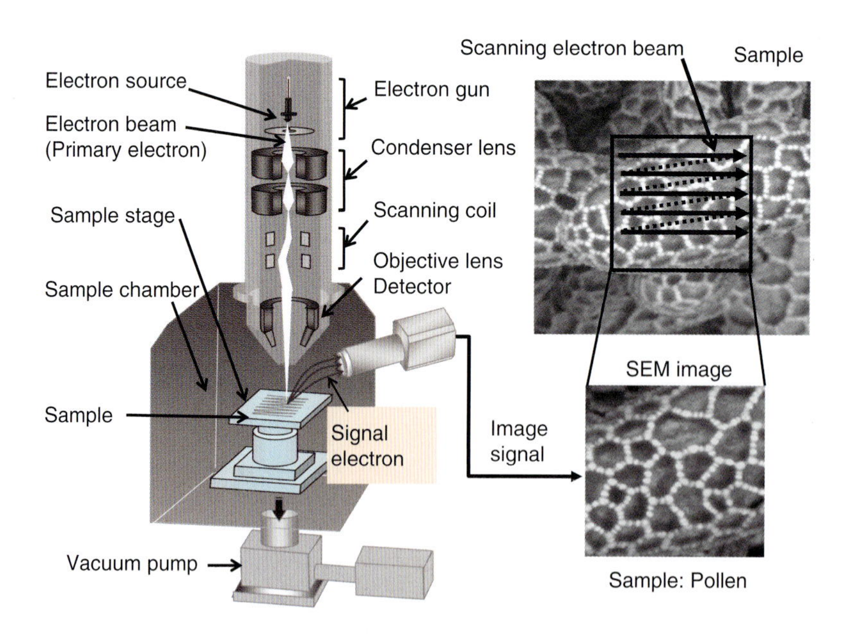

Figure 3.1 Structure of SEM and principle of SEM image formation. From Sato M, Koubunshi, 63, p648–651, 2014 (Sato 2014).

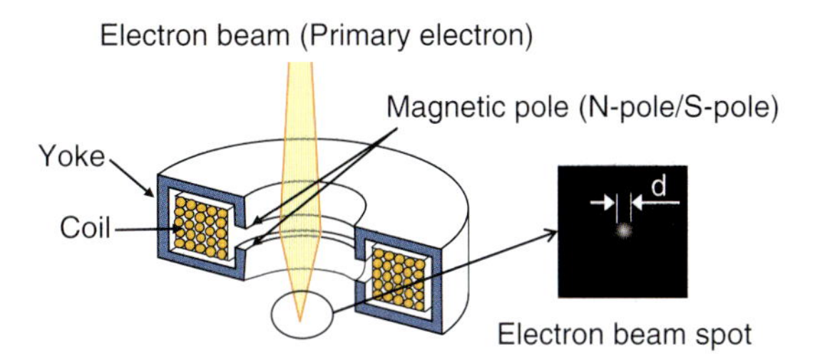

Figure 3.2 Principle of electron lens. From Sato M, Koubunshi, 63, p648–651, 2014 (Sato 2014).

the objective lens are irradiated on to the sample as a spot. The smaller the diameter of the spot (*d*), the higher is the resolution.

3.2.2 Mechanism of SEM Resolution

3.2.2.1 Resolution and Convergent Angle

Figure 3.3 shows the relationship between the primary electron convergence angle, α, and resolution, R. Resolution R shows a V-shape variation against α and α (= α_{opt}) at which R is minimized (= R_{min}) is the operating point of SEM. The V-shape characteristic shown in Figure 3.3 is formed of R_1, the component that becomes smaller with an increase in α, and R_2, the component that becomes larger with an increase in α. R1 is determined by the brightness of the electron source and wavelength of electron λ ($\propto V_{acc}^{-1/2}$, where V_{acc} is the accelerating voltage). With an increase in the brightness, R_1 shifts downward and finally reaches the diffraction limit ($0.61\lambda/\alpha$). At the diffraction limit, resolution can be improved by increasing the accelerating voltage, resulting in a shorter wavelength. R_2 depends on the characteristics of the electron lens. For example, the electron beam trajectory accurately focuses on one point in the central part of the lens, as shown in Figure 3.4a. When α is increased to expand the focusing area to the outer part of the lens, as shown in Figure 3.4b, the outer trajectory is defocused, resulting in a larger spot. This phenomenon is known as spherical aberration. In order to reduce this aberration, α needs to be increased only in the central part of the lens, which can be helped by shortening the focal length as shown in Figure 3.4c.

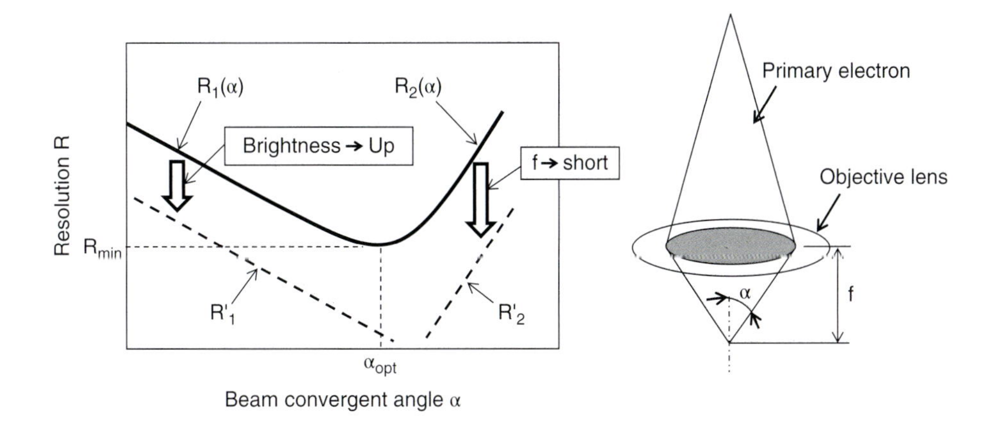

Figure 3.3 Relationship between beam convergence angle and resolution.

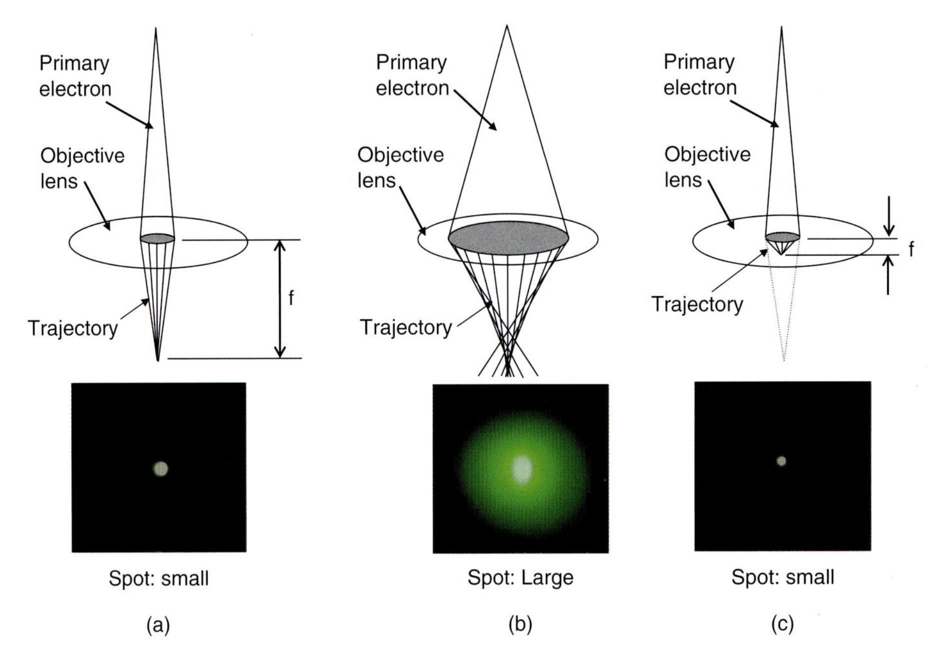

Figure 3.4 Focusing area of objective lens and spot size on sample. (a) Trajectory with small focusing area. (b) Trajectory with large focusing area. (c) Trajectory with shortened focal length.

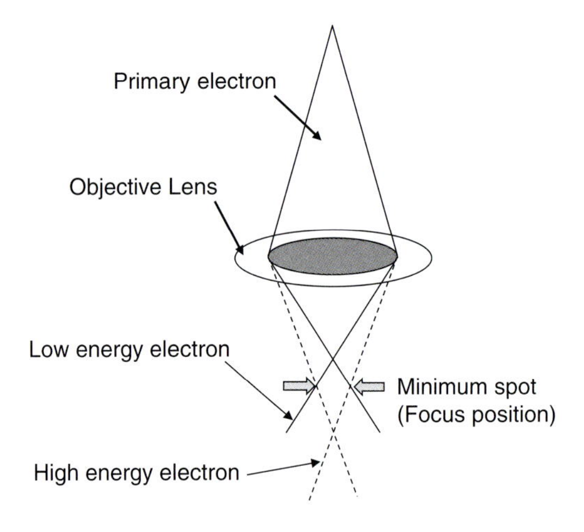

Figure 3.5 Primary electron of different energies and their focusing trajectories.

There is another factor causing degradation of resolution. That is the energy spread of primary electrons (ΔV) determined by the electron source. As shown in Figure 3.5, electrons of different energy focus at different points. This phenomenon is referred to as chromatic aberration. Broadening of the spot is proportional to $(\Delta V/V_{acc})f\alpha$. The lower the accelerating voltage, the larger is the chromatic aberration. Chromatic aberration can be reduced by using the FE electron source with small ΔV and shortening the focal length (f) of the objective lens.

3.2.2.2 Objective Lens Systems and Detector Configuration

Figure 3.6 shows typical objective lens systems and detector geometry. Figure 3.6a shows an out-lens system used for incipient SEM. In an out-lens system, the detector is located below the objective lens. Figure 3.6b shows an in-lens system in which a small sample is located between the magnetic poles (in the magnetic field) to enable a short focal length. Figure 3.6c shows a semi in-lens system (also called a snorkel lens system). In a semi in-lens system, a magnetic field is generated below the objective lens and a short focal length similar to the in-lens system is available for a large sample located beneath the objective lens. Since the sample is located in the magnetic field in both the in-lens system and semi in-lens system, most of the electron signal is attracted by the magnetic field and travels upwards beyond the objective lens. Therefore, the main detectors are located above the objective lens. In the case of the semi in-lens system, a detector is also located below the objective lens and is used when the distance between the objective lens and the sample (working distance, WD) is long.

3.2.2.3 Resolution Characteristic of Each System

Figure 3.7 illustrates resolution characteristics for each combination of electron source and objective lens system. Line (a) shows the diffraction limits of the electron, curves (b) reflect

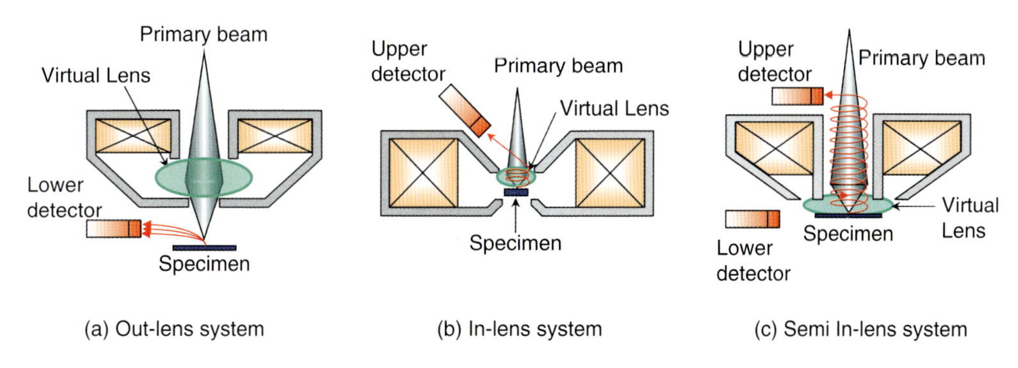

Figure 3.6 Objective lens system and detector geometry.

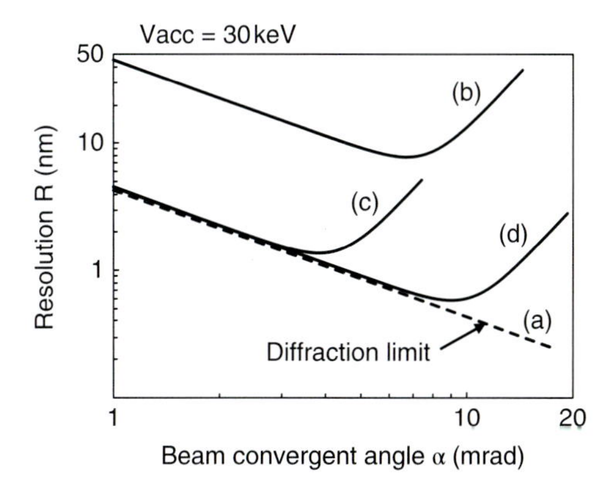

Figure 3.7 Resolution characteristic of each SEM. (a) Diffraction limit of electron. (b) W-filament and out-lens system. (c) FE electron source and out-lens system. (d) FE electron source and in-lens system.

the combination of W-filament and out-lens system, curve (c) reflects the FE electron source and out-lens system and curve (d) reflects the FE electron source and in-lens system. Historically, resolution of the SEM was improved from (b) to (c) to (d). Subsequent to this, α_{opt} has been increased to achieve higher resolution in a number of diverse ways.

3.3 COMMERCIALIZATION OF FE-SEM AND THE IMPACT OF ITS APPLICATION

3.3.1 Commercialization of FE-SEM

Figure 3.8 shows typical electron sources for SEM. With a W-filament electron source, thermionic electrons are emitted from a heated filament. On the other hand, the cold emission FE electron source operates at room temperature. An intense electric field is generated at the end of the sharp cathode to emit electrons by a tunneling effect. In the FE electron source, electrons are extracted by the intense electric field and provide about a thousand times higher brightness than the W-filament. With this high brightness, the resolution of the SEM reaches the diffraction limit determined by the wavelength of electrons. The diffraction limit is calculated as $0.61\sqrt{(1.5/V_{acc})}/\alpha$, where α is the optimum convergence angle of the beam determined by inherent aberrations of the objective lens and the acceleration voltage, etc. If optimum α is 0.01 rad at an accelerating voltage of 30 keV, the diffraction limit is 0.43 nm. In addition, the cathode operates at room temperature, resulting in a very small energy spread (ΔV) of 0.2–0.3 eV, which is about one-tenth that of the W-filament. The high brightness and small energy spread of the FE electron source improves the resolution of SEM tremendously.

Hitachi Ltd developed the first commercial FE-SEM, model HFS-2, in 1972 with the technical guidance of Dr A.V. Crewe *et al.* HFS-2 drastically improved the resolution of SEM from around 10 nm to 3 nm (Komoda and Saito 1972). However, the early FE electron gun had a problem in maintaining stable electron emission over a prolonged period of time. Field emission is, in a manner, a discharging phenomenon in vacuum and to sustain stable discharge, an ultrahigh vacuum of 10^{-8} Pa is required. Figure 3.9 shows the structure of the FE electron gun. Extraction voltage (V_1) is applied between the cathode and anode to cause electron emission. The biggest challenge was that electrons emitted out of the cathode hit the anode to knock out gas molecules. These gas molecules degrade the vacuum around the cathode, resulting in unstable electron emission. At the time, there was insufficient technology to prevent outgassing from the anode. A decade of research followed

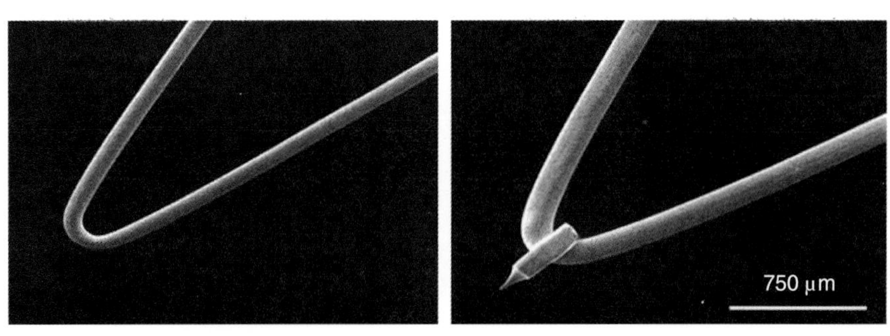

Figure 3.8 Electron source (cathode) for SEM.

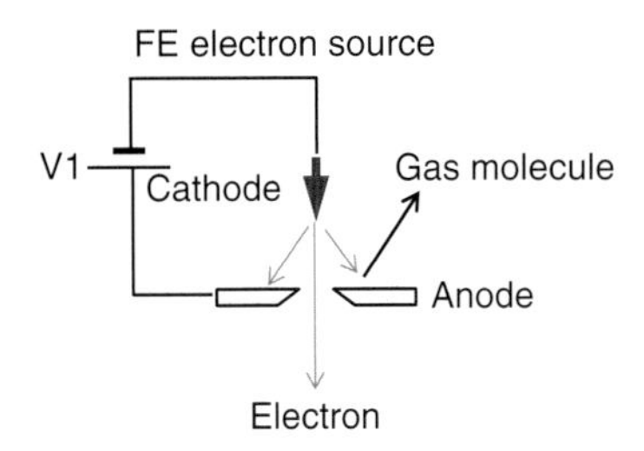

Figure 3.9 Structure of FE electron gun and out-gassing from anode.

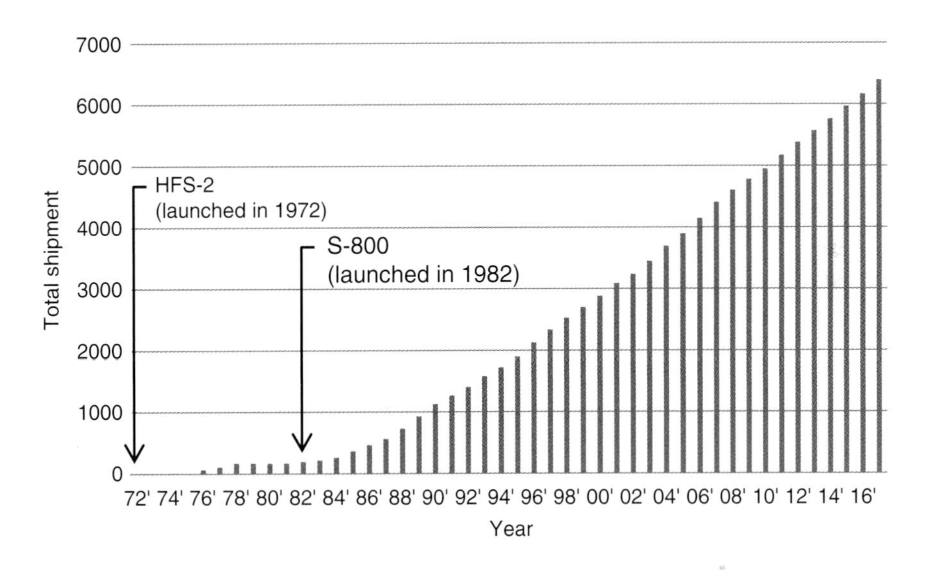

Figure 3.10 Shipment record of Hitachi FE-SEM (1971~2017).

the development of HFS-2, resulting in a new technology to directly heat the anode inside the electron gun and succeeded in significantly reducing outgassing from the anode. In 1982 this technology was adopted in the S-800 FE-SEM (2 nm resolution at 30 keV) to substantially improve the stability and reliability of FE operation (electron emission). In addition, an MPU (microprocessor unit) was employed on the S-800 to automate the delicate adjustment of extraction voltage (V_1), resulting in quantum improvement in operability (Saito *et al.* 1982). Figure 3.10 shows the shipment record of Hitachi FE-SEM after development of HFS-2. This clearly indicates rapid market penetration of FE-SEM in the wake of the S-800 launch. FE technology has brought about substantial results to date, and the original development of HFS-2 was certified as an IEEE (Institute of the Electrical and Electronics Engineers, Inc.) milestone in 2012.

3.3.2 Impact of FE-SEM on Applications

Practical applications of the FE electron source have greatly improved the resolution of SEM. Figure 3.11 shows images of W-filament SEM and those of FE-SEM (both with an out-lens system). While there is not much difference between the two at a magnification of 10 000 times, the difference becomes obvious at a magnification of 100 000 times, reflecting

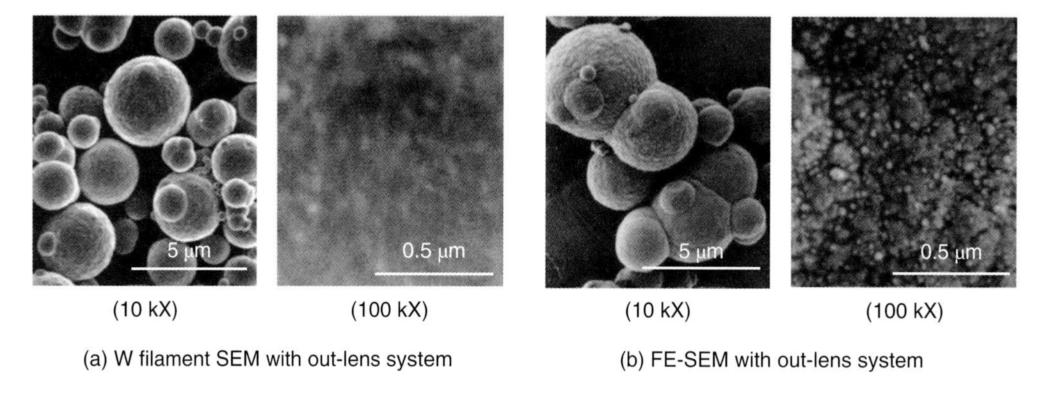

Figure 3.11 Resolution difference between W-filament SEM and FE-SEM.

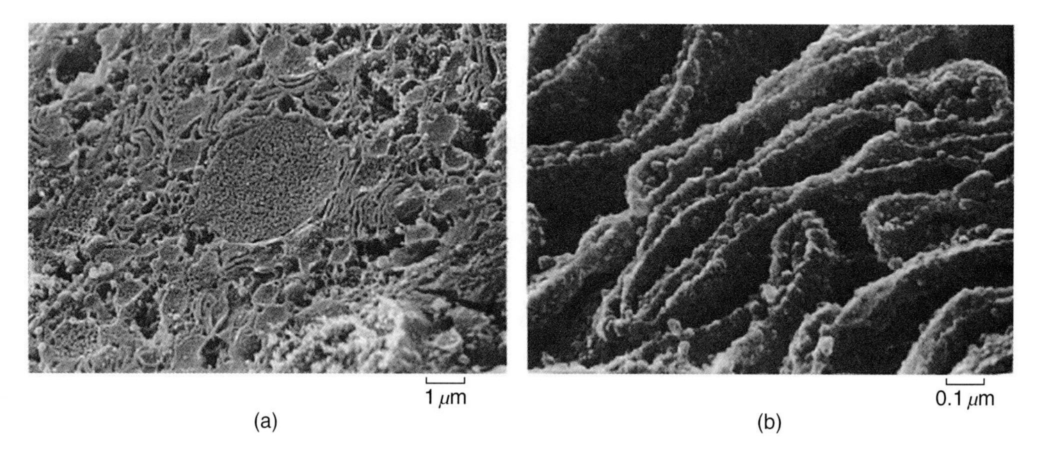

Figure 3.12 SEM image of freeze-fractured liver cell taken with FE-SEM (Hitachi HFS-2). Accelerating voltage: 25 keV, tilting angle: 45°. Images courtesy of Emer. Prof. Keiichi Tanaka, Tottori University, Japan.

the difference in resolution. Although the practical limit of SEM magnification had been around 20 000 times until FE-SEM was introduced, FE-SEM provides sharp images even at a magnification of 100 000 times and the morphology of organisms, which was blurred due to insufficient resolution, has been continuously unveiled (Tanaka 1981, 1989b; Tanaka and Mitsushima 1989). Figure 3.12 shows an example FE-SEM image of a freeze-fractured liver cell. A cross-section of a round nucleus can be identified in the central part of Figure 3.12a. Also, a granular structure deemed to be a number of ribosomes was confirmed in the surrounding endoplasmic reticulum (ER) (Figure 3.12b).

3.4 DEVELOPMENT OF IN-LENS FE-SEM AND ITS IMPACT

3.4.1 Development of the In-Lens FESEM

The development of S-800 has brought SEM resolution to 2 nm, but further improvement of resolution was considered theoretically impossible at that time (around 1980) because of primary electrons scattering inside the sample causing emissions of signal electrons from a large area of the sample. On the other hand, Koike *et al.* incorporated an SEM function in the TEM (transmission electron microscope) in 1970 and the SEM images showing much higher resolution than conventional SEM attracted a great deal of interest (Koike 1976; Koike *et al.* 1971). While the objective lens of TEM is an in-lens system with a short focal length, the SEM of the day employed an out-lens system with a long focal length, resulting in a large spherical aberration. Koike pointed out that the spherical aberration of SEM is a major limiting factor for resolution. Prof. Tanaka, School of Medicine, Tottori University, looked at the experiment of Koike *et al.* and considered it possible to exceed the resolution of 2 nm, previously thought to be the limit of SEM, by combining an in-lens system and an FE electron source. He proposed the instrument development be given to Hitachi in 1982 and the development of the in-lens FE-SEM started in 1983 with the participation of TEM engineers and SEM engineers.

In order to realize ultrahigh resolution SEM, it is essential not only to reduce the spot size of the primary electron beam but also to prevent contamination during observation at a high magnification. Residual hydrocarbon gas molecules in the specimen chamber adhere to the sample by electron beam irradiation and prevent faithful observation of sample surface. This phenomenon is referred to as contamination. Figure 3.13a shows an example of contamination. Clean vacuum with little hydrocarbon gas molecules is mandatory to prevent contamination and obtain a faithful image of the sample surface, as shown in Figure 3.13b. Since the vacuum pump technology of the day (oil-diffusion pump) used oil, it was difficult to achieve a clean vacuum. By good fortune, the oil-free turbo molecular pump (TMP) was developed by Seiko Seiki Co., Ltd around this time. Hitachi decided to employ this oil-free TMP for the in-lens FESEM. The first TMP was developed by Pfeiffer Co. in 1958, but this was not oil-free. Subsequently, Leybold Co. developed an oil-free TMP with magnetic suspension in 1976 (Sawada 1995). It also achieved low vibration as a result of the magnetic suspension design. The introduction of this oil-free TMP resulted in clean vacuum and enabled stable ultrahigh resolution imaging with little contamination. The first functional in-lens FE-SEM was completed in 1985 and the world's highest resolution of 0.5 nm was achieved (Tanaka 1989a, 2008, 2009a,b). The instrument was delivered to Tottori University and named UHS-T1 (ultra-high resolution SEM Tottori 1).

After eagerly waiting for the completion, Tanaka utilized UHS-T1 to observe the human immunodeficiency virus (HIV) from which infection led to acquired immune deficiency syndrome (AIDS) HIV virus clearly resolved by FE-SEM (Tanaka 1989c). As HIV had become a major societal issue at that time, this news rapidly spread around the world. He also imaged bacteriophage (T2) with vivid clarity in 1988. Figure 3.14a shows the feet of bacteriophage clearly imaged by UHS-T1, which could not be observed using conventional

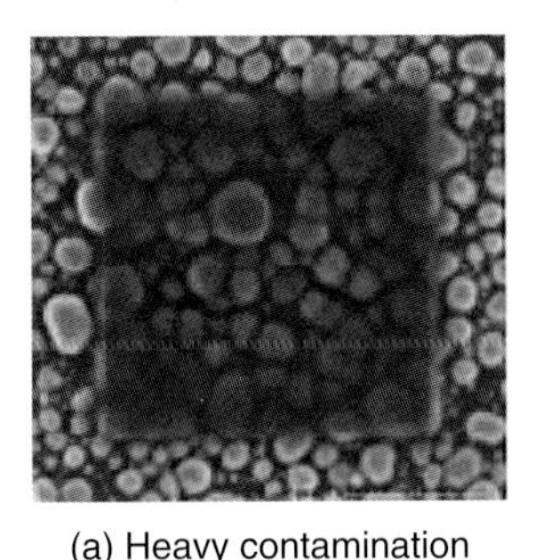
(a) Heavy contamination

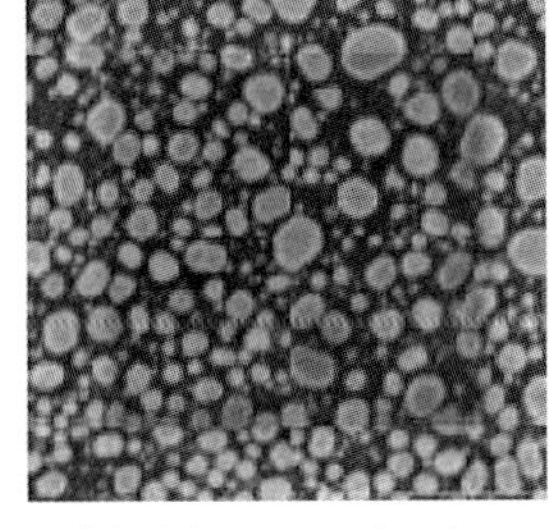
(b) Light contamination

Figure 3.13 Contamination on sample surface (sample: gold particles on carbon).

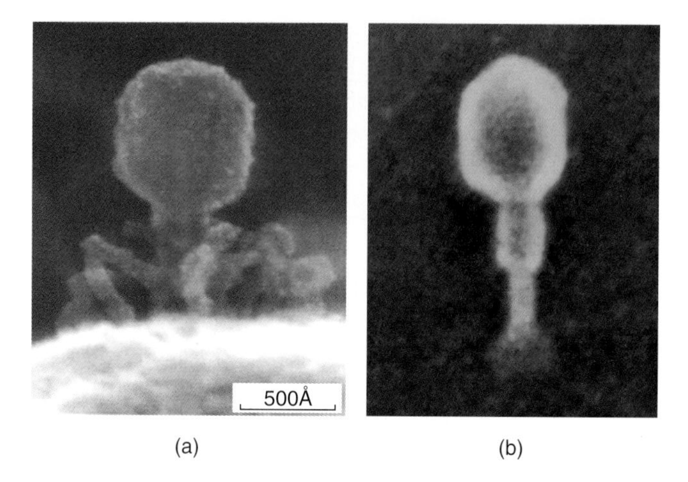

Figure 3.14 SEM images of T_2 bacteriophage (a) taken with in-lens FE-SEM (Hitachi UHS-T1) and (b) taken with conventional FE-SEM (Hitachi HFS-2). From Inada *et al.*, Hitachi's development of cold-field emission scanning transmission electron microscope, Advances in imaging and electron physics, Academic press, 2009, vol 159, pp123–186, Copyright (2017), with permission from Elsevier. Images courtesy of Emer. Prof. Keiichi Tanaka, Tottori University, Japan, and the late Prof. Albert Victor Crewe, Chicago Univ., US.

FE-SEM (Figure 3.14b) (Tanaka 1989c; Nakadera *et al.* 1991; Tanaka and Yamagata 1992; Inada *et al.* 2009). Technologies developed for UHS-T1 were employed in commercialized in-lens FE-SEM, the S-900, in 1986 (Nagatani and Sato 1986; Nagatani *et al.* 1987).

3.4.2 Adoption of the High Sensitivity Backscatter Electron Detector (YAG Detector)

The S-900 also employed a high sensitivity backscattered detector with YAG (yttrium aluminum garnet). This detector was developed by Autrata (Autrata *et al.* 1978; Autrata 1992) and provided high resolution backscattered electron image in combination with in-lens FE-SEM. Taking advantage of this performance, Walther and Müller *et al.* at the Federal Institute of Technology, Zurich (ETH) developed a method to observe a freeze fracture face of a high pressure frozen sample (Walther *et al.* 1995; Walther and Müller 1997). Figure 3.15 shows a YAG-BSE image of high pressure freeze fractured rat pancreas tissue. The fractured surface of membrane is clearly observed at an accelerating voltage of 10 keV and this structure was confirmed to be consistent with a freeze replica image by TEM (Walther 2003).

3.4.3 Expansion of a Low Accelerating Voltage Application

Prior to the widespread use of FE-SEM, samples were usually coated with evaporated metals to prevent charging caused by electron beam irradiation and observed at an accelerating voltage of around 20 keV. At a high accelerating voltage, however, primary electrons penetrate deep beneath the surface of the sample and the contrast of surface morphology is degraded. Low accelerating voltage below several kilovolts is required to observe surface morphology with good contrast. Figure 3.16 shows secondary electron images of a diatom imaged at an accelerating voltage of 15 keV and 1 keV. Surface structure

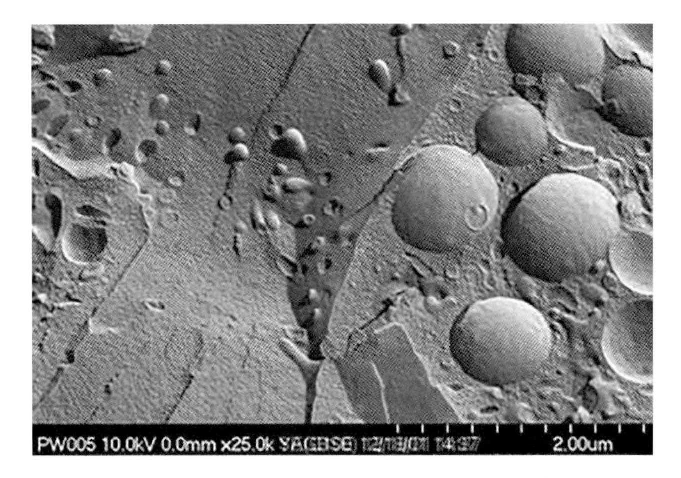

Figure 3.15 YAG-BSE image in the cryo-SEM of a high pressure frozen tissue. Image courtesy of Prof. Paul Walther, Ulm University, Germany.

Figure 3.16 Comparison of contrast between different accelerating voltages. (Sample: Diatom).

located in the center of the diatom is transparent and the structure cannot be observed at an accelerating voltage of 15 keV (Figure 3.16a) whereas the micromorphology of the diatom surface can be clearly observed at an accelerating voltage of 1 keV (Figure 3.16b). In addition, a low accelerating voltage suppresses charging caused by electron beam irradiation and allows imaging without coating metals. This also gives an advantage of observing a true surface without particle deposition. While chromatic aberration degrades resolution at a low accelerating voltage, in-lens FE-SEM ensures sufficient resolution even at a low accelerating voltage. Prof. Osumi of Japan Women's University looked at the performance of in-lens FE-SEM at a low accelerating voltage and proposed the development of S-900LV (LV, low voltage), which has improved performance at a low accelerating voltage with a shorter focal length than the S-900 (Nagatani *et al.* 1990; Sato *et al.* 1990). The S-900LV (commercial model SEM S-900H) enabled sub-2 nm resolution at a low accelerating voltage of 3 keV to allow faithful observation of the cell surface (Nagatani *et al.* 1990; Sato *et al.* 1989, 1990). Osumi utilized S-900LV to reveal

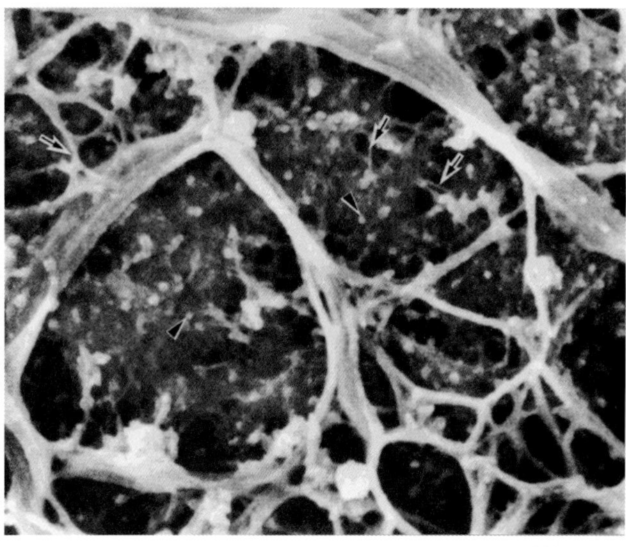

Figure 3.17 Uncoated surface structure of regenerated protoplast in fission yeast revealed by in-lens FE-SEM (Hitachi S-900LV). From Osumi, M. *et al.*, Ultrahigh resolution low-voltage SEM reveals ultrastructure of the glucan network formation from fission yeast protoplast, *J. Electron Microsc.*, 1995, 44,198–206. Copyright (2015), with permission from Oxford University Press (Japanese Society of Microscopy). Images courtesy of Emer. Prof. Masako Osumi, Japan Women's Univ., Japan.

the cell-wall constituent regenerated on the protoplast surface of fission yeast without coating (Figure 3.17 and see Chapter 16 by Osumi in Volume II). The cell-wall component, a glucan network regenerated on the protoplast surface, was clearly observed in Figure 3.17. The arrow indicates the microfilament constituting this fiber. This result from SEM images correlates well with the TEM images, in which there are microfilaments 2 nm in diameter on the protoplast surface and the filaments are considered to be synthesized twisting and branching (Naito *et al.* 1991). Also, the three-dimensional conformation of the glucan fiber as a cell wall skeleton located on the surface was revealed by this result (Osumi *et al.* 1995). This technique was applied to analyze mutant cells related by synthesis of cell wall components, and its roles were investigated (Konomi *et al.* 2003).

3.5 INTRODUCTION OF SEMI IN-LENS FE-SEM

3.5.1 Objective Lens of the Semi In-Lens System

In the development of in-lens FE-SEM, ultrahigh resolution imaging was attained by limiting sample size. However, there was a growing demand in the semiconductor industry to observe patterns printed on a 6-inch silicon wafer at the same resolution attainable with in-lens systems; the semi in-lens system objective lens has addressed this need. The prototype of a semi in-lens was proposed by Mulvey *et al.* in 1973 under the name of snorkel lens (Mulvey and Newman 1973), but it was not commercialized at that time. The first commercial product with a snorkel lens system was model S-6100 CDSEM (critical dimension SEM) (Otaka *et al.* 1989), launched in 1989. S-6100 was a dedicated instrument to inspect the pattern dimension of semiconductor devices and employed a snorkel lens system to adapt to shrinking semiconductor devices. The snorkel lens system indicated in Figure 3.6c was then applied to FE-SEM and model S-4500 was developed in 1992 (Sato *et al.* 1993, 1994). The snorkel lens system has allowed ultrahigh resolution imaging of a large sample at the same resolution as an in-lens system and dramatically expanded ultrahigh resolution imaging applications. With the snorkel lens system, the sample is located in the magnetic field of the objective lens. As the in-lens system already existed, the term semi in-lens system has become widely used.

3.5.2 Advance of the Signal Detection System

3.5.2.1 Improved Sensitivity of the Signal Detection System

In in-lens and semi in-lens systems, secondary electrons emitted from the sample are attracted by the magnetic field and travel upwards beyond the objective lens and thus the secondary electron detector is located above the objective lens. Figure 3.18 shows the structure of a secondary electron detector and the path of the electron. This detector is named an Everhart–Thornley (ET) detector after its two inventors (Everhart and Thornley 1960). The ET detector accelerates secondary electrons to collide with a scintillator and the resultant photons are detected by a photomultiplier. However, since a high voltage of 10 keV is applied to the scintillator to accelerate the secondary electrons, the path of the primary electrons is changed by the electric field. The ET detector is located above the objective lens in the in-lens and semi in-lens systems and the change in the path deflects the primary electron from the center of the objective lens, resulting in degradation of resolution. Thus in early in-lens SEM, there was a trade-off between the effect on the primary electron (resolution degradation) and detection efficiency of the secondary electron and it was necessary to suppress some of the signal detection efficiency. It was the development of the ExB (E cross B) system that solved this problem by applying electric field (E) and magnetic field (B) orthogonally to control the trajectory of the primary electrons and secondary electrons (Sato *et al.* 1993). The original concept was invented by Kuroda of Hitachi Central Research Laboratory *et al.* in 1993 (Kuroda *et al.* 1993). Then the system was improved to control the primary electron trajectory more precisely (Sato *et al.* 2000) and was employed on model S-4500 FE-SEM. Figure 3.19 shows the structure of the ExB system and electron trajectory. A force due to the electric field and that due to the magnetic field are balanced out to direct the primary electrons straight through the ExB system. Secondary electrons moving in a direction opposite to the primary electrons are

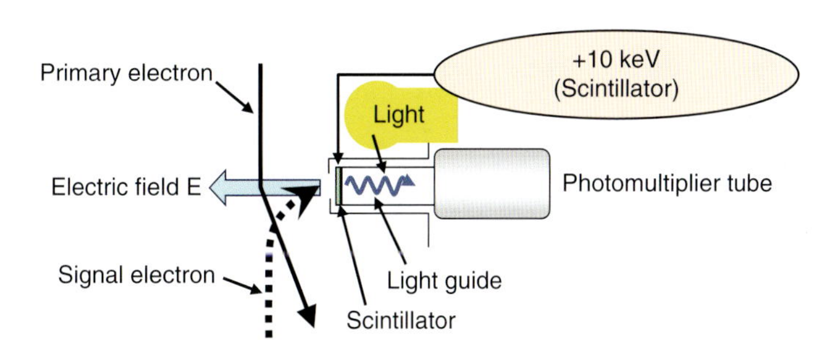

Figure 3.18 Structure of Everhart–Thornley detector and trajectory of electrons. From Sato M, Koubunshi, 63, p648-651, 2014 (Sato 2014).

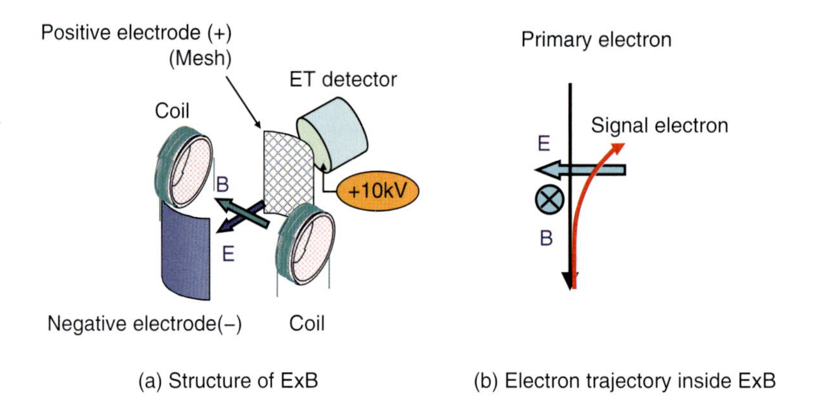

Figure 3.19 Structure of ExB and trajectory of electron passing through ExB. From Sato M, Koubunshi, 63, p648–651, 2014 (Sato 2014).

attracted strongly toward the detector by the force due to the magnetic field, which acts oppositely from the primary electron, according to the Lorentz force law. By this method, high sensitivity secondary electron detection was established by the ExB system without affecting the primary electron trajectory.

3.5.2.2 Evolution of Signal Discrimination

The ExB system has allowed high-efficiency detection of a low energy secondary electron to provide sharp images with a high signal-to-noise ratio. On the other hand, further investigation revealed that backscattered electrons with energy equivalent to that of primary electrons provides individual contrast depending on the ejection angle. A new technology was developed to separate and individually detect secondary electron (SE) and backscattered electrons with different ejection angles (BSE-H with a high angle, BSE-L with a low angle), as shown in Figure 3.20. The technology was employed on model S-5200 in-lens FE-SEM in 2000 (Sato *et al.* 2001) and enabled high sensitivity detection of backscattered electrons at a low accelerating voltage, which previously had proven very difficult. Also, it has a beneficial effect on avoiding the negative impact of charging. Figure 3.21 shows SEM images of each signal at an accelerating voltage of 1.5 keV, where the sample is ceramic with nickel particles. Figure 3.21a, b, and c represent SE, BSE-L, and BSE-H, respectively, and these signals are detected above the objective lens, while Figure 3.21d represents a signal

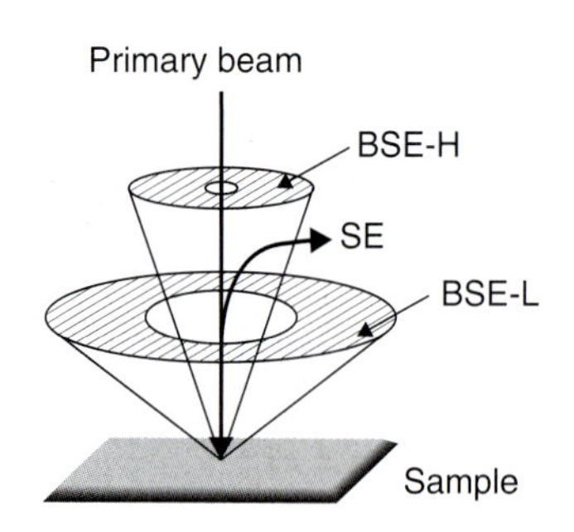

Figure 3.20 Various signals available on SEM with signal discrimination in-lens FE-SEM (Hitachi S-5200).

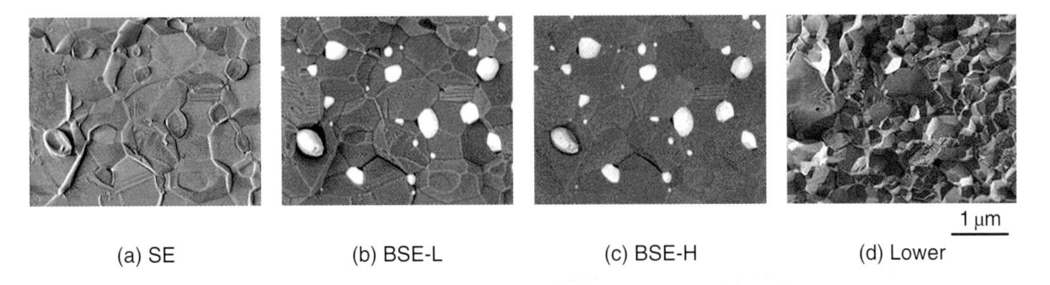

Figure 3.21 Contrast difference by signal discrimination (sample: Al_2O_3/Ni ceramic composites, accelerating voltage: 1.5 keV). From Sato M, Koubunshi, 63, p648–651, 2014 (Sato 2014). Images courtesy of Dr Tohru Sekino, ISIR, Osaka University, Japan.

detected by the lower detector in the semi in-lens system. In the semi in-lens system, most of the secondary electrons are detected above the objective lens and the backscattered electrons ejected at a low angle are the dominant signal for the lower detector. The contrast varies with different detected signal electrons for the same sample indicated in Figure 3.21. The SE signal (Figure 3.21a) is influenced by charging to give an impression that surface topography has collapsed. The BSE-L signal (Figure 3.21b) shows emphasized surface topography and bright nickel particles with a strong compositional contrast. The BSE-H signal (Figure 3.21c) also shows strong compositional contrast of nickel particles but less surface topography. A lower detector image (Figure 3.21d) shows strong surface topography but little compositional contrast of nickel particles. In this way, the detected signal can select information to be emphasized or suppressed. Such signal discrimination technology has also been employed on semi in-lens FE-SEM. In addition to the difference in ejection angle, nowadays signals can be discriminated by energy difference, and precisely optimized information can be obtained from a sample according to the application.

3.5.3 Applications

Improved sensitivity of signal detection and signal discrimination technology have enabled low damage observation with less primary electron irradiation and contrast optimization by signal control of the secondary and backscattered electrons. The following images show applications utilizing these signal detection technologies.

Figure 3.22 shows a secondary electron image of a V-shaped stereocilia bundle on a sensory hair cell taken with a semi in-lens FE-SEM (S-4800). High-efficiency secondary electron detection technology reduced charging and electron beam damage to allow observation of three rows of the stereocilia (Figure 3.22a). Horizontal tip links between stereocilia are clearly imaged at a higher magnification of 150 000 times (Figure 3.22b) (Rzadzinska and Jones 2006; Rzadzinska *et al.* 2004; Mogensen *et al.* 2007; Hertzano *et al.* 2008).

Low damage observation at low accelerating voltage is also beneficial for cryo electron microscopy. Figure 3.23 shows cryo SEM images of a mint leaf taken with semi in-lens FE-SEM (S-4700) with Leica EM VCT100 system. Trichomes and essential oil-accumulating granular trichomes of a mint leaf were observed at an accelerating voltage of 0.5 keV at −125 °C (Samuels and McFarlane 2012).

Figure 3.24 shows a high resolution cryo SEM image of an actin filament taken with an in-lens FE-SEM (S-5200). The observed actin fiber agrees rather well with the model obtained by cryo TEM data after averaging (McGough *et al.* 1997) and the native structure of one helical filament about 10 nm in width was clearly imaged (Walther 2008).

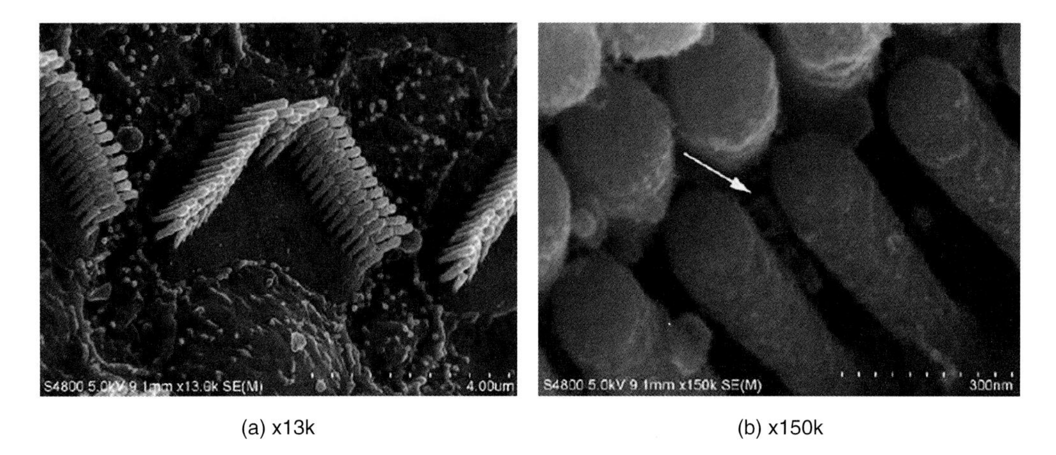

Figure 3.22 V-shaped Stereocilia bundle on the apical surface of the normal outer hair at postnatal day 21 taken with semi in-lens FE-SEM (Hitachi S-4800) at an accelerating voltage of 5 keV. From Razadzinska, A. and Jones, C., Microscopy and the genetics of deafness, *Hitachi E.M. News*, 1, 19–21, 2006 (Rzadzinska *et al.* 2004) and Hertzano R et al., PLoS Genet, 4(10): e1000207, 2008 (Hertzano *et al.* 2008). Images courtesy of Dr Agnieszka Rzadzinska-Prosser, The Wellcome Trust Sanger Institute, UK. (Current address: Sawston medical center, 10 London RD, Swanston)

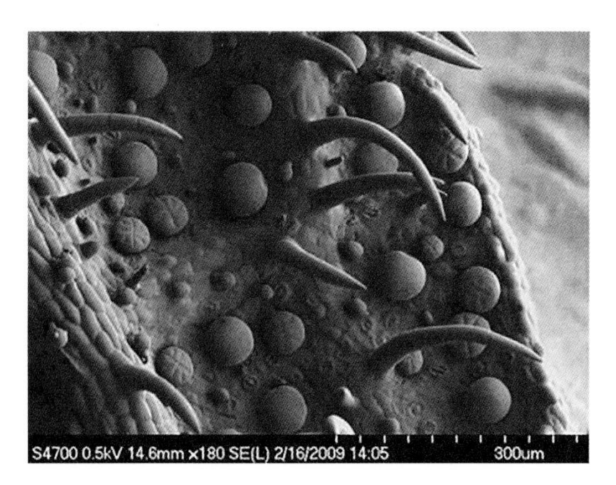

Figure 3.23 Cryo-SEM of trichomes and essential oil-accumulating glandular trichomes of a mint leaf. Mounted on a Leica EM VCT100 sample stage. Frozen and fractured with the Leica EM VCT100 loading box. Transferred with the Leica EM VCT100 shuttle to a semi in-lens FE-SEM (Hitachi S-4700) with a EM VCT100 Cryo-Stage. Images courtesy of Dr. Lacey Samuels and Dr. Heather E. McFarlane, Dept. Botany, Univ. British Columbia, Canada.

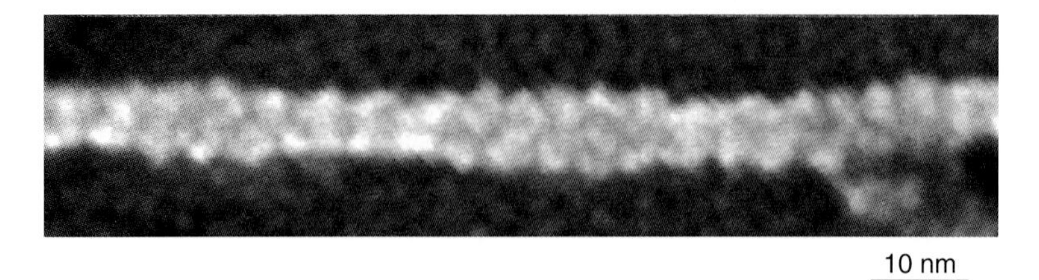

Figure 3.24 Cryo-SEM of a single actin filament taken with in-lens FE-SEM (Hitachi S-5200). Image courtesy of Prof. Paul Walther, University of Ulm, Germany.

Figure 3.25 shows immuno-gold stained SEM images of *Staphylococcus aureus*. High sensitive backscattered electron detection technology to convert secondary electrons at low accelerating voltage also allowed imaging of fine gold particles on a sample surface with sharp contrast (Yamaguchi *et al.* 2010) the same as imaging with YAG scintillators (Erlandsen *et al.* 2003).

Figure 3.26 shows a backscattered electron image of a resin-embedded section with inverted contrast. The section of rat cerebellar cortex was prepared by rapid freezing-freeze substitution and collected on a glass slide (Tohyama *et al.*, unpublished data). The visualized result is highly consistent with TEM observation. Nucleus, mitochondria, nerve axon and vesicles are identified and the membrane structure can be clearly confirmed (Dan *et al.* 2013).

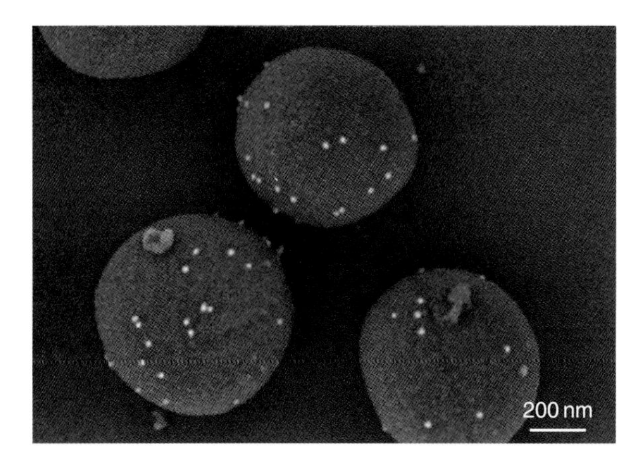

Figure 3.25 Immuno-gold stained SEM of *Staphylococcus aureus* taken with semi in-lens FE-SEM (Hitachi SU8020) at an accelerating voltage of 1.5 keV, LABSE+SE. Image courtesy of Dr Masashi Yamaguchi, Chiba University, Japan.

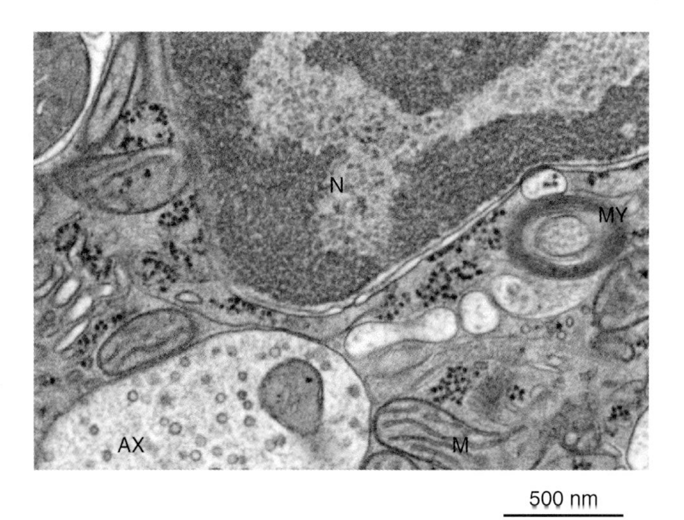

Figure 3.26 Ultrathin section of rat cerebellar cortex substituted with osmium tetroxide aceton after He rapid freezing on glass slide taken with semi in-lens FE-SEM (Hitachi SU8020) at an accelerating voltage of 1.5 keV, LABSE N, nucleus; M, mitochondrion; MY, myelin; AX, axon. From Dan et al., J. Electr. Microsc. Technol. Med. Biol., 27(2): 94, 2013 (Dan *et al.* 2013). Image courtesy of Mr Kinji Ishida and Dr Koujiro Tohyama, Iwate Medical University, Japan.

3.6 RESOLUTION IMPROVEMENT BY THE DECELERATION METHOD

3.6.1 Retarding Method and Boosting Method

There is a technique called beam deceleration to decelerate primary electrons before interaction with the sample. There are two kinds of deceleration method: the retarding method (Ezumi *et al.* 1996; Pease 1969) and the boosting method (Frosien *et al.* 1989; Zach 1989). Both methods improve resolution at a low accelerating voltage and enable ultralow accelerating voltage imaging below 100 eV.

Figure 3.27a shows the principle of the retarding method. In the retarding method, decelerating voltage ($-V_d$) is applied on the sample to decelerate the primary electron. For instance, voltage of the primary electron irradiated on the sample (landing voltage, *Vi*) is 100 eV ($Vi = V_{acc} - V_d = 100$ eV) where the accelerating voltage of the electron gun is 2 keV ($V_{acc} = 2$ keV) and the decelerating voltage applied to the sample is −1.9 keV ($-V_d$=−1.9 keV). Here the sample needs to be flat so as not to disturb the electric field between the sample and the objective lens.

Figure 3.27b shows the principle of the boosting method. In the boosting method, the primary electrons accelerated in the electron gun are further accelerated by an electrode located in the electron path. Then the electron is decelerated to the original accelerating voltage at the exit of the objective lens and irradiated on the sample. The boosting method is suited to the out-lens system in which the accelerating electrode can be placed in the magnetic field and offers the advantage of having no constraints on sample shape due to no electric field being generated between the sample and the objective lens. On the other hand, as the boosting method is applied to an out-lens system, the resolution is inferior compared to semi in-lens or in-lens systems, especially at a high accelerating voltage where the boosting method is ineffective (Frosien *et al.* 1989; Zach 1989).

3.6.2 Resolution Improvement Effect by the Deceleration Method

The maximum effect of the deceleration method is the reduction of chromatic aberration at a low accelerating voltage. Chromatic aberration is proportional to $\Delta V/V_{acc}$ where the

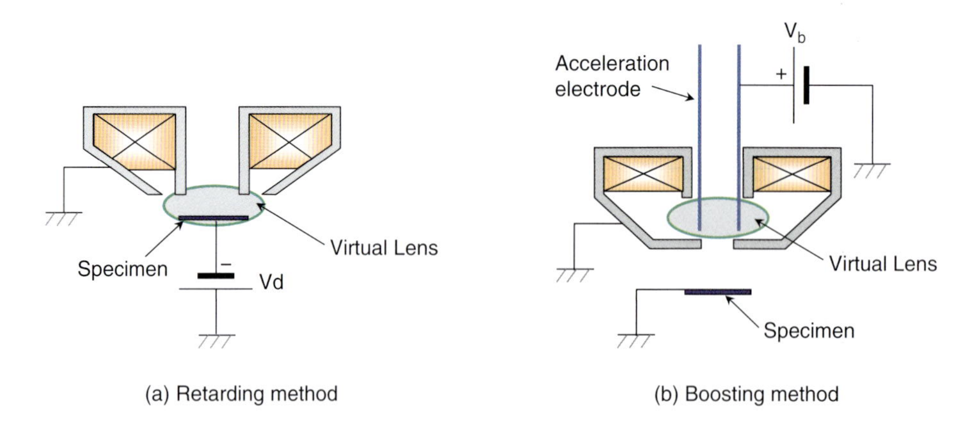

Figure 3.27 Optics for deceleration methods.

(a) Without deceleration
Accelerating voltage : 500 eV

(b) With deceleration by retarding method
Landing voltage : 500 eV

Figure 3.28 Resolution improvement by deceleration technique (sample: mesoporous silica, landing voltage: 500 eV, magnification: ×150 000). From Endo, A. *et al.*, Direct observation of surface structure of mesoporous silica with low acceleration voltage FE-SEM, in *Colloids and Surface A: Physicochemical Engineering Aspects*, 2010, vol. 357, pp. 11–16. Copyright (2015), with permission from Elsevier. Images courtesy of Dr. Akira Endo, AIST, Japan.

accelerating voltage of the primary electron is V_{acc} when it passes through the objective lens (the magnetic field of the objective lens, to be precise). For example, when a primary electron accelerated to 100 eV in the FE electron gun ($\Delta V = 0.3$ eV) focused at the objective lens, $\Delta V/V_{acc}$ is 3×10^{-3} ($\Delta V/V_{acc} = 0.3/100 = 3 \times 10^{-3}$). When a primary electron accelerated to 2 keV is decelerated to 100 eV by the deceleration method ($V_d = 1.9$ keV), then $\Delta V/V_{acc}$ is 1.5×10^{-4} ($\Delta V/V_{acc} \fallingdotseq 0.3/2000 = 1.5\times10^{-4}$), reducing the chromatic aberration to one-twentieth. Whilst the deceleration method has the effect of reducing spherical aberration, the main effect is the reduction of chromatic aberration at a low accelerating voltage.

3.6.3 Expanded Applications with Ultralow Accelerating Voltage

The deceleration method has expanded applications at an ultralow accelerating voltage below 1 keV. The advantage of an ultralow accelerating voltage is to be able to observe samples sensitive to an electron beam or image top-most surface at high resolution. Figure 3.28 shows an example application to mesoporous silica. While the structure of mesoporous silica is blurred at an accelerating voltage of 500 eV without deceleration (Figure 3.28a), the resolution is much improved to resolve micropore below 10 nm at a landing voltage of 500 eV with deceleration by the retarding method (Figure 3.28b) (Endo *et al.* 2010).

3.7 POPULARIZATION OF A SCHOTTKY EMISSION ELECTRON SOURCE AND PROGRESS OF A COLD FE ELECTRIC GUN

3.7.1 Popularization of a Schottky Emission Electron Source

In addition to the aforementioned FE electron source and W-filament electron source, the Schottky emission (SE) electron source is also widely used. The SE electron source was developed by Swanson *et al.* in the 1970s (Swanson and Crouser 1969; Swanson and Martine 1975). The Schottky emission (SE) and field emission (FE) are technical terminology used in the studies of science. However, SE is also called "Schottky FE" and "thermal type FE" and is often confused with the FE emission source. Therefore, the FE electron source, which operates at room temperature, is called cold FE (CFE) to distinguish it from the SE electron source. The SE electron source has a brightness close to the CFE electron source and requires less ultrahigh vacuum than a CFE electron source. Since the SE electron source provides a large beam current and superior current stability, it is better suited for analytical applications requiring a large beam current. On the other hand, the energy spread (ΔV) of the SE electron source is more than double that of the CFE electron source, and hence high resolution imaging performance of the SE electron source at low accelerating voltage is not as good as that of the CFE electron source.

3.7.2 Progress of the CFE Electron Gun

In recent years, the NEG (non-evaporable getter) pump has been incorporated into the electron gun to operate the electron source under extreme high vacuum (10^{-9} Pa) (Kasuya *et al.* 2010). Figure 3.29 shows the temporal change in emission current of the CFE electron source. Absorbed gas molecules on the cathode are removed by flashing (instantaneous heating) to clean the surface of the cathode to ensure stable electron emission. The optimum condition at which the brightness of the electron source is highest and the current fluctuation, so-called "FE noise", is smallest occurs immediately after flashing.

With a conventional CFE electron source, gas molecules begin to be absorbed to the cathode right after flashing, resulting in decay of the emission current (Figure 3.29). SEM observation is conducted only within a stable condition after an emission decay, which means that the high brightness condition that occurs just after flashing is not available for SEM observation due to the large variation of emission current.

By operating the CFE electron source under extreme high vacuum, the high brightness conditions that result after flashing can be maintained for a long time, as shown in Figure 3.29 (dotted line) and a SEM observation can be conducted in this condition with a higher beam current and superior current stability.

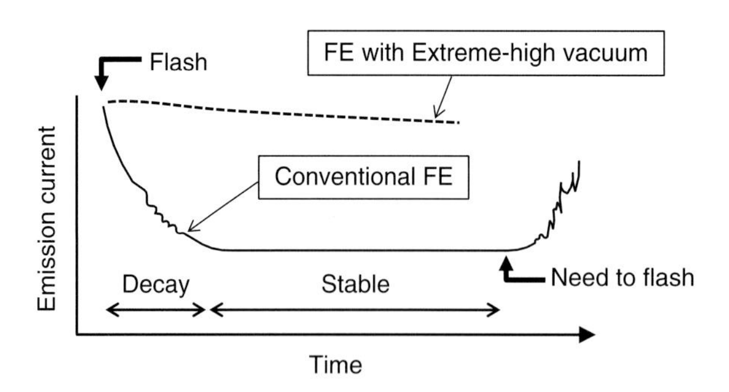

Figure 3.29 Emission current variation on cold FE electron sources.

3.8 ADVENT OF TRULY "EASY TO USE" FE-SEM

3.8.1 Outline

FE-SEM has become widely used as one of the crucial analysis tools and the user base is rapidly expanding. On the other hand, some users are finding difficulty in using FE-SEM effectively. In order to get the maximum out of FE-SEM and obtain SEM images with good contrast and sharpness, appropriate settings such as accelerating voltage, beam current, detector, etc., are essential. Although these operations are not too difficult for expert users, this may not be the case for novice SEM users. A new user interface, EM Wizard, was therefore developed for novice users to allow them to operate the FE-SEM easily and obtain good images without having to be an expert. EM Wizard was introduced on SU5000 Schottky FE-SEM launched in 2014.

3.8.2 No Expertise Necessary

In EM Wizard, the desired image property is selected for the intended use rather than setting each of the SEM conditions (Konishi *et al.* 2013). Figure 3.30 shows the observation mode selecting window. Five typical imaging modes are available, such as standard imaging, surface profile imaging, elemental distribution imaging, etc., and an example image is shown for each mode with characteristic contrast. In addition, the balance of image information such as spatial resolution, surface topography information, elemental information, etc., is displayed as a cobweb chart to visually illustrate what is emphasized

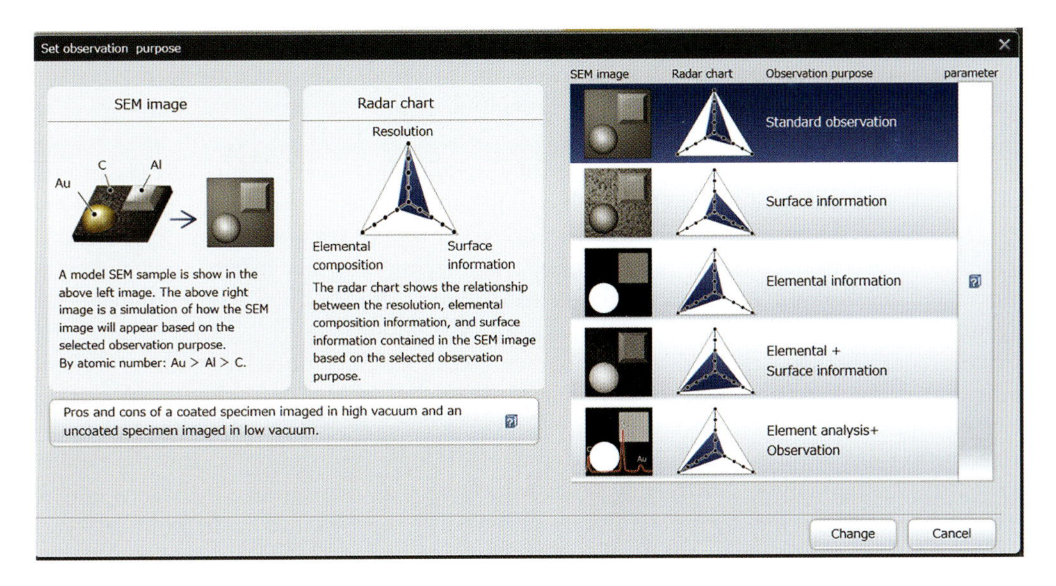

Figure 3.30 Purpose selecting window in EM Wizard.

and what is lost. Expert users operate the SEM with an in-depth understanding of the relationship between image information and the required SEM conditions. EM Wizard automatically sets appropriate SEM conditions for the selected image property, and the optimum SEM conditions can be set without the need for in-depth knowledge of SEM operation.

3.8.3 Simplified Beam Alignment

EM Wizard also provides an automatic precise beam alignment function (Sato *et al.* 2013), auto calibration (Shigeto *et al.* 2014), across all SEM conditions for each intended use. Since sample shape affects the precision of automatic beam alignment, a special dedicated sample is used for reliable alignment. Once the SEM conditions are set, the alignment value is simultaneously set for auto calibration and the beam alignment can be significantly simplified.

3.8.4 Coaching for Skill Improvement

As a result of progress in operability and performance of the SEM, more people are showing an interest in SEM and a desire to attain higher skill levels than ever before. Responding to the situation, a variety of learning support functions such as "Advanced operation guide", "Application assistance", "Astigmatism correction simulator", etc., are available in EM Wizard to allow users to improve their skills by themselves, even in the working environment, without the need for expert supervision.

3.9 CLOSING REMARKS

In recent years, sample processing systems such as manipulators and cutting devices like microtomes are utilized with SEM for the elucidation of internal structures. A cross-section of a resin block or a resin section are imaged as shown in Figure 3.26, and these images of serial sections are collected to determine the three-dimensional structure of cells and tissues (Denk and Horstmann 2004) in the biological field. There is also an increased use of FIB-SEM, an integrated system of FIB and SEM widely used in the material science field. FIB-SEM allows consecutive sectioning and imaging of a specific site at even smaller intervals. The resultant serial SEM images are combined to produce a three-dimensional reconstruction of the structure of a sample (Knott *et al.* 2008).

Recently a FIB-SEM with a unique orthogonal layout was introduced for a high precision three-dimensional structural analysis (Figure 3.31) (Man *et al.* 2013). The ideal orthogonal column layout, in which SEM and FIB are at 90° angles to each other, eliminates cross-section image shrinkage due to a low incident angle electron beam and the field of view shift that normally occurs during serial sectioning with conventional FIB-SEM configurations. In addition, the layout enables optimum SEM performance at the point of beam coincidence (Sonomura *et al.* 2013). Figure 3.32 shows the reconstructed structure of the proximal tubule of a rat kidney. The structure of the microvillus present in the proximal tubule can be clearly observed in three dimensions (Ichikawa *et al.*, unpublished data).

In this way, new applications can also trigger technological advancements of instruments. It is hoped that instrument technology evolves in concert with application technology to further unravel vital processes and structures in the life sciences.

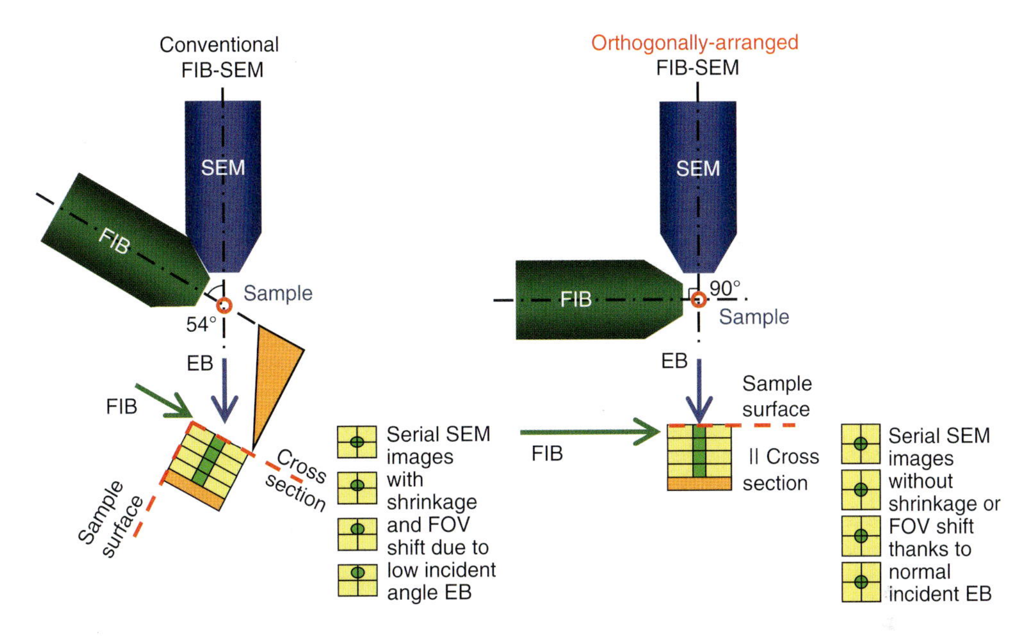

Figure 3.31 Conventional FIB-SEM and orthogonally-arranged FIB-SEM (orthogonal column layout of SEM and FIB).

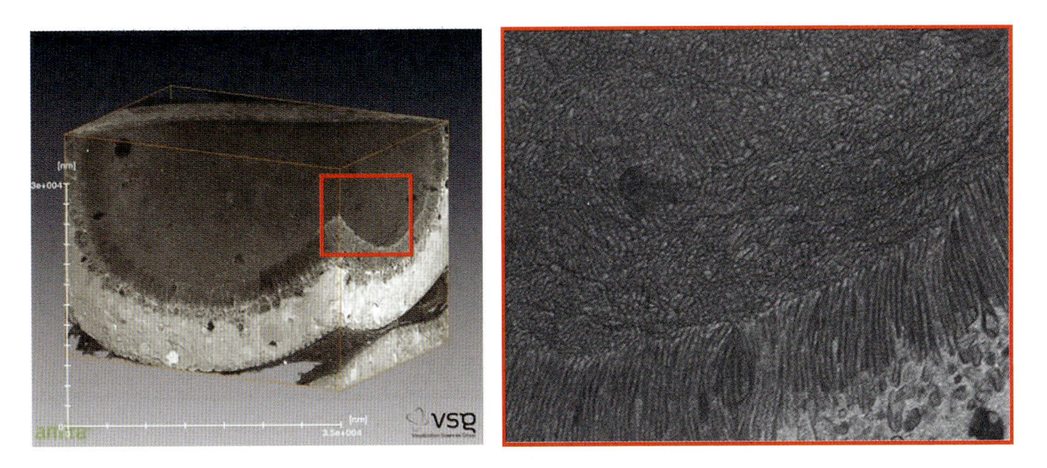

Figure 3.32 3D reconstruction of proximal tubule of rat kidney revealed by orthogonally-arranged FIB-SEM (Hitachi MI4000L) (K. Ichimura *et al.*, unpublished data). FIB condition, accelerating voltage: 30 keV, slice step: 50 nm. SEM condition, accelerating voltage: 3 keV, signal: BSE, image number: 800. Images courtesy of Dr Koichiro Ichimura, Juntendo University School of Medicine, Japan.

ACKNOWLEDGEMENTS

Application images generously provided by the following researchers are included in this article. The authors express their sincere thanks to Emer. Prof. Keiichi Tanaka of Tottori University for Figures 3.12 and 3.14; the late Prof. Albert Victor Crewe of Chicago University for Figure 3.14; Prof. Paul Walther of Zentrale Einrichtung Elektronenmikroskopie,

University Ulm for Figures 3.15 and 3.24; Emer. Prof. Masako Osumi of Japan Women's University for Figure 3.17; Prof. Tohru Sekino of ISIR, Osaka University for Figure 3.21; Dr Agnieszka Rzadzinska-Prosser of The Wellcome Trust Sanger Institute for Figure 3.22; Dr Heather E. McFarlane of Department of Botany, University of British Columbia for Figure 3.23; Dr Masashi Yamaguchi of Medical Mycology Research Center, Chiba University for Figure 3.25; Prof. Koujiro Tohyama and Mr Kinji Ishida of EMBIR, Iwate Medical University for Figure 3.26 with permission from Japanese Society of Electron Microscopy Technology for Medicine and Biology; Dr Akira Endo of National Institute of Advanced Industrial Science and Technology for Figure 3.28; and Dr Koichiro Ichimura of Juntendo University School of Medicine for Figure 3.32.

REFERENCES

Autrata, R. (1992) Single crystal detector suitable for high resolution scanning electron microscopy. *EMSA Bull.*, 22, 54–58.

Autrata, R., Schauer, P., Kuapil, J.S. and Kuapil, J. (1978) A single crystal of YAG – new fast scintillator in SEM. *J. Phys. E, Sci. Instrum.*, 11, 707–708.

Crewe, A.V., Wall, J. and Welter, L.M. (1968) A high-resolution scanning transmission electron microscope. *J. Appl. Phys.*, 39, 5861–5868.

Dan, Y., Nakazawa, E., Ogashiwa, T., Ishida, K., Hanasaka, T., and Tohyama, K. (2013) Observation of back-scattered electron images of biological ultrathin sections under lower accelerating voltage (in Japanese). *J. Electr. Microsc. Technol. Med. Biol.*, 27 (2), 94.

Denk, W. and Horstmann, H. (2004) Serial block-face scanning electron microscopy to reconstruct three-dimensional tissue nanostructure. *PLoS Biol.*, 2, e329.

Endo, A., Yamada, M., Kataoka, S., Sano, T., Inagi, Y. and Miyaki, A. (2010) Direct observation of surface structure of mesoporous silica with low acceleration voltage FE-SEM. *Colloids and Surfaces A: Physicochemical and Engineering Aspects*, 357, 11–16.

Erlandsen, S., Chen, Y., Frethem, C., Detry, J. and Wells, C. (2003) High-resolution backscatter electron imaging of colloidal gold in LVSEM. *J. Microsc.*, 211, 212–218.

Everhart, T.E. and Thornley, R.F.M. (1960) Wide-band detector for micro-ampere low energy electron current. *J. Sci. Instrum.*, 37, 246–248.

Ezumi, M., Otaka, T., Mori, H., Todokoro, H. and Ose, Y. (1996) Development of critical dimension measurement scanning electron microscope for ULSI (S-8000 series), in *Proceedings of SPIE 2725, Metrology, Inspection, and Process Control for Microlithography* X, pp. 105–113.

Frosien, J., Plies, E. and Anger, K. (1989) Compound magnetic and electrostatic lenses for low voltage applications. *J. Vac. Sci. Technol. B7*, 6, 1874.

Hertzano, R., Shalit, E., Rzadzinska, A.K., Dror, A.A., Song, L., Ron, U., Tan, J.T., Shitrit, A.S., Fuchs, H., Hasson, T., Ben-Tal, N., Sweeney, H.L., Angelis, M.H., Steel, K.P., Avraham, K.B. (2008) A Myo6 of Mutation Destroys Coordination between the Myosin Heads, Revealing New Functions of Myosin VI in the Stereocilia of Mammalian Inner Ear Hair Cells, *PLoS Genetics*, 4 (10): e1000207.

Inada, H., Kakibayashi, H., Isakozawa, S., Hashimoto, T., Yaguchi, T., Nakamura, K. (2009) Hitachi's development of colf-field emission scanning electron microscopes, *Advanced in Imaging and Electron Physics* (ed. Peter Hawkes) Academic Press, *Bulington*, 159, 124–187.

Kasuya, K., Katagiri, S. and Ohshima, T. (2010) Stabilization of a tungsten <310> cold field emitter. *J. Vac. Sci. Technol. B*, 28, 55–60.

Knott, G., Marchman, H., Wall, D. and Lich, B. (2008) Serial section scanning electron microscopy of adult brain tissue using focused ion beam milling. *J. Neurosci.*, 28, 2959–2964.

Koike, H. (1976) Analytical electron microscopy (in Japanese). *ZAIRYOKAGAKU*, 13, 217–227.

Koike, H., Ueno, K. and Suzuki, M. (1971) Scanning device combined with conventional electron microscope. *Proc. EMSA*, 29, 28–29.

Komoda, T. and Saito, S. (1972) Experimental resolution limit in the secondary electron mode for a field emission source scanning electron microscope, in *Scanning Electron Microscopy/1972* (eds O. Johari and I. Corvin), IIT Research Institute, Chicago, IL, pp. 129–136.

Konishi, Y., Sato, M., Takano, M., Tamayama, N., Nishimura, M., Watanabe, S. and Konomi, M. (2013) Charged-particle-beam device, specimen-observation system, and operation program, Patent WO 2013/137466A1 (JP5348152).

Konomi, M., Fujimoto, K., Toda, T. and Osumi, M. (2003) Characterization and behavior of alpha-glucan synthase in *Schizosaccharomyces pombe*. *Yeast*, 18, 433–444.

Kuroda, K., Todokoro, H., Yoneda, S., Saito, S. and Otaka, T. (1993) Hitachi, Ltd: Electron beam device (in Japanese), Japan Patent JP1797109.

Man, X., Umemoto, A., Asahara, T., Suzuki, H., Hasuda, M. and Fujii, T. (2013) An introduction of real time 3D analytical FIB-SEM system "MI4000L" (in Japanese), in *Proceedings of the 33rd NANOTS 2013*, Osaka, pp. 103–108.

McGough, A., Pope, B., Chiu, W. and Weed, A. (1997) Cofilin changes the twist of F-actin: Implications for actin filament dynamics and cellular function. *J. Cell Biol.*, 138 (4), 771–781.

Mogensen, M.M., Rzadzinska, A. and Steel, K.P. (2007) The deaf mouse mutant whirler suggests a role for whirling in actin filament dynamics and stereocilia development. *Cell Motil. Cytoskeleton*, 64 (7), 496–508.

Mulvey, T. and Newman, C.D. (1973) New electron optical system for SFM and SEM, in *Scanning Electron Microscopy – Systems and Applications* (ed. W.C. Nixon), Institute of Physics London Conference Series 18, p. 16.

Nagatani, T. and Sato, M. (1986) Instrumentation for ultra high resolution scanning electron microscopy, in *Proceedings of the 12th ICEM Mtg* (ed. T. Imura), pp. 2101–2104.

Nagatani, T., Saito, S., Sato, M. and Yamada M. (1987) Development of an ultra high resolution scanning electron microscope by means of a field emission source and in-lens system. *Scanning Microsc.*, 1 (3), 901–909.

Nagatani, T., Sato, M. and Osumi, M. (1990) Development of an ultra-high-resolution low voltage (LV) SEM with an optimized "in-lens" design, in *Proceedings of the 12th International Congress on Electron Microscopy*, Seattle, pp. 388–389.

Naito, N., Yamada, N., Kobori, H. and Osumi, M. (1991) Contrast enhancement by ruthenium tetroxide for observation for the ultrastructure of yeast cells. *J. Electron Microsc.*, 40 (6), 416–419.

Nakadera, T., Mitsushima, A. and Tanaka, K. (1991) Application of high-resolution scanning electron microscopy to biological macromolecules. *J. Microsc.*, 163 (Pt 1), 43–50.

Osumi, M., Yamada, N., Yaguchi, H., Kobori, H., Nagatani, T. and Sato, M. (1995) Ultrahigh-resolution low-voltage SEM reveals ultrastructure of the glucan network formation from fission yeast protoplast. *J. Electron Microsc.*, 44 (4), 198–206.

Otaka, T., Sato, M., Ohtsuka, S., Shimizu, M., Nishioka, T., Arima, J. and Yamada, M. (1989) Development of high resolution C-D measurement SEM (Model S-6100) (in Japanese), in *Proceedings of the 109th Workshop in the 132th Committee on the Application of Charged Particle Beam to the Industry*, Japan Society for the Promotion of Science, pp. 159–163.

Pease, R.F.W. (1969) Low-voltage scanning electron microscopy, in *Record of IEEE 9th Symposium on Electron, Ion, and Laser Beam Technology* (ed. R.F.W. Pease), San Francisco Press, Inc., pp. 176–187.

Rzadzinska, A. and Jones, C. (2006) Microscopy and the genetics of deafness. *Hitachi E.M. News*, 1, 19–21.

Rzadzinska, A.K., Schneider, M.E., Davies, C., Riordan, G.P. and Kachar, B. (2004) An actin molecular treadmill and myosins maintain stereocilia functional architecture and self renewal. *J. Cell Biol.*, 164 (6), 887–897.

Saito, S., Nakaizumi, Y., Yamada, M. and Nagatani, T. (1982) A field emission SEM controlled by microprocessor EMI.I. *Deutsche Gesellschaft fur Electronenmikroscopiee*, V, 379–380.

Samuels, L. and McFarlane, H.E. (2012) Plant cell wall secretion and lipid traffic at membrane contact sites of the cell cortex. *Protoplasma*, 249 (1), 19–23.

Sato, M. (2014) History of SEM performance improvement (in Japanese), *Koubunshi*, 63, 648–651.

Sato, M., Kageyama, K. and Nakagawa, M. (2001) Development of a high resolution in-lens PC-FESEM S-5200. *Recent Res. Devel. Vacuum Sci. and Technol.*, 3, 263–270.

Sato, M., Nakaizumi, Y., Yamada, M. and Nagatani, T. (1990) Development of a low accelerating voltage SEM (S-900H). *Hitachi Instrument News Electron Microscopy Edition*, 19, 45–49.

Sato, M., Otaka, M., Esumi, M., Takane, J. and Yoshida, M. (2013) Adjustment method of charged particle beam and charged particle beam device (in Japanese), Japan Patent JP5348152.

Sato, M., Otsuka, S., Miyamoto, R. and Osumi, M. (1989) Development of low-voltage and high resolution SEM. Instrumentation (in Japanese). *Biomed. SEM*, 18, 1–3.

Sato, M., Tamochi, R., Suzuki, T. and Goto, M. (1994) Effect of detector geometries on SEM images. *Hitachi Review, Ed. Hitachi Ltd*, Tokyo, 43 (4). 191–194.

Sato, M., Todokoro, H. and Kageyama, K.: (1993) A snorkel-type conical objective lens with E cross B field for detecting secondary electrons, in *Proceedings of SPIE2014, Charged-Particle Optics*, pp. 17–23.

Sato, M., Todokoro, H. and Otaka, T. (2000) *Hitachi, Ltd: Scanning electron microscope, Japan Patent* JP3081393.

Sawada, M. (1995) The history, the present and the future of the molecular pump and dry pump (in Japanese). *Turbomachinery*, 23 (11), 630–634.

Shigeto, K., Sato, M., Saito, T., Hosoya, K., Takahoko, Y. and Ando, T. (2014) Charged particle radiation device and adjustment method for charged particle radiation device, Patent WO 2014/199709A1 (JP5464534).

Sonomura, T., Furuta, T., Nakatani, I., Yamamoto, Y., Unzai, T., Matsuda, W., Iwai, H., Yamanaka, A., Uemura, M. and Kaneko, T. (2013) Correlative analysis of immunoreactivity in confocal laser-scanning microscopy and scanning electron microscopy with focused ion beam milling. *Frontiers Neural Circuits*, 7, e26.

Swanson, L.W. and Crouser, L.C. (1969) Angular confinement of filed electron and ion emission. *J. Appl. Phys.*, 40, 4741–4749.

Swanson, L.W. and Martine, N.A. (1975) Field electron cathode stability studies: Zirconium/tungsten thermal-field cathode. *J. Appl. Phys.*, 46, 2029–2050.

Tanaka, K. (1981) Demonstration of intracellular structure by high resolution scanning electron microscopy. *Scan Electron Microsc.*, Pt 2, 1–8.

Tanaka, K. (1989a) *Cho-mikuro sekai e no chosen (in Japanese)*, Iwanami Shinsho, Tokyo, 213 pp.

Tanaka, K. (1989b) High resolution scanning electron microscopy of the cell. *Biol. Cell*, 65 (2), 88–98.

Tanaka, K. (1989c) Ultrahigh resolution scanning electron microscopy of biological materials. *J. Electron Microsc.*, 38 (Suppl.), S105–109.

Tanaka, K. (2008) Scanning electron microscope of the dream and tapirus, Part 1 (in Japanese). *Microscopia*, 25 (4), 8–9.

Tanaka, K. (2009a) Scanning electron microscope of the dream and tapirus, Part 2 (in Japanese). *Microscopia*, 26 (1), 4–5.

Tanaka, K. (2009b) Scanning electron microscope of the dream and tapirus, Part 3 (in Japanese), *Microscopia*, 26 (2),4–5.

Tanaka, K. and Mitsushima, A. (1989) A preparation method for observing intracellular structures by scanning electron microscopy *J. Microsc.*, 133 (Pt 2), 213–222.

Tanaka, K. and Yamagata, N. (1992) Ultrahigh-resolution scanning electron microscopy of biological materials. *Arch. Histol. Cytol.*, 55 (Suppl.), 5–15.

Walther, P. (2003) Recent progress in freeze-fracturing of high-pressure frozen samples. *J. Microsc.*, 212, 34–43.

Walther, P. (2008) High-resolution cryo-SEM allows direct identification of F-actin at the inner nuclear membrane of *Xenopus oocytes* by virtue of its structural features. *J. Microsc.*, 232 (2), 379–385.

Walther, P. and Müller, M. (1997) Biological ultrastructure as revealed by high resolution cryo-SEM of blockfaces after cryo-sectioning. *J. Microsc.*, 196, 279–287.

Walther, P., *et al.* (1995) Double layer coating for high resolution low temperature SEM. *J. Microsc.*, 179, 229–237.

Yamaguchi, M., Ikeda, R., Nishimura, M. and Kawamoto, S. (2010) Localization by scanning immunoelectron microscopy of triosephosphate isomerase, the molecules responsible for contact-mediated killing of *Cryptococcus*, on the surface of *Staphylococcus*. *Microbiol. Immunol.*, 54, 368–370.

Zach, J. (1989) Design of a high-resolution low-voltage SEM. *Optik*, 83, 30–40.

4

A History of JEOL Field Emission Scanning Electron Microscopes with Reference to Biological Applications

Kazumichi Ogura[1] **and Andrew Yarwood**[2]

[1]*JEOL Ltd, Akishima, Tokyo, Japan*
[2]*Andrew Yarwood, JEOL (UK) Ltd, Welwyn Garden City, Hertfordshire, UK*

4.1 THE FIRST JEOL SCANNING ELECTRON MICROSCOPES

JEOL has been developing scanning electron microscopes for many years and is one of the original companies to sell the first commercial SEMs. As part of this development, SEMs for biological applications are very much a part of the history of JEOL SEMs.

The first SEMs used a tungsten thermionic filament source rather than a field emission source. The field emission gun was described by Crewe *et al.* in 1968 and was patented by V.J. Coates and L.M. Welter in 1972. JEOL started development of their own cold field emission gun in 1972 some 7 years after the first commercial SEM.

In fact JEOL developed an electron probe micro analyser (EPMA) named the JXA-1 in 1958. The first commercial EPMA, the JXA-3 was completed in 1963. Three units of the JXA-3 were delivered to Osaka University, Hokkaido University in Sapporo and the NKK Corporation, which was the second largest steel company (now JFE Steel Corporation). The electron optics of these EPMAs was basically the same as that of a conventional SEM. However, at that time the electron optics of the EPMA was designed to make a narrow electron probe, which was used to excite characteristic X-rays from specimens. In any event the EPMA was not the instrument to make a structural observation of specimens.

Following the EPMA, JEOL started SEM development in 1963 and the first commercial type SEM, the JSM-1, was introduced in 1965 and completed in 1966 (Figure 4.1).

Biological Field Emission Scanning Electron Microscopy, First Edition.
Edited by Roland A. Fleck and Bruno M. Humbel.

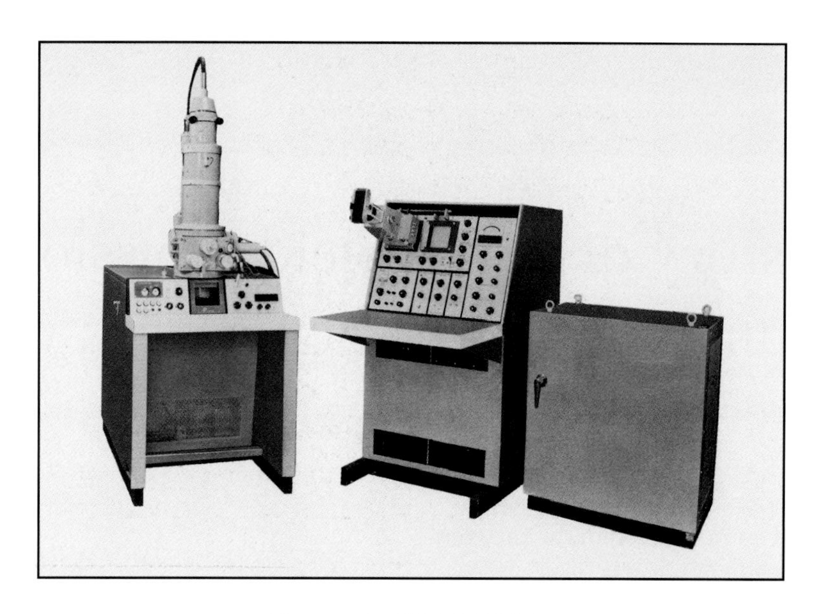

Figure 4.1 JSM-1 (1966).

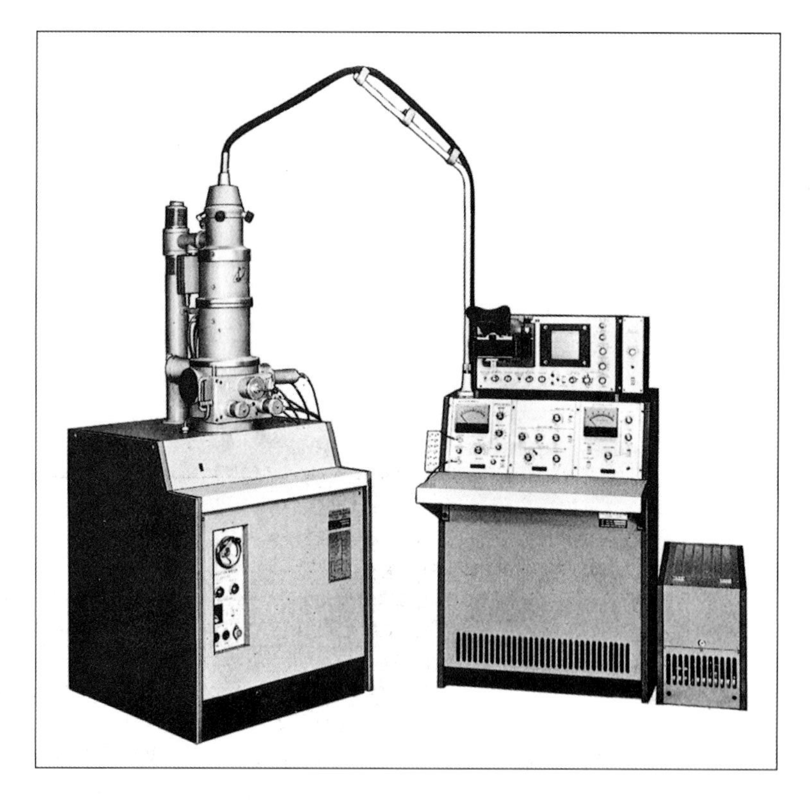

Figure 4.2 JSM-2 (1967).

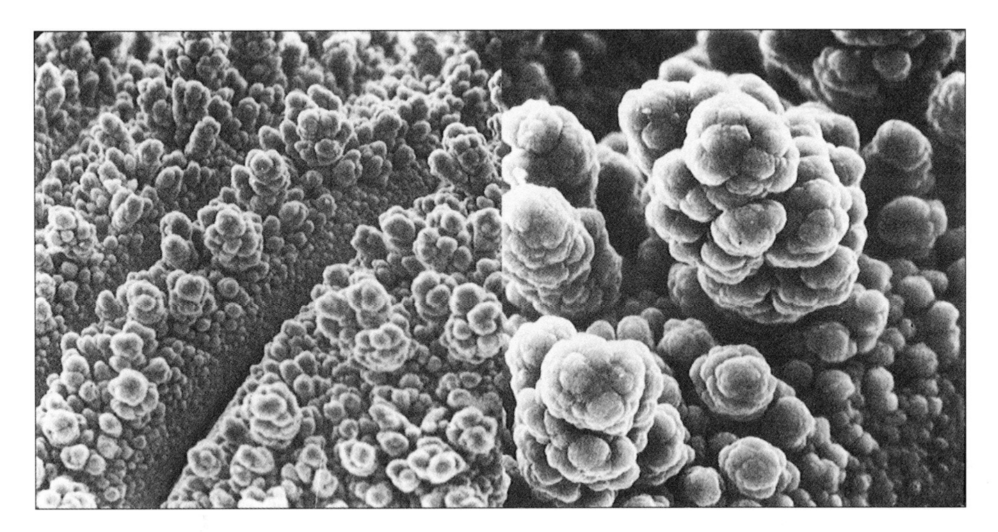

Figure 4.3 Secondary electron images from the top of a feather-like dendritic crystal of graphite. The left and right images were taken at SEM magnifications of 240× and 800× respectively (from I. Nakada and Y. Tamai, JSM-2 scanning electron microscope, *JEOL News*, 5 (9), 3–5, 1967).

It is interesting to note that 1965 was the same year as the Cambridge Instrument Company introduced their first SEM, the Stereoscan Series 1.

JEOL accelerated SEM development and introduced a multipurpose SEM, the JSM-2, in 1967 (Figure 4.2). The JSM-2 came with secondary electron imaging, backscattered electron imaging and absorbed current imaging as standard. Detectors for scanning transmission electron imaging and cathodoluminescence imaging were available as options (see Figure 4.3).

Three years later in 1970 JEOL introduced an analytical SEM, the JSM-U3, which had a five-axis goniometer stage and was capable of attaching two channels of JEOL spectrometers for the established technique of wavelength dispersive spectroscopy (WDS) (Figure 4.4).

4.2 THE FIRST CRYO-SEM

In the same year JEOL started development of the Cryo-SEM, which allowed observation of frozen specimens. The first Cryo-SEM was developed in collaboration with Prof. T. Nei of The Institute of Low Temperature Science at Hokkaido University and Prof. J. Tokunaga of The Laboratory of Electron Microscopy at Kyushu Dental College. The completed Cryo-unit (SMU3-CRU) was mounted on to a JSM-U3. This was in fact the first Cryo-SEM in the world. The two groups succeeded in the observation of frozen moulds and insect larvae in 1972 (see Figure 4.5).

The specimens were uncoated and the cryo-SEM images were taken at less than 5 kV accelerating voltage to reduce the effects of specimen charge-up. Specimen charge-up is frequently seen when imaging non-conductive samples and as a result is one of the major problems experienced when using SEMs to observe frozen hydrated specimens. The use of low accelerating voltage is one of several methods described in 1969 by Boyde and Wood to reduce specimen charge-up.

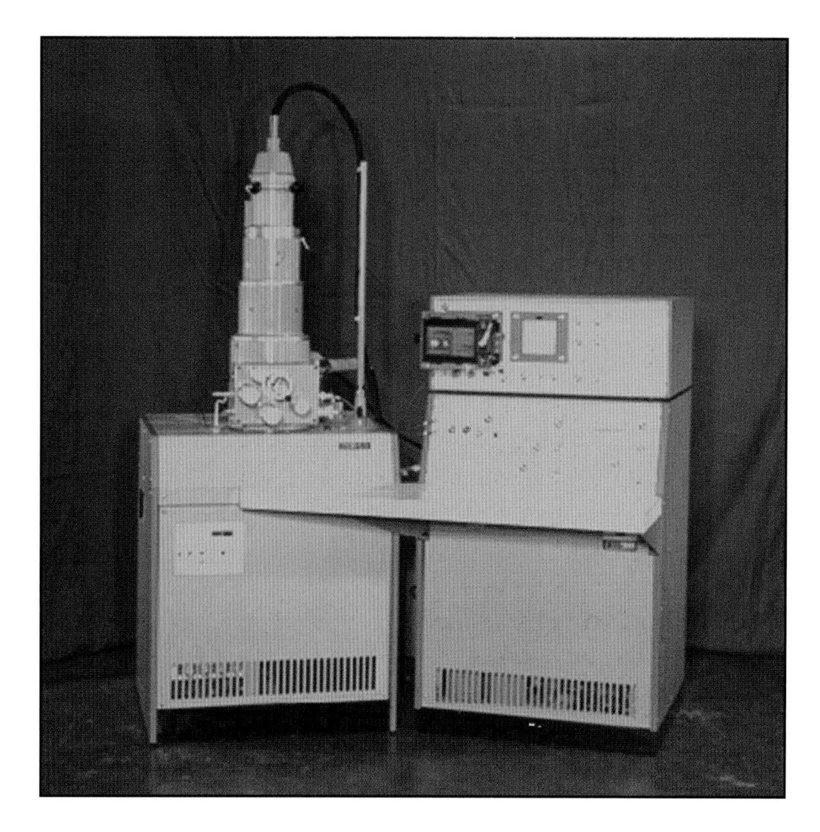

Figure 4.4 JSM-U3 (1970).

Figure 4.5 Cryo-SEM images of the frozen hydrated conidial heads of *Aspergillus niger*. The left and right images were taken at SEM magnifications of 300× and 670× respectively (from J. Tokunaga and M. Tokunaga, Cryo-scanning microscopy of coniospore formations in *Aspergillus niger*, *JEOL News*, 11, 51–55, 1973).

4.3 DEVELOPMENT OF JEOL FIELD EMISSION SEMS

Commercial SEMs at this time primarily used tungsten hairpin filaments to generate an electron source of approximately 15–20 μm. The source can then be focussed to an electron probe of just a few nanometres in diameter and scanned across the specimen surface in order to obtain high resolution images.

Unfortunately, when the electron probe is focussed to the small diameter required for high resolution imaging the number of electrons in the probe is also reduced, leading to images with a poor signal to noise ratio. The solution to this problem is to reduce the diameter of the electron source, which in turn increases the density of electrons in the focussed probe at the specimen surface. Field emission electron sources provide the solution to this problem.

Table 4.1 shows that field emission sources will create an electron source approximately 1000 times smaller than a conventional thermionic source with at least 2 orders of magnitude greater brightness.

JEOL started development of a field emission SEM in 1972. The result of this was the first JEOL commercially available multipurpose FE-SEM, the JFSM-30, which was introduced to the market in 1974 (Figure 4.6).

The JFSM-30 had a very unique electron optics system. It had three different lens modes. Firstly, there was the normal mode, where both the condenser lens (CL) and the objective lens (OL) were working rather like a conventional SEM. Secondly, in the high resolution mode, which was for high magnification imaging, the CL was not used. The OL was used to focus the fine electron probe directly on to the specimen surface. Finally, the bright mode was used for low magnification imaging with a large depth of field. In this case only the CL was working. Interestingly, a development of this third mode is used in many later JEOL field emission SEMs for low magnification and large depth of field imaging.

The vacuum system of the JFSM-30 was a totally dry oil free system. It consisted of two sputter ion pumps on the field emission gun, titanium sublimation pumps on the lens and specimen chambers and a liquid nitrogen sorption pump on the specimen pre-evacuation chamber. Liquid nitrogen sorption pumps are ideal for cryo-SEM as they are extremely effective at pumping water vapour. Unfortunately they have to be regenerated by heating and can only be used when enough liquid nitrogen is available to cool the pump sufficiently for efficient operation.

Table 4.1 This table shows the differences between the thermionic electron sources and the two main types of field emission electron sources

	Thermionic emission		Field emission	
Electron source	Tungsten	Lanthanum hexaboride (LaB_6)	Cold (W310)	Schottky (W100/ZrO)
Electron source size	15–20 μm	10 μm	5–10 nm	15–20 nm
Work function	4.5 eV	2.7 eV	4.3 eV	2.6–2.8 eV
Brightness ($A/cm^2\ rad^2$)	10^5	10^6	10^8	10^8
Energy spread (eV)	3–4	2–3	0.3	0.7–1
Lifetime	~50 hours	~1000 hours	Several years	~3 years
Cathode temperature (K)	2800	1900	300	1800
Current fluctuation (/hour)	<1%	<2%	~5%	<1%

Figure 4.6 JFSM-30 (1974).

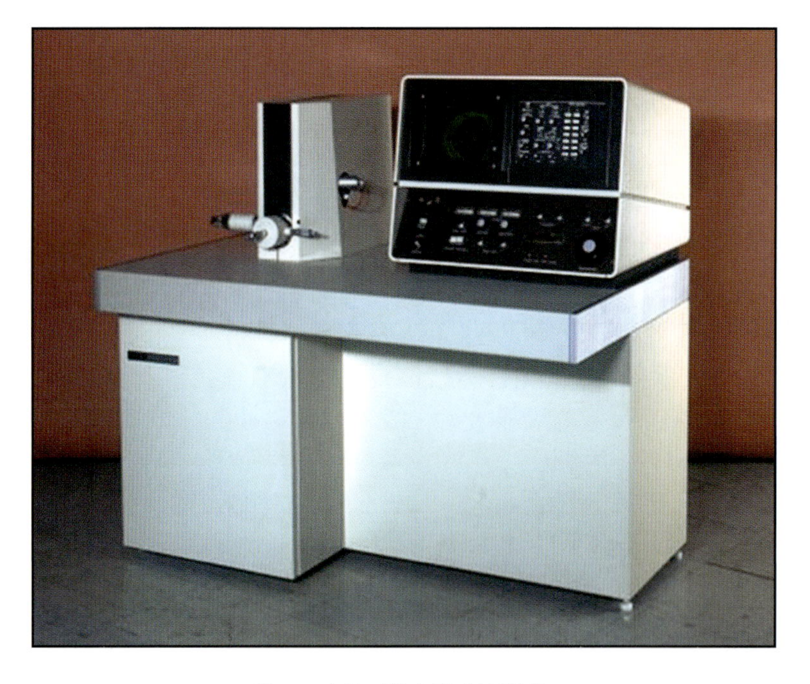

Figure 4.7 JSM-F15 (1976).

Two units were delivered to Rolls-Royce in the United Kingdom and one was delivered to Yale University in the United States. The other two units were delivered to Keio University and Canon Inc. in Japan.

In 1978 JEOL announced two single purpose field emission SEMs, the JSM-F7 and JSM-F15 (Figure 4.7). They each had a fixed accelerating voltage, 7 kilovolts (kV) in the

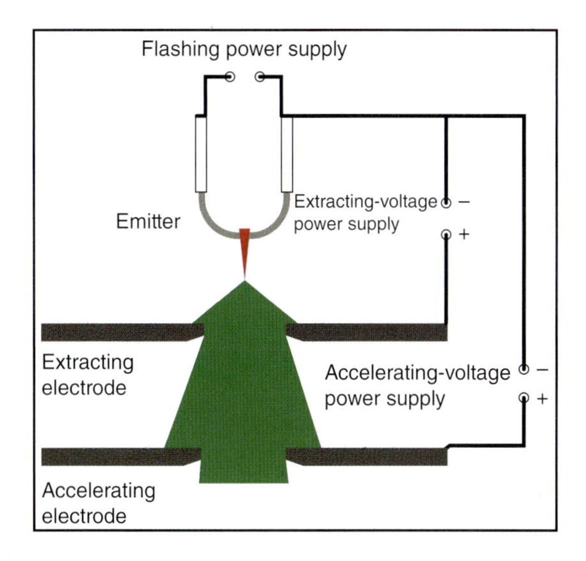

Figure 4.8 Schematic of a cold field emission source.

case of the F7 and 15 kV for the F15. The two instruments were designed only for imaging. On both models JEOL experimented with automation of the control of the FE-SEM, with features such as auto pumping, auto emission, etc.

The next milestone for JEOL was 1982 when the very successful JSM-840 series was introduced. These multipurpose SEMs were provided with a tungsten hair-pin filament or LaB_6 electron guns. JEOL successfully sold more than one thousand JSM-840 series units within three years from 1983. During the big boom years of the JSM-840, JEOL started development of field emission versions of this instrument using cold field emission as the electron source (Figure 4.8). The result of this development was announced in the form of the JSM-840F, which was guaranteed to resolve 1.8 nm at 35 kV accelerating voltage.

The JSM-840 series SEMs use a conventional SEM objective lens which is designed to create a strong magnetic field inside the conical tip of the lens. This has the advantage of providing a non-magnetic environment for the specimen. However, once the scanning probe exits from the OL final aperture it will defocus noticeably as there is no focussing field in the few millimetres between the OL and the specimen surface. This results in a loss of resolution that is proportional to the distance of the sample from the OL final aperture.

4.4 IN-LENS FIELD EMISSION SEM DEVELOPMENT

To overcome this limitation it is possible to position the specimen inside the electromagnetic lens in the physical gap between the two lens pole-pieces. In this case the scanning probe will be focussed all the way to the specimen surface, thus reducing aberrations and maintaining the highest resolution condition for SEM imaging. Transmission electron microscopes (TEMs) are designed with in-lens OL systems (Figure 4.9) and as a result the design is easily accessible for JEOL to improve FE-SEM performance by at least 3 times at 30 kV. With this in mind JEOL started the development of an ultrahigh resolution field emission SEM in 1985.

Using an in-lens type objective lens JEOL ultimately developed the JSM-890 in 1987, which was introduced to the market in 1988 (Figure 4.10). The objective lens design was

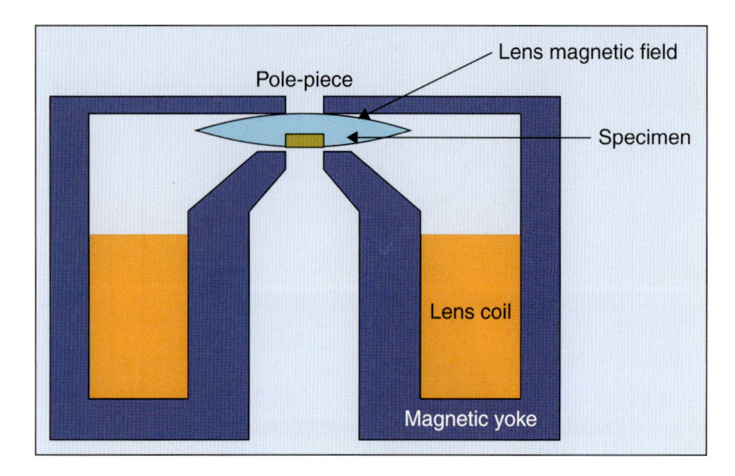

Figure 4.9 A schematic of an in-lens type objective lens showing the specimen fully immersed in the objective lens magnetic field.

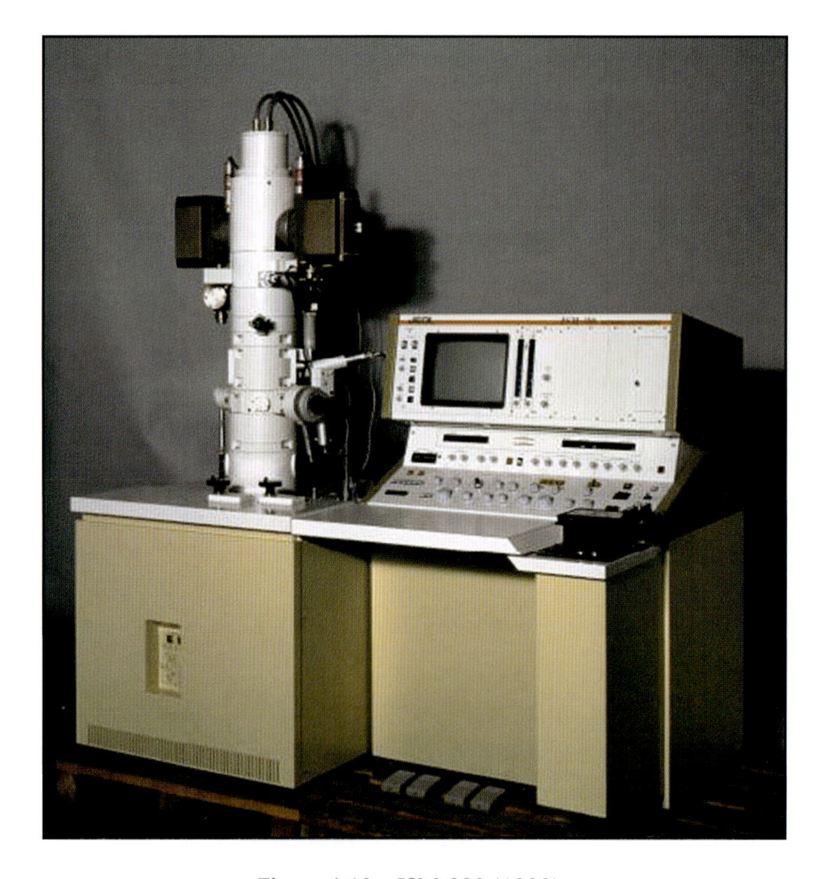

Figure 4.10 JSM-890 (1988).

taken directly from a 200 kV transmission electron microscope (TEM). The JSM-890 used a cold field emission source owing to the smaller electron source and reduced energy spread when compared to a Schottky field emission source (see Table 4.1). This combination gave the JSM-890 a subnanometre resolution at 30 kV, which was significantly better than the conventional lens on the JSM-840F (1.8 nm at 35 kV). As a result the JSM-890 set new performance standards for FE-SEMs.

The secondary electrons (SE) were imaged using an Everhart Thornley type SE detector positioned above the objective lens pole-piece. The SE image resolution of the JSM-890 was 0.6 nm at 30 kV. However, owing to the TEM style goniometer stage the specimen size has to be much smaller than in a typical SEM, with maximum dimensions in *X*, *Y* and *Z* of just 4 mm × 8 mm × 2 mm. The small specimen is positioned inside the strong magnetic field of the OL, in the same situation as in a TEM.

In the late 1970s and early 1980s sample preparation methods for biological soft tissues, microbiological, bacterial and other samples were drastically improved due to the development of new methods of fixation, critical point drying and freeze fracture preparation. Many anatomical and microbiological researchers were very much interested in the in-lens FE-SEM to observe the fine surface details and internal ultrastructure of their biological tissues. With this interest in high resolution imaging of biological and other beam sensitive materials there was a necessity to use low accelerating voltage, such as 1 kilovolt, as this greatly improved the resolution of small surface structures whilst reducing damage from the intense focussed electron probe in the field emission SEM.

The next development was in fact a conventional FE-SEM as the JSM-890 was something of a specialised instrument with a limited market. Standard cylinder or pin-type SEM stubs could not be used and the TEM style specimen holder and goniometer could only take thin narrow specimens.

The JSM-6400F was released in 1993 using updated optics from the highly successful JSM-840 series fitted to a larger specimen chamber. The JSM-6400F also used the recently introduced digital operating system and frame-store from the JSM-6400 series SEMs. The built-in frame-store digitises the SEM image signals directly and produces digital images as standard. It could also be used in combination with, or as an alternative to, the conventional photo-CRT and film camera assembly. As this instrument was primarily designed for the semiconductor industry a smaller version, the JSM-6300F, was rapidly introduced to satisfy the requirements of the more general and biological markets.

In 1994 the JSM-890 was also replaced with an updated digital version called the JSM-6000F. Although this instrument was an outstanding performer, the limitations of sample size and lack of a full cryo-SEM capability was too limiting for many biologists.

4.5 INTRODUCTION OF THE JEOL SEMI IN-LENS

To overcome the specimen size limitation of the JSM-6000F and to improve the image resolution at low kV, JEOL developed a new type of objective lens. The new type of lens looks similar to a conventional SEM objective lens. However, the two OL pole-pieces are modified to allow the magnetic field to escape from the OL into the SEM specimen chamber. The two pole-pieces are positioned in such a way that the external magnetic field is still strongly focussed but extends away from the conical tip of the objective lens for several millimetres. Normal bulk samples can then be placed close to the OL to take advantage of the

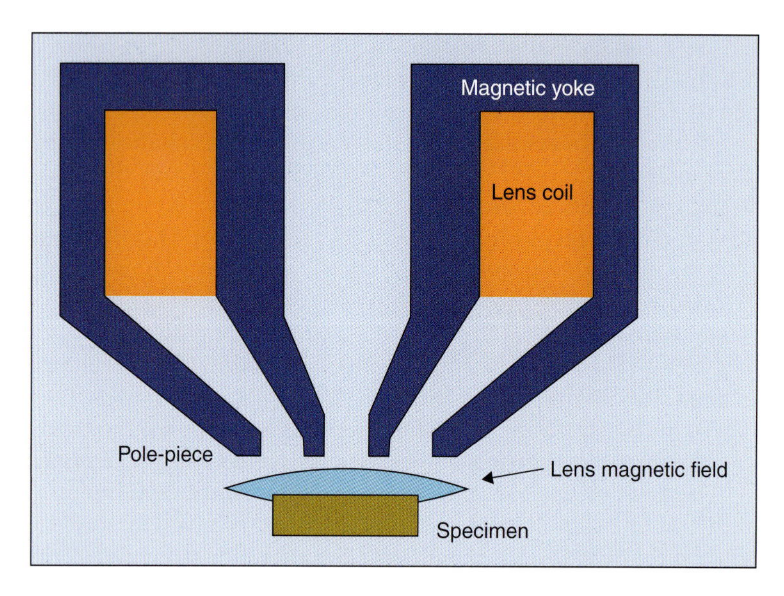

Figure 4.11 A schematic of a semi in-lens type objective lens. Although the field is external to the lens the specimen can still be immersed in the magnetic field.

fully focussed electron beam created by this external magnetic field. This new lens design was called the semi in-lens type objective lens (Figure 4.11).

The first semi in-lens type FE-SEM from JEOL was the JSM-6320F, which was introduced to the market in 1995, just one year after the JSM-6000F. The specimen is positioned below the OL just as in a normal FE-SEM. However, due to the semi in-lens design the specimen is still immersed in the strong magnetic field of the OL which results in greatly improved resolution, particularly at low accelerating voltages.

The high performance of a cold field emission source at low accelerating voltages, coupled with the semi in-lens design, was a popular imaging tool for biologists. The semi in-lens design also allowed users to install cryo-SEM stages to take full advantage of the high performance with minimal beam damage at low accelerating voltages. The performance of the JSM-6320F was guaranteed at 1.2 nm at 15 kilovolts and 2.5 nm at just 1 kilovolt.

With both in-lens and semi in-lens field emission SEMs the strong magnetic field traps the secondary electrons (Figure 4.12). If the specimen is immersed in the magnetic field the SE signal cannot be collected by a standard Everhart Thornley SE detector, which is normally positioned in the specimen chamber of the SEM.

Due to their low energy the trapped secondary electrons will spiral up in the magnetic field through the upper pole-piece of the objective lens. The secondary electron detector and its collector assembly are placed off-axis above the pole-piece final aperture (Figure 4.13). As a result both in-lens and semi in-lens FE-SEM designs are extremely efficient at collecting a high proportion of the pure secondary electron signals generated at the specimen surface.

The semi in-lens design results in a very high quality image of the specimen surface with the appearance of being illuminated from the observer's viewing point. This effect is very useful when looking at highly topographical surfaces such as those frequently encountered when observing freeze fractured hydrated cryo-SEM samples (see Figure 4.14).

For the next few years development of field emission SEMs was dominated by the rapid evolution of computer technology. Although the JSM-6320F was an excellent performer,

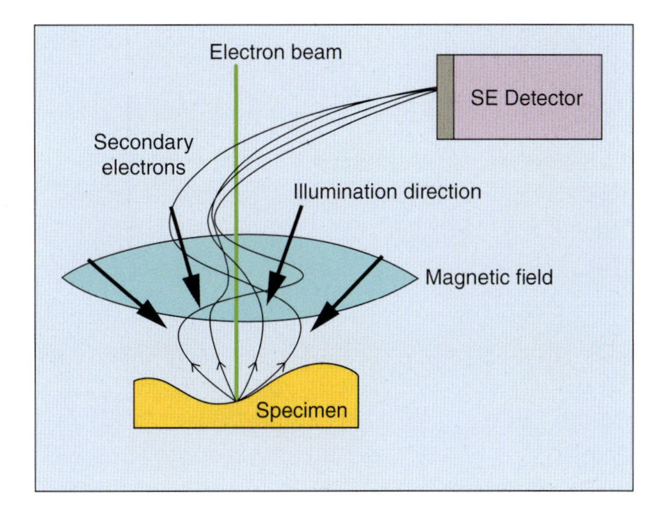

Figure 4.12 Diagram showing how the magnetic field traps secondary electrons. Once they are above the magnetic lens field the trapped electrons are collected by an off-axis in-lens SE detector. The specimen is therefore viewed as if illuminated from directly above rather than from the side, as seen with a conventional in chamber SE detector.

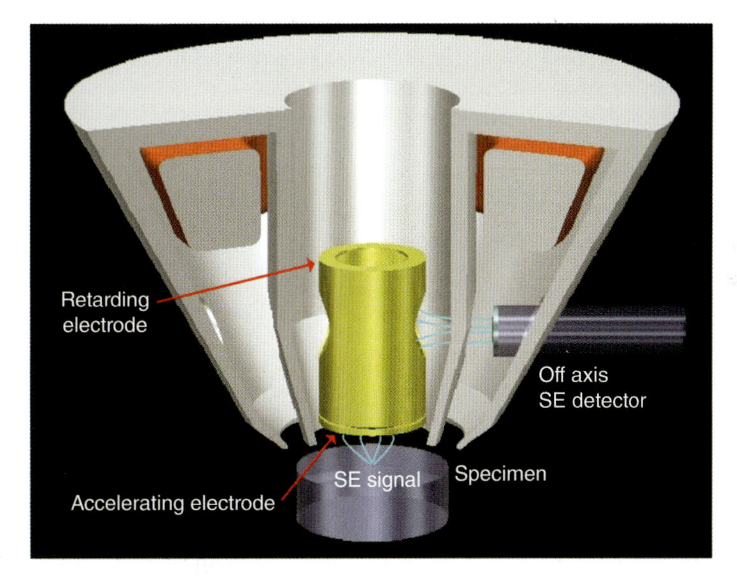

Figure 4.13 Schematic showing the relative positions of the specimen and off-axis SE detector in a semi in-lens FE-SEM pole-piece. In order to improve SE collection efficiency later versions of the semi in-lens design (from 1999) included an electrode assembly positioned inside the objective lens.

it was deemed old fashioned with its rotary controls and lack of a modern graphical user interface (GUI). JEOL developed a very reliable Windows™ like GUI and dedicated computer control system for its FE-SEMs and released the JSM-6340F semi in-lens design in 1997 to replace the JSM-6320F. A more conventional FE-SEM, the JSM-6330F replaced the JSM-6300F series soon after in 1998.

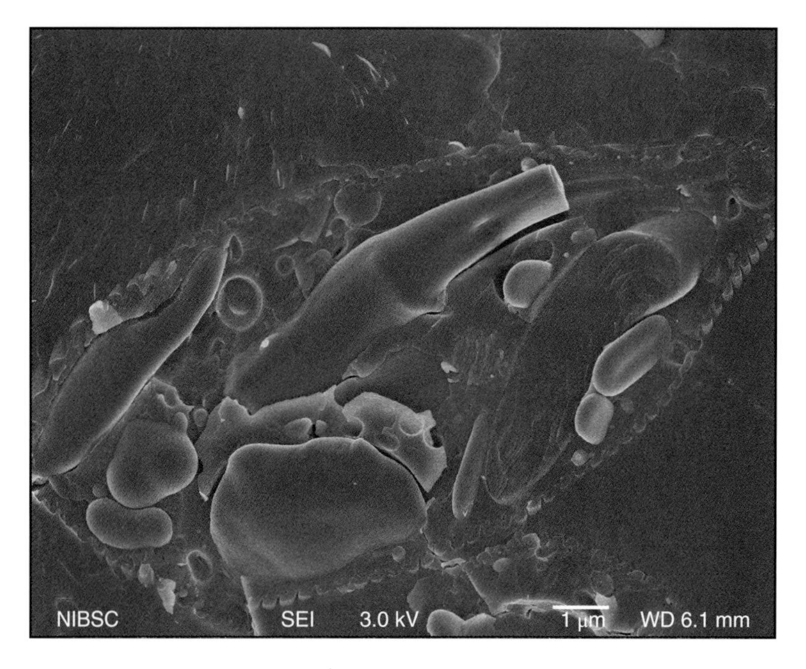

Figure 4.14 *Euglena gracilis* Klebs CCAP 1224/5Z specimen imaged on a JSM-7401F using a Baltec cryo-SEM cold stage. The algal cell suspension was high pressure frozen, freeze-fractured and lightly etched to enhance surface detail. The fracture surface was electron beam coated with carbon/platinum. This typical image from a semi in-lens field emission SEM shows how the top-down illumination effect enables the observer to see detail on all sides of three-dimensional structures.

Optically the JSM-6340F was very similar to its predecessor. When this type of instrument was sold to biologists and food technologists a high end cryo-stage and preparation system was often fitted to take advantage of the high performance at low accelerating voltages of a cold field emission SEM on their beam sensitive samples.

However, computer technology was developing rapidly at this time, with Microsoft Windows™ operating systems becoming the dominant and preferred option for many scientific instrument users. From 1999 onwards JEOL developed a completely new range of field emission SEMs starting with two models, the JSM-6700F and the JSM-6500F.

4.6 EVOLUTION OF THE JEOL SEMI IN-LENS FE-SEMs

The JSM-6500F, completed in 2000, used a conventional objective lens and was capable of 5 nm resolution at 1 kilovolt accelerating voltage. More significantly, the JSM-6500F had a Schottky type thermal field emission gun, which had been developed for JEOL's state of the art Auger SEM system. Although not ideal for biological applications the JSM-6500F worked well, with high current/high resolution operation for high speed imaging and analysis. Later on we will see how the excellent JEOL Schottky field emission gun would develop into a very desirable source for all applications due to its excellent stability and relatively long lifetime when compared with other Schottky electron sources.

Of the two new instruments the JSM-6700F (1999) proved to be more popular with biological SEM users with a guaranteed resolution at 1 kilovolt of just 2.2 nm and 1.0 nm at 15 kilovolts. The combination of JEOL's conical electrode type cold field emission source with a strongly excited semi in-lens objective lens design meant that operators could obtain fine surface details from biological samples at low accelerating voltages. To further improve SE collection efficiency an accelerating and retarding electrode assembly was designed to fit inside the semi in-lens pole-piece. This assembly accelerates the secondary electrons into the semi in-lens bore. The retarding electrode then decelerates the electrons at the correct position for efficient collection of the secondary electrons by the off-axis SE detector.

Although the preceding JEOL semi in-lens designs offered similar features, the Microsoft Windows™ GUI, larger specimen chamber and airlock, alignment free optics and improved stage design made the JSM-6700F a much more desirable instrument in the biological field emission market.

The JSM-6700F also introduced a new condenser lens design uniquely combined with the aperture angle control lens in order to optimise the electron probe diameter to the objective lens aperture at any given probe current setting. For the operator this means that the resolution of the JSM-6700F does not change significantly when using an incident probe current from a few picoamperes up to 0.5 nanoamperes or greater. For cryo FE-SEM studies this is ideal as the operator can adjust the incident electron beam current to limit beam-induced damage or improve signal to noise as required without affecting the resolution of small structures on the specimen surface.

At the same time new FE-SEM cryo stages, for example the Gatan Alto 2500, became available to take advantage of the performance of the JSM-6700F (Figure 4.15). The ease of use of this type of instrument combination along with the high resolution performance at low accelerating voltages set new standards for biological imaging. The ability to use an SEM to observe small structures such as large membrane proteins and viruses became a reality for many laboratories (see Figure 4.16).

It is interesting to note that the development of immunocytochemistry, using immunogold labels in particular, led to the use of field emission SEMs to identify the gold markers on the

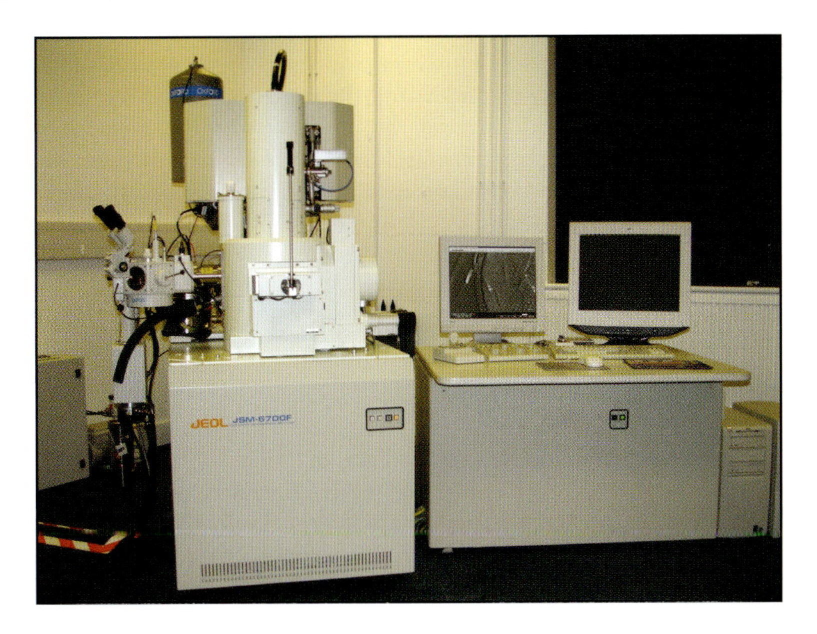

Figure 4.15 A JSM-6700F (2000) fitted with a Gatan Alto 2500 cryo system and an Oxford EDS detector for X-ray microanalysis.

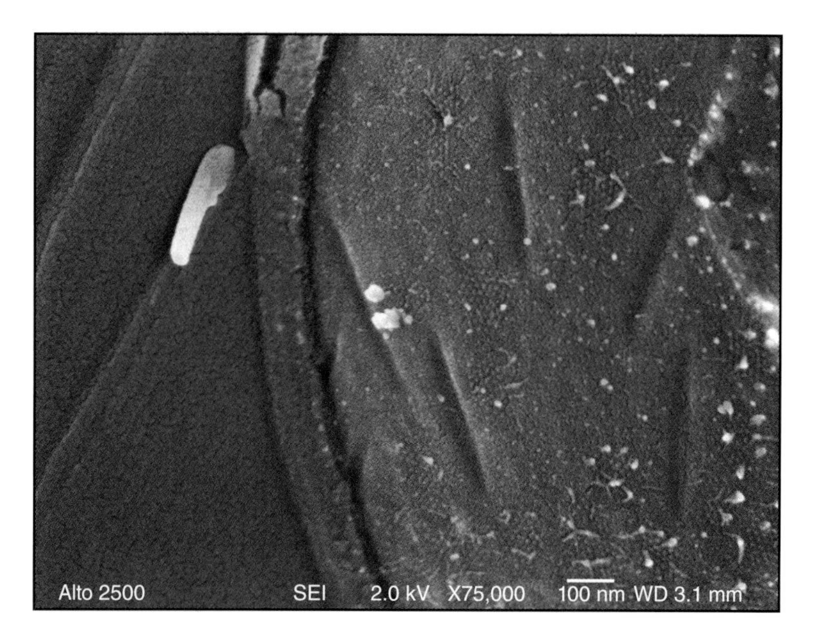

Figure 4.16 Cryo FE-SEM image of a yeast cell clearly showing cell membrane protein arrays (~10 nm). The yeast sample was prepared using a Gatan Alto 2500 cryo system, coated with platinum and imaged using a JSM-6700F field emission SEM.

surface of the samples in addition to the normal TEM studies on ultrathin resin sections as described by Hermann, Schwarz and Müller (1991) for example. Immunogold markers had been observed for several years, normally at around 6 to 8 kilovolts, using a conventional backscattered electron detector and mixing the signal with the secondary electron signal to show the location of the immunogold markers. Typically gold markers of 10 nm diameter or greater were used for this technique. However, 5 nm gold markers were somewhat harder to image due to their transparency at >5 kV and the presence of thick conductive coatings to prevent charging at these rather high accelerating voltages for biological sample imaging.

This problem and similar problems in observing fine surface structures, nanoparticles, etc., in other sciences using both secondary and backscattered electrons led to the rapid development of field emission SEMs.

4.7 DEVELOPMENT OF BEAM DECELERATION AND THE JEOL ENERGY FILTER

As a result, JEOL released the JSM-7400F in 2002, just 2 years after the JSM-6700F. With this instrument two new technologies were introduced.

Firstly, in-lens energy filtering in the form of the JEOL r-filter was provided as standard (Figure 4.17). The JSM-6700F has a basic form of energy filtering, but it has limited control and is set up as standard to reduce the effect of specimen charge-up rather than provide additional information. With the JSM-7400F it is possible to adjust the in-lens r-filter to collect just secondary or high angle backscattered electrons for example. The JEOL r-filter is also designed to filter imaging electron energies without deflecting the primary electron beam. The r-filter proves to be a great advantage when working at very short working

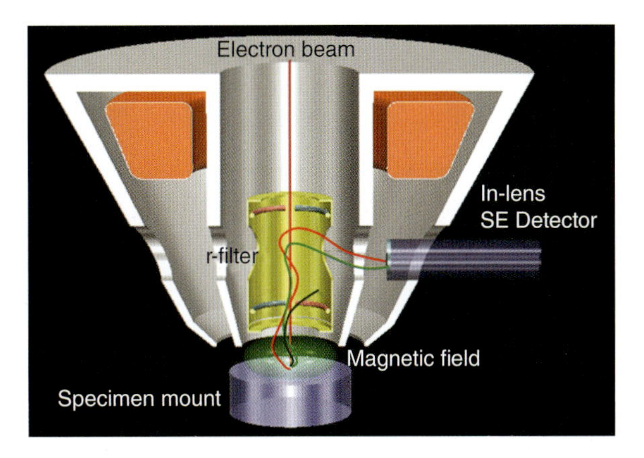

Figure 4.17 Schematic of the JEOL r-filter showing possible secondary SE trajectories for high (red), medium (green) and low (dark blue) energy secondary electrons.

distances for high resolution imaging as it reduces the need for a retractable backscattered electron detector.

Secondly, the JSM-7400F introduced beam deceleration as a standard fitment. Using beam deceleration a guaranteed resolution of 1.5 nm at 1 kilovolt was possible with the JSM-7400F. This is an improvement of more than 30% in low accelerating voltage resolution, and the JSM-7400F is also able to operate at 100 volts accelerating voltage, which overcomes one of the perceived disadvantages of a cold field emission equipped SEM.

Beam deceleration works simply by applying a negative voltage to the specimen mount. Sample biasing as it is known is not a new technique, as positive and negative values of a few hundred volts have been used for many years to reduce specimen charge-up when imaging non-conducting or poorly conducting specimens. The difference with the JSM-7400F is that it can apply between 100 volts and 2 kilovolts of negative bias to the specimen.

The high negative bias will reduce the energy of the incident electrons, thus reducing the landing voltage of the electrons on the specimen surface whilst retaining the resolution of the FE-SEM at the accelerating voltage set in the electron gun. For example, if the JSM-7400F is used at 3 kilovolts accelerating voltage, an image resolution of 1.5 nm is achievable. If we then apply a negative bias to the specimen of 2 kilovolts, the incident electrons will be decelerated to just 1 kilovolt landing voltage, that is

$$\text{Landing voltage} = \text{gun voltage} - \text{specimen bias voltage} \tag{4.1}$$

In this example the image resolution will be same as that achieved at 3 kV, which is 1.5 nm.

The beam deceleration effect is thought to continue into the specimen surface, thus reducing the damaging effect of high energy electrons even more. Secondary electron emission is also increased due to the incident electron specimen interactions being closer to the surface. A useful observation is that beam induced contamination at the specimen surface is also reduced when using beam deceleration. These effects all help to protect the specimen surface from beam induced damage, which is why JEOL developed what is known as the gentle beam mode (see Figure 4.18). An interesting article (Bouwer *et al.*, 2017) about the use of beam deceleration to improve the performance of field emission SEMs for a serial block face SEM clearly explains the physics behind beam deceleration and how it can be

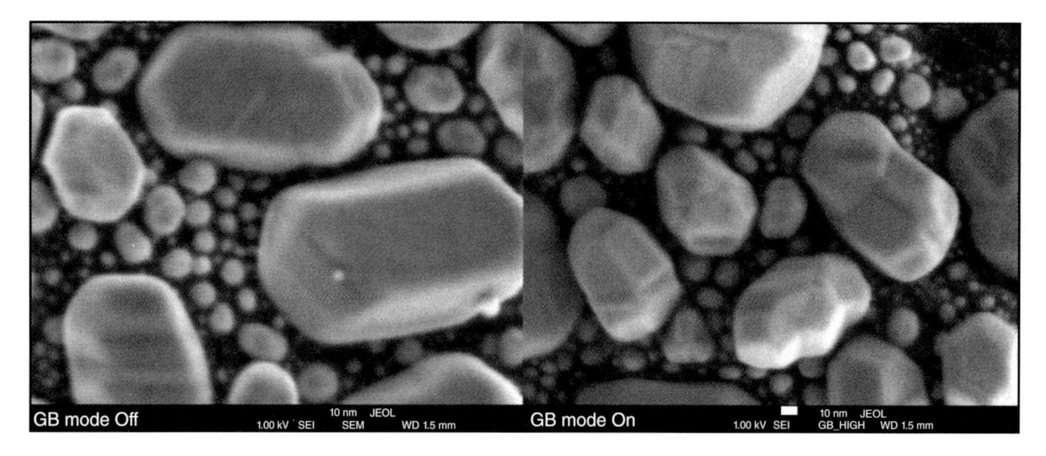

Figure 4.18 Image showing how the JEOL gentle beam mode improves imaging at low landing voltages. With the GB mode on (right hand image) the resolution of the small particles is improved and surface topography is significantly enhanced. Gold on carbon test specimen imaged on a JSM-7500F. Micron bar: 10 nm.

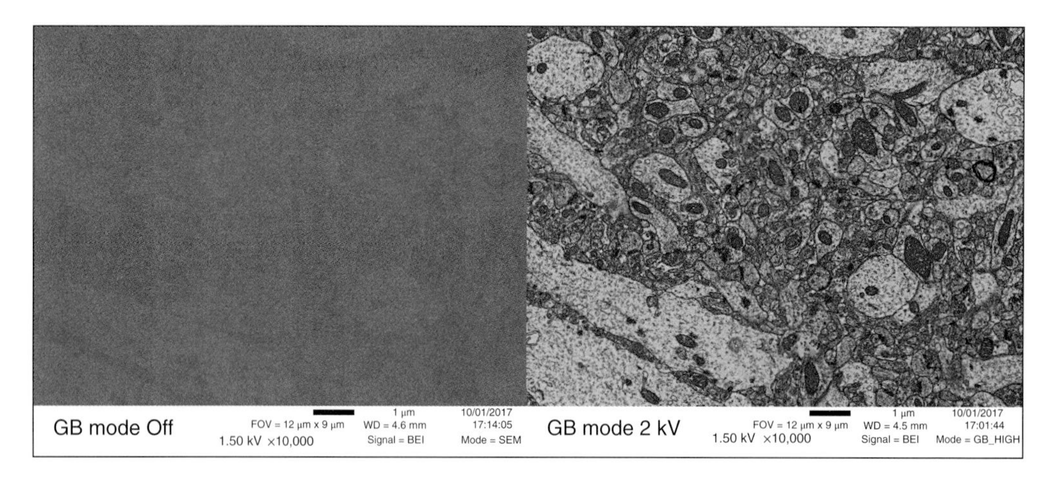

Figure 4.19 These backscattered electron images show the increase in signal achieved by using specimen biasing at low landing voltages. With a 2 kilovolt specimen bias in GB mode (right hand image), the signal to noise ratio and resolution of the backscattered electron image is greatly improved. Ultrathin section of brain tissue mounted on a silicon wafer. JSM-7610F. Micron bar: 1 μm.

used to greatly improve low accelerating voltage performance, particularly when imaging the backscattered electron signal (Figure 4.19).

JEOL's gentle beam mode is invaluable when imaging beam-sensitive samples such as biological specimens or more importantly fully hydrated cryo specimens on a cryo-SEM stage (Figure 4.20).

The JSM-7401F also introduced a feature known as the SE enhancer, which is extremely useful for cryo FE-SEM (Figure 4.21). When operating the instrument in the gentle beam mode, secondary electrons created at the specimen surface will be accelerated by the specimen bias. Many of the accelerated secondary electrons will pass through the objective lens aperture and will impact on the inner surfaces of the in-lens r-filter assembly. This will in

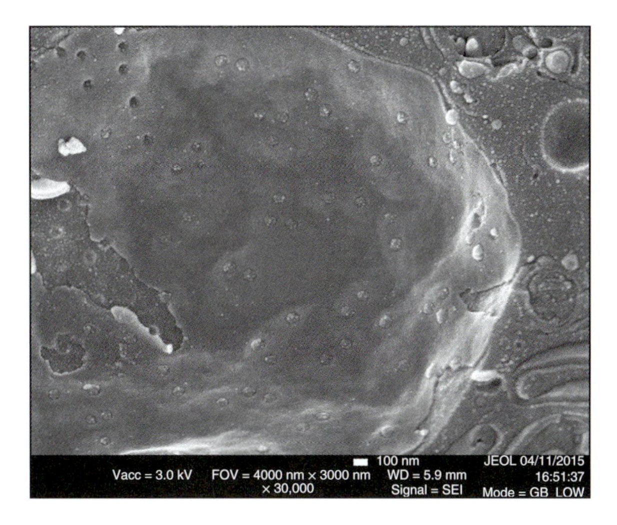

Figure 4.20 This semi in-lens FE-SEM image of the inner surface of a nuclear envelope in a frozen gut specimen clearly shows the basket structure of the nuclear pore complexes. Gentle beam was used with a primary voltage of 4 kV and a beam deceleration of 1 kV, resulting in a landing voltage of 3 kV. These settings help to prevent damage to the fine protein structures whilst maintaining good spatial resolution and image contrast across the highly topographic specimen. This image was taken using a JSM-7610F and a Quorum PP3010T cryo-SEM stage. The fracture surface was sputter-coated with gold/palladium. Micron bar: 100 nm.

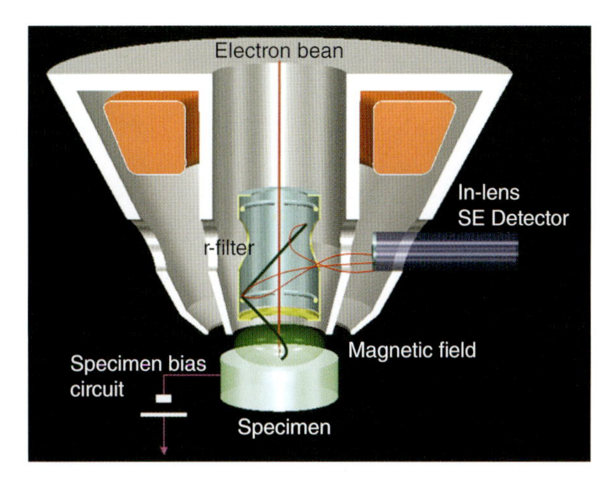

Figure 4.21 Schematic of the JEOL SE enhancer. The specimen bias accelerates secondary electrons (dark green) into the r-filter. This generates many more secondary electrons (red), which are collected simultaneously by the in-lens SE detector.

turn generate more secondary electrons (SE III type), which will be collected by the in-lens SE detector. As a result the SE signal from the specimen is amplified when using the gentle beam mode and hence the term SE enhancer. For cryo FE-SEM operation this SE enhancement is extremely useful as it means that the signal to noise ratio can be increased even though the operator may have reduced the incident electron probe current in order to reduce beam-induced damage of the cryo specimen.

In line with the development of the JSM-7400F, the preceding JSM-6500F was significantly updated in 2003 and developed into the JSM-7000F. Just like its predecessor, the JSM-7000F used a conventional objective lens with a Schottky field emission gun. The improvements, however, resulted in a large performance boost, particularly at a low accelerating voltage, where 3 nm was now guaranteed at 1 kV without beam deceleration technology. This helped to establish the JSM-7000F as an excellent general purpose and analytical field emission SEM platform.

4.8 A UNIQUE ABERRATION CORRECTED FE-SEM

In 2003 JEOL also developed a novel new type of field emission SEM called the JSM-7700F (Figure 4.22).

This instrument offered a state of the art performance of just 1 nm at 1 kV and an incredible 0.6 nm resolution at 5 kV accelerating voltage. The JSM-7700F had an aberration corrector fitted just before the semi in-lens objective lens to make pre-set corrections for chromatic aberration and allow automatic correction of the spherical aberration. A cold field emission source and a modern cantilever type TEM stage contributed to the performance of the instrument.

Although imaging on biological samples was impressive, the sample size limitations (5 mm × 18 mm × 8 mm) and the necessity to learn how to operate an aberration corrected FE-SEM did not ultimately result in success in the biological field.

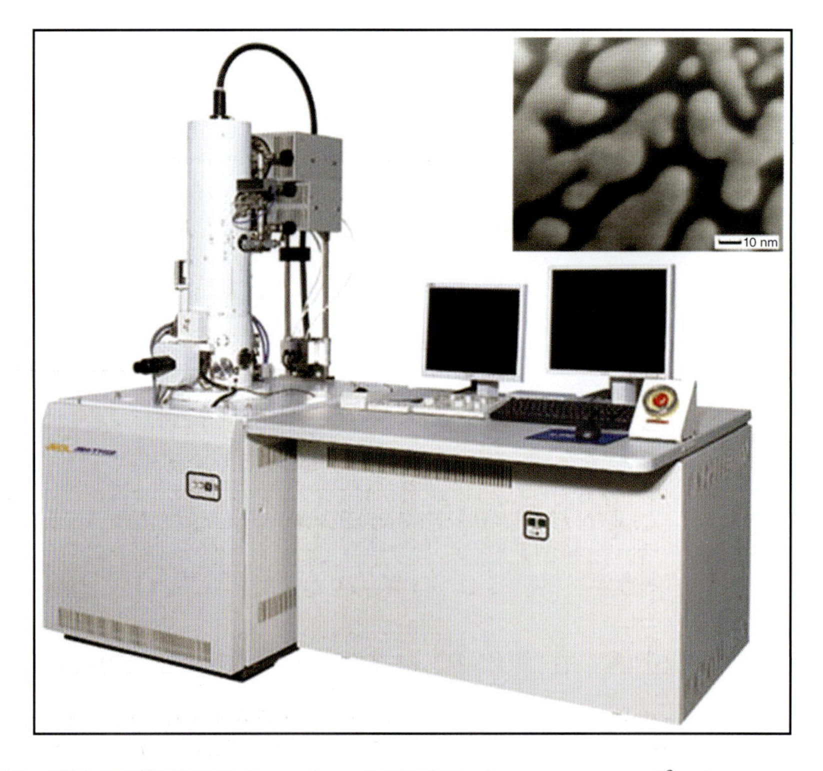

Figure 4.22 JSM-7700F (2003). Image inset 1 000 000× instrument magnification on an evaporated gold test specimen. Micron bar: 10 nm.

4.9 ONGOING SEMI IN-LENS FE-SEM DEVELOPMENT

For the next few years JEOL worked on improvements to the existing FE-SEM technology, developing the JSM-7401F in 2004. The JSM-7401F had an improved r-filter for in-lens energy filtering, but was otherwise a routine systems update, which offered the same 1.5 nm performance at 1 kV as the JSM-7400F. When coupled with a cryo stage the JSM-7401F also offered all the performance and advantages of using the gentle beam system on fully hydrated cryo specimens and has produced many state of the art biological cryo field emission SEM images (Figure 4.23).

The year 2006 saw the development of the JSM-7500F. This instrument was a natural evolution of the JSM-7401F cold field emission gun semi in-lens technology with the gentle beam system. Improvement in electronics and the objective lens design resulted in a resolution improvement down to 1.4 nm at 1 kV. A new Windows™ based GUI was also introduced in order to future proof JEOL FE-SEMs and provide a faster and better operating experience for the users. The JSM-7500F has proved to be very popular for the imaging of nano sized structures in the material sciences, as well as maintaining popularity as a high performance cryo FE-SEM (Figure 4.24).

Interestingly, 2006 also saw an operating system upgrade to the JSM-6700F to create the JSM-6701F as there was still an interest in a basic semi in-lens field emission SEM. In 2007 JEOL developed the JSM-7001F series from the JSM-7000F, with options for low vacuum operation and in-lens secondary electron detection using annular rather than off-axis SE detectors. The new instrument offered better resolution at 1 kV (2.5 nm) when fitted with the in-lens detector system, but was otherwise a natural upgrade to the previous model.

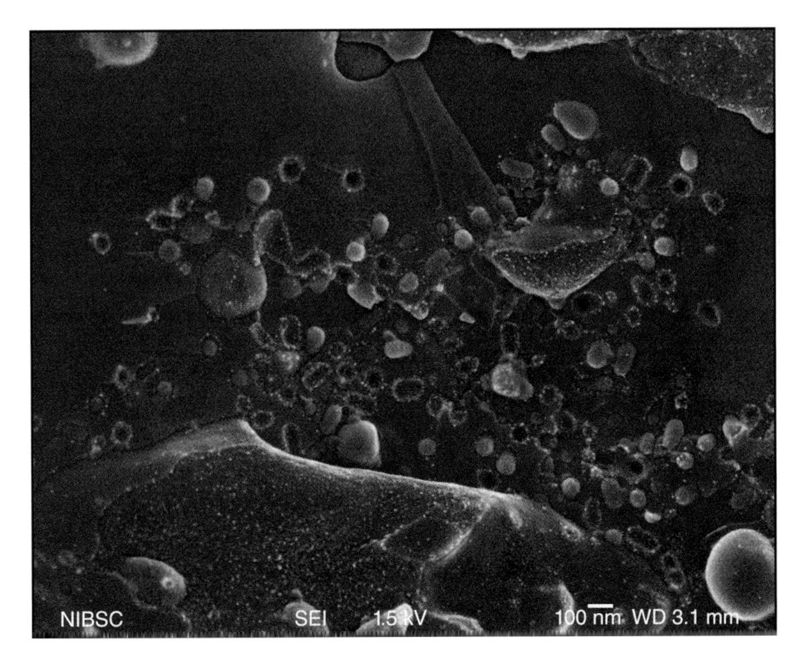

Figure 4.23 Cultured influenza viruses prepared by high pressure freezing and imaged in a JSM-7401F using a Baltec cryo-SEM cold stage. Both gentle beam and the SE enhancer were used at 1.5 kV to clearly show the viral proteins with no evidence of beam induced damage. (Image courtesy of NIBSC from the NIBSC JEOL Leica 2013 cryo workshop.)

Figure 4.24 Cryo FE-SEM image of a detergent specimen prepared and imaged using a GATAN Alto 2500 cryo system attached to a JSM-7500F microscope. Individual membrane layers can be clearly seen in this image. No beam deceleration was used. (From Gatan JEOL Workshop 2007; sample courtesy of Marilyn Carey.)

The JEOL Schottky field emission design was proving to be very successful so the next field emission instrument to be developed was basically a Schottky FE-SEM based on a semi in-lens high performance platform. This design became the JSM-7600F (2008). Whilst offering the advantages of the JSM-7500F type instrument, the JSM-7600F also offered high current and high stability with a 1.5 nm performance at 1 kV. The ability of the gentle beam mode to reduce aberrations and improve performance is in part responsible for improving the performance of the JSM-7600F so that it is almost identical to the JSM-7500F with its highly coherent cold field emission source. Later developments do in fact improve the performance of the JSM-7600F even more.

However, there was a requirement for more flexible instrumentation and higher performance, as well as strong competition from other field emission SEMs with Schottky type field emission sources. One fact is that the semi in-lens design has some sample limitations due to the strong magnetic field, which in part means that the JSM-7500/7600F platform is unlikely to be sufficiently flexible to satisfy many customer requirements for multitechnique high performance FE-SEMs.

4.10 JEOL INTRODUCES THE SUPER HYBRID LENS

After a few years of development JEOL designed a new range of field emission SEMs that use what is known as the super hybrid lens technology. The first version of this new instrument was introduced in 2011 as the JSM-7800F (Figure 4.25).

The JSM-7800F includes the JEOL gentle beam technology and improves on low kV performance by offering 1.2 nm at 1 kV with a Schottky field emission source for high current and high stability. Low vacuum technology is also available with the JSM-7800F,

Figure 4.25 JSM-7800F (2011).

which also offers even more flexibility should it be required. It is interesting to note that this resolution is obtained at a 2 mm working distance, which is quite relaxed when compared to other instruments. In fact, with the new super hybrid lens there is no reason to work any closer than a 2 mm working distance from the objective lens.

In order to do this JEOL developed a new lens design that combines the performance of a semi in-lens design with in-lens electrostatic technology and an annular through-the-lens electron detector assembly for energy filtering. The electrostatic voltages are used to accelerate the primary beam electrons through the objective lens and then decelerate them back to the correct voltage just before exiting the pole piece into the specimen chamber. The new lens design also shields the specimen from the strong electromagnetic field at the tip of the objective lens pole-piece.

The annular electron detector technology used is of a design that is seen frequently in many modern field emission SEMs. The annular scintillator has a grid that can be used to filter the electrons quite simply by changing the bias voltage of the grid in front of it. With no bias the detector is passive and will collect all electrons that impact on its surface, especially in the gentle beam mode.

When imaging normally, positive biasing is useful for collecting more secondary electrons, whereas a negative bias will tend to repel the low energy secondary electrons, which will leave a backscattered electron signal.

If an off-axis secondary electron detector is also fitted the user can collect the secondary electrons even when the energy filter (detector grid bias) is set to collect backscattered electrons.

With the new super hybrid lens technology and a JEOL Schottky electron source the JSM-7800F is proving to be a successful instrument and is popular for all applications including biology and cryo-SEM. At the same time, the JSM-7001F was updated and introduced as the JSM-7100F. Improvements in electron optical design and electronics helped the JSM-7100F to improve low voltage performance by 20% when the in-lens detector is fitted. As mentioned earlier, the JSM-7600F was not forgotten in the JEOL line-up and was in fact upgraded in 2013 and released as the JSM-7610F.

An improved SE enhancer with the option to create even more secondary electron signals and improvements to the r-filter amongst other updates has led to an instrument that can resolve an impressive 1.3 nm at 1 kV when using the gentle beam mode. It is interesting that this is very close to the performance of the base version of the JSM-7800F. As a result customers now have the option to choose the well-established JSM-7610F semi in-lens technology for their biological and cryo FE-SEM applications or go for the high end option with more features in the form of the JSM-7800F.

4.11 DEVELOPMENT OF THE JEOL GENTLE BEAM SYSTEM

However, it was to be expected that JEOL would take advantage of the performance potential of the JSM-7800F platform. In 2014 the JSM-7800F Prime with the JEOL Schottky

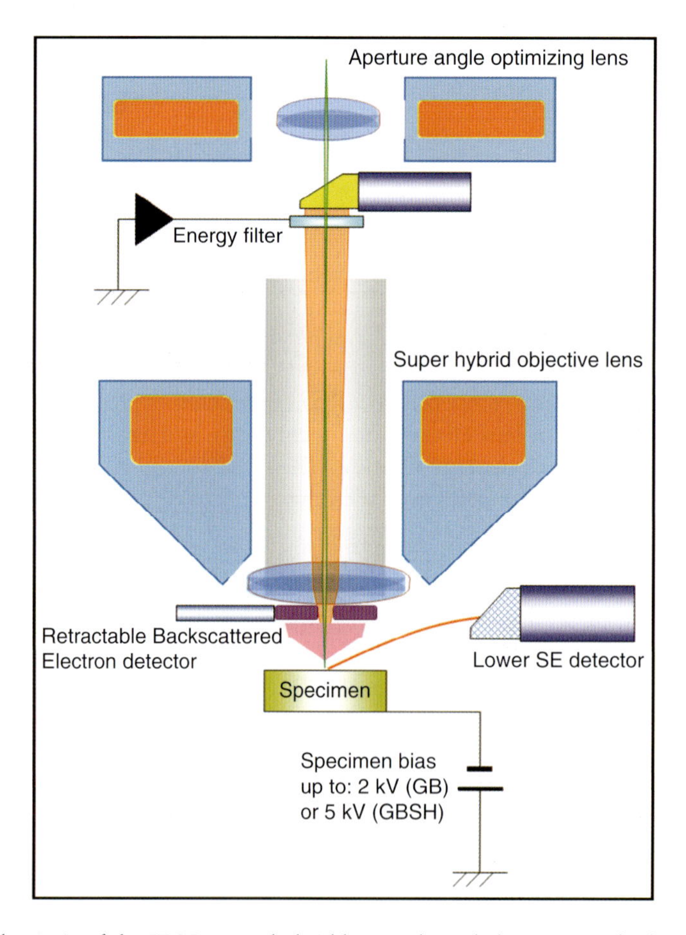

Figure 4.26 Schematic of the JEOL super hybrid lens and gentle beam super high system as fitted to the JSM-7800F series field emission SEMs.

Plus field emission source was developed and offered subnanometre performance (0.7 nm) at just 1 kV. In order to achieve this performance JEOL developed a new version of the gentle beam system called gentle beam super high or GBSH for short (Figure 4.26).

There are two differences in the GBSH design when compared to the existing gentle beam technology.

The main difference is that up to 5 kilovolts of negative bias can be set on the specimen, which is a significant increase on the 2 kV available with the standard gentle beam system. When coupled with the super hybrid lens technology it is clear why the resolution of the JSM-7800F Prime is significantly better.

The second difference is the design of the specimen holder. Specimen biasing up to 2 kilovolts works surprisingly well on a variety of specimen holders and will also work well on many stage attachments such as a cryo-stage as long as the stage has been designed to be compatible with specimen biasing technology. In the case of GBSH, it is much more difficult to contain and control stage biasing values of up to 5 kilovolts. As a result specially designed specimen holders must be used to take advantage of the gentle beam super high mode on the JSM-7800F Prime (Figure 4.27). The actual operation of the instrument is, however, no different to working as normal on the instrument and the results are generally worth it.

Ongoing collaboration work with the JSM-7800F Prime in the field of biological cryo FE-SEM should hopefully yield some excellent and interesting results in the next few years. JEOL is continuing to develop field-emission SEMs especially for the high resolution low voltage imaging applications. These developments and frequent collaborations with customers will ensure that JEOL can provide some of the best state of the art imaging platforms for biological, correlative, cryo FE-SEM and similar applications for many years to come. A summary of the developments given in this chapter are shown in Table 4.2.

Figure 4.27 Backscattered electron image of the surface of a daisy pollen grain. The sample is uncoated and was imaged at just 50 volts on a JSM-7800F Prime using the gentle beam super high mode.

Table 4.2 A summary of the SEM and field emission SEM developments as presented in this chapter

Year	Instrument	Features
1958	JXA-1	JEOL's first Electron Probe Micro Analyser.
1963	JXA-3	JEOL's first commercial EPMA.
1965–1966	JSM-1	JEOL's first commercial SEM.
1967	JSM-2	A development of the JSM-1 with STEM and CL options.
1970	JSM-U3	Analytical SEM with a 5-axis stage and a 2 channel WDS.
1970–1972	SMU3-CRU	Development of the first cryo-stage option for the JSM-U3. It was used to observe frozen hydrated specimens at low voltage.
1974	JFSM-30	JEOL's first commercial multi-purpose field emission SEM.
1978	JSM-F7 & JSM-F15	2 fully automated field emission SEM's designed for imaging using 7 and 15 kilovolt fixed accelerating voltages respectively.
1982	JSM-840	Introduction of the very successful JSM-840 series SEMs.
1988	JSM-840F	Cold field emission version of the JSM-840
1988	JSM-890	High resolution in-lens type FESEM with a TEM goniometer.
1993	JSM-6400F & JSM-6300F	Introduction of a new digitally controlled field emission SEM with large (6400F) or standard specimen chambers (6300F).
1994	JSM-6000F	Digitally controlled version of the in-lens JSM-890 FESEM.
1995	JSM-6320F	JEOL introduces its first Semi in-lens cold field emission SEM.
1997	JSM-6340F	Computer controlled version of the semi in-lens JSM-6320F with a new Windows™ style graphical user interface.
1998	JSM-6330F	Updated JSM-6300F with the new JSM-6340F interface.
1999	JSM-6700F	Introduction of the newly designed high performance semi in-lens cold FESEM with Windows™ PC control.
2000	JSM-6500F	JEOL introduces a new high current Schottky field emission SEM with a conventional lens and Windows™ PC control.
2002	JSM-7400F	JEOL releases Gentle Beam (beam deceleration) and r-filter (electron signal energy filter) in the semi in-lens cold FESEM.
2003	JSM-7000F	Updated JSM-6500F with improved low voltage performance.
2003	JSM-7700F	JEOL develops the aberration corrected semi in-lens FESEM.
2004	JSM-7401F	Introduction of the SE enhancer with an improved r-filter.
2006	JSM-6701F	Operating system and electronic upgrade of the JSM-6700F.
2006	JSM-7500F	A development of the JSM-7401F with improved performance
2007	JSM-7001F	Upgraded JSM-7000F with a new hybrid lens, in-lens detectors and low vacuum options.
2008	JSM-7600F	JEOL introduces a new high stability, high current Schottky field emission gun version of the semi in-lens FESEM.
2011	JSM-7100F	An upgrade of the JSM-7001F to improve performance.
2011	JSM-7800F	Introduction of the new Super Hybrid Lens gives conventional lens SEM operation with very high performance imaging.
2013	JSM-7610F	An upgrade to the JSM-7600F greatly improves performance.
2014	JSM-7800F Prime	JEOL introduces Gentle Beam Super High beam deceleration technology to the JSM-7800F for sub nanometre performance.
2016	JSM-7200F	An upgrade of the JSM-7100F on a new electronics platform with improved hybrid lens optics.
2017	JIB-4700F	JSM-7200F based FIB SEM for correlative cryo-FESEM.
2017	JSM-7900F	Upgraded JSM-7800F Prime with easy to use Neo Engine.

In 2017 the JSM-7800F Prime was updated with a new platform from the JSM-7200F to enhance operation and automation. This instrument was released as the JSM-7900F which introduced the Neo Engine. This development is designed to make it easy for all users to obtain the very highest performance from the instrument using new automation algorithms associated with the new firmware.

Similar upgrades were also added to the JEOL Focussed Ion Beam range of FESEM's to create the JIB-4700F which was introduced at the same time as the new JSM-7900F. The JIB-4700F is designed to be a very stable high performance FIB-SEM, ideally suited to correlative cryo TEM and FESEM applications.

Ongoing collaboration work with JEOL FESEMs and FIB-SEMs in the field of biological cryo FE-SEM should hopefully yield some excellent and interesting results in the next few years. JEOL is continuing to develop field-emission SEMs especially for the high resolution low voltage imaging applications. These developments and frequent collaborations with customers will ensure that JEOL can provide some of the best state of the art imaging platforms for biological, correlative, cryo FE-SEM and similar applications for many years to come.

4.12 CONCLUSION

JEOL has always been passionately involved in field emission SEM development, design and manufacturing to provide reliable and state of the art tools to the many institutions worldwide who use this type of technology as part of their daily research. Biological and other sciences observing beam sensitive materials benefit greatly from this continued and ongoing development. JEOL will continue to collaborate with customers and users to ensure that this ongoing development will continue to provide novel and high performance SEMs to the scientific community.

REFERENCES

Bouwer, J.C., Deerinck, T.J., Bushong, E., Astakhov, V., Ramachandra, R., Peltier, S.T. and Ellisman, M.H. (2017) Deceleration of probe beam by stage bias potential improves resolution of serial block-face scanning electron microscopic images. *Advanced Structural and Chemical Imaging*, 2 (1), 11; http://doi.org/10.1186/s40679-016-0025-y.

Boyde, A. and Wood, C. (1969) Preparation of animal tissues for surface-scanning electron microscopy. *Journal of Microscopy*, 90, 221–249.

Crewe, A.V., Eggenberger, D.N., Wall, J. and Welter, L.M. (1968) Electron gun using a field emission source. *Review of Scientific Instruments*, 39 (4), 576–583.

Hermann, R., Schwarz, H. and Müller, M. (1991) High precision immunoscanning electron microscopy using Fab fragments coupled to ultra-small colloidal gold. *J. Struct. Biol.*, 107, 38–47.

Nakada, I. and Tamai, Y. (1967) JSM-2 scanning electron microscope. *JEOL News*, 5 (9), 3–5.

Tokunaga, J. and Tokunaga, M. (1973) Cryo-scanning microscopy of coniospore formations in *Aspergillus niger*. *JEOL News*, 11, 51–55.

5

TESCAN Approaches to Biological Field Emission Scanning Electron Microscopy

Jaroslav Jiruše, Vratislav Košťál and Bohumila Lencová
TESCAN, Brno, Czech Republic

5.1 HISTORICAL INTRODUCTION

TESCAN Company was established in 1991 by a group of five former employees of Tesla Brno, a Czechoslovak enterprise with a 40 year tradition of producing electron microscopes. In a little more than twenty years, TESCAN has grown into a large company with hundreds of employees and a number of branches around the world.

TESCAN began with the development and production of digital electronics for upgrading older SEMs and also produced detectors and attachments for SEMs. TESCAN then progressed to build its own instruments. Notable milestones of TESCAN as a developer and manufacturer of electron microscopes include: the first fully digital tungsten Proxima SEM, which was introduced in 1996, followed by the first generation of the highly successful VEGA SEMs with a thermoemission electron gun in 1999 and by its variable-pressure version in 2000, for which a new type of secondary electron detector in the variable pressure environment was developed (Jacka *et al.* 2003). In 2004, TESCAN introduced the unique four-lens Wide Field Optics™ with intermediate lens (IML), initially with a fish eye mode providing macro imaging, later replaced by the wide field mode, which enabled undistorted imaging of a specimen with ultra-low magnification as low as nearly 1:1.

At the Microscopy Conference 2005 in Davos, TESCAN exhibited the new MIRA FE-SEM equipped with a Schottky field emission gun, utilizing In-Flight Beam Tracing™ to optimize the adjustment of the electron optical system. The MIRA series was gradually extended with a low vacuum option and with in-beam detectors and beam deceleration

Biological Field Emission Scanning Electron Microscopy, First Edition.
Edited by Roland A. Fleck and Bruno M. Humbel.

technology to become an all-round high performance analytical FE-SEM. Live stereoscopic 3D SEM imaging was introduced in 2007.

The portfolio of TESCAN was extended with a combined FIB-SEM system LYRA, where an Orsay Physics gallium FIB was integrated with the VEGA column in 2007 and with the MIRA column in 2008. Integration of the unique time-of-flight secondary ion mass spectrometry (TOF-SIMS) setup and scanning probe microscopy (SPM), as well as a number of detectors, manipulators, gas injection systems and analyzers, was exhibited at the Microscopy and Microanalysis Conference in 2011 in Nashville, making it a truly multifunctional laboratory of its own (Jiruše *et al.* 2012b; Whitby *et al.* 2012; Zadražil *et al.* 2012). It was extended in 2014 by integrating the confocal Raman microscopy (Jiruše *et al.* 2014a), thus bringing electron, ion and photon beams into a single instrument. A further unique addition to the combined FIB-SEMs in 2011 was then the FERA Plasma FIB (Jiruše *et al.* 2012a; Hrnčíř *et al.* 2012), using a focused beam of xenon ions (instead of gallium) and thereby increasing the sputtering rate up to 50 times.

Finally in 2013, MAIA, the ultra-high resolution FE-SEM equipped with an immersion lens, was introduced at the Microscopy and Microanalysis Conference in Indianapolis (Jiruše *et al.* 2013). This new SEM column was used also in other two FIB-SEM instruments, GAIA (Jiruše *et al.* 2014c) in 2014 and XEIA (Jiruše *et al.* 2015) in 2015. All FE-SEMs can be equipped with a cathode lens (beam deceleration method, BDM) with optimized resolution for low voltage applications (Jiruše *et al.* 2014b).

For its rapid R&D and business progress, TESCAN was twice awarded the Czech company-of-the-year prize. In 2013 TESCAN as SEM manufacturer and Orsay Physics as FIB manufacturer merged to form TESCAN Orsay Holding. Several daughter companies together with a number of distributors constitute the world-wide sales and service network.

5.2 BIOLOGICAL SAMPLES IN SEM

SEM is a routine analytical tool for many biological applications. Scientists choose SEM for its high-magnification imaging with nanometer-scale resolution, but many microscopists often appreciate other features of SEMs as well. In general, imaging of biological samples in SEM requires different approaches compared to imaging of conducting material samples. Biological samples contain a significant amount of water, which makes them largely incompatible with the high vacuum environment inside the microscope (Schatten and Pawley 2008; Stokes 2008). In terms of atomic composition, they predominantly contain light elements, such as carbon, hydrogen, nitrogen and oxygen. This results in a large interaction volume of the primary beam, excessive damage to sensitive biological molecules and low emission of electrons. In addition, biological samples are usually not conductive, so charging of the sample surface often makes observation difficult. Besides, the excessive drying causes degradation of internal sample structures, collapsing membranes and overall damaging or destroying the sample. Many strategies have been developed to deal with biological samples and each of these strategies is specific to the particular task. Since the variety of applications is enormous, one needs to carefully take into account all aspects of a sample and the information being sought before selecting a particular technique. This includes selection of the appropriate sample preparation as well as imaging conditions. While some users may be interested in observing very small details, others may need to visualize complex topography of large volumes. In the following sections, our aim is to provide the reader with an overview and variability of the diverse approaches being used by TESCAN for imaging organic and biological materials.

5.2.1 Why FE-SEM?

Schottky field-emission SEM has many important advantages compared to thermoemission tungsten SEM. The source brightness is 10 000 times higher, the size of the virtual source is 2000 times smaller and the energy spread is 3 times smaller, all together leading to much better resolution.

Not only is the best possible resolution improved threefold (i.e. typically 1 nm for FE-SEM compared to 3 nm for tungsten SEM at 30 keV), but the overall dependence of resolution on acceleration voltage is also substantially better. At low acceleration voltages, for example of 1 keV, the difference in resolution is better by one order of magnitude (1.5 nm versus 15 nm). Moreover, since angular intensity of the tungsten emitter decreases with decreasing acceleration voltage, both beam current and the resulting signal-to-noise ratio are also decreasing. As a practical consequence, in order to gather sufficient signal from a weakly contrasting specimen, an increase of the beam current is required, which leads to further deterioration of the resolution. Thus for biological specimens of low contrast, which require the use of low voltages, it is practically impossible to get good quality images with fine detail for tungsten SEM and the use of FE-SEM is essential. Its price is higher than for tungsten SEM because ultra-high vacuum of 10^{-7} Pa or better is needed in the gun chamber, but the quality of images (given by their resolution and signal-to-noise ratio) at low voltages is invaluable.

Thermoemission LaB_6 or CeB_6 sources do provide enhanced performance compared to a tungsten source, leading to a SEM resolution of 2 nm at 30 keV, but the price of LaB_6 SEM is higher because of the higher required vacuum, and the low voltage performance is still worse than with FE-SEM.

5.2.2 SEM Optics and Displaying Modes

Compared to photonic microscopy, scanning electron microscopy offers an unprecedented depth of focus and a dynamic range of magnifications. This is advantageous in displaying larger samples with complex topography, such as fossils, bones, plants or even small animals. Moreover, users often want to switch between high and low magnifications easily for complex navigation and quick analysis without complicated and time consuming steps. The ideal SEM should be able to handle all these tasks with minimum effort from the operator. This typically means that most of the alignment steps should be done automatically and the microscope should be easily controlled via a user-friendly software interface.

The electron optical system consists of a gun, one or two condenser lenses, the intermediate lens (IML) and the objective lens (OL). The condenser lenses demagnify the primary source and regulate the beam current. Two magnetic lenses, IML and OL, and the deflection/scanning system project the electron beam on to the specimen surface and scan over it. The addition of the TESCAN proprietary intermediate lens extends the range of SEM display modes as shown in Figure 5.1. Basically there are four different display modes in which the IML, OL and scanning coils can work.

For high resolution imaging (resolution mode) at medium to low beam currents and short working distance the electron beam is focused by OL only. In the "so-called" depth mode, both OL and IML are activated; thus the aperture angle of the beam in the objective lens is decreased and the object is imaged with an increased depth of focus. This is particularly useful for imaging highly topographic samples (see Figure 5.2) as well as 3D stereo imaging.

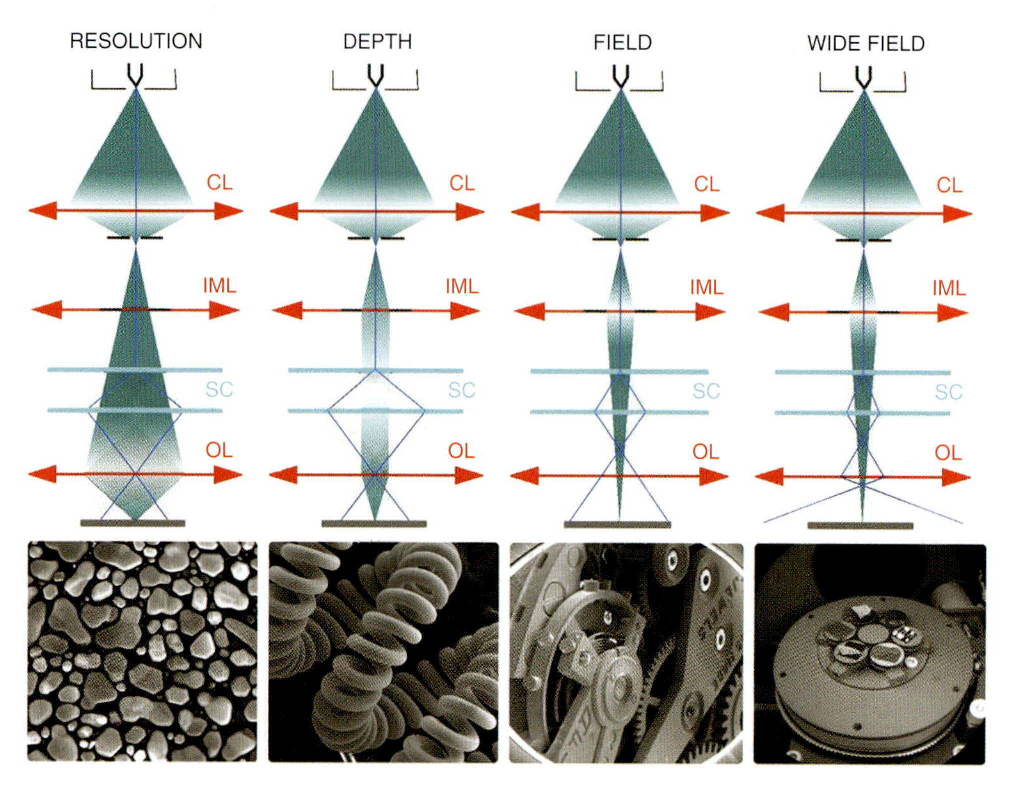

Figure 5.1 Display modes used in TESCAN SEMs. CL denotes condenser lens (there are two condenser lenses for thermoemission tungsten or LaB_6 SEM), IML intermediate lens, OL objective lens and SC scanning coils. Beam focusing and scanning is shown schematically.

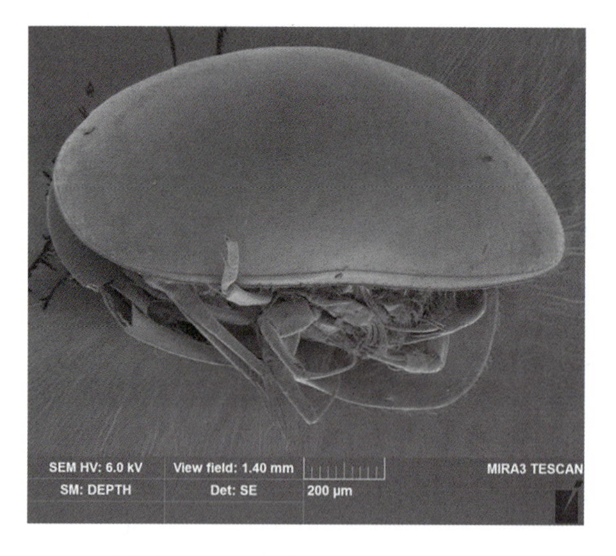

Figure 5.2 The image of a seed shrimp (*Ostracoda*). Large depth of focus is achieved by incorporating an intermediate lens into the column.

A large field of view is achieved in the "so-called" field mode when the beam is focused only by IML. The depth of focus is naturally even more increased in this mode. The spot size is larger compared to resolution or depth modes but it still allows imaging up to a hundred thousand times magnification.

An extremely large field of view (more than 100 mm) can be reached in the wide field mode, unique to TESCAN microscopes, when both OL and IML lenses are used, resulting in magnifications down to 1×. Originally, this method of imaging was developed as the world's first fish eye mode in SEMs. Later on, a more sophisticated method of scanning correction was developed to compensate for the "fish eye" distortion. The main advantage of the wide field mode is that the live image allows simple orientation and navigation with large samples (see Figure 5.3). Together with the other modes it enables easy and continuous magnification adjustment over the whole range from 1 × to 1 000 000 ×.

An alternative approach of obtaining large panoramic images is provided by a software module, the so-called image snapper, which acquires multiple smaller images and then stitches them into one larger mosaic image (see Figure 5.4). This post-processing method may create a very large image, but one has to be aware of its limitations. Small inaccuracies of stage movement and image distortion together with uneven brightness and contrast of individual images may lead to visible disturbing stitching artifacts. Thus, whenever possible, we recommend using live, real-time and free of stitching artifacts wide field optics imaging.

The ultra-fast scanning system with a minimum dwell time of 20 ns also allows compensation of static and dynamic image distortions. The computer-driven scanning coils also compensate for image rotation if the working distance or the magnification changes. Another important feature of the electron optics is its capability to easily produce three dimensional images. A light stereomicroscope is a commonly used tool for many biologists interested in observations of complex, topographic samples such as small insects and fossils. A similar 3D effect can be obtained with the SEM as well. The traditional way of generating a 3D stereoscopic image was by tilting the sample. However, this is time consuming and it

Figure 5.3 Mole jaws imaged in the wide field mode. The image was captured as one field with mixed SE and BE signals. (Sample courtesy of James Weaver, Harvard University.)

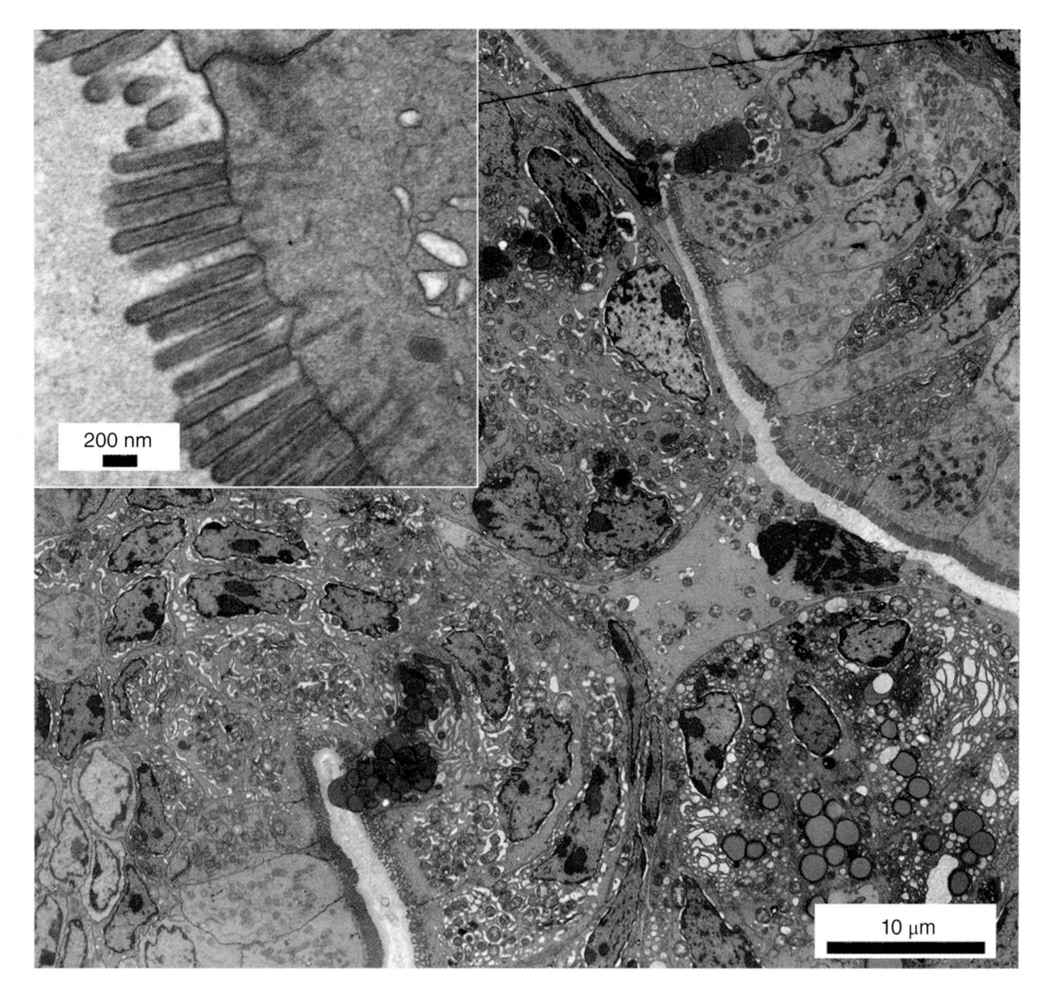

Figure 5.4 A panoramic view of rat intestine using a bright field transmitted electron detector. The picture is composed of 49 stitched images (tiles) with total 42 000 × 42 000 pixels with a pixel size of 1.2 nm. One tile is also shown.

cannot be used for observing in real time, for example manipulation of objects on the sample surface. An innovative method uses two separate 2D images of the same object taken at different incidence angles (two scans) and encoded in different colors (typically chromatically opposite ones, e.g. red and cyan; see Figure 5.5). Then they are superimposed into a single 3D anaglyph image which is viewed using 3D anaglyph glasses. Thus the tilting of the stage is no longer necessary and real time observation is possible, including automatic software synchronization of images and correction of distortion and astigmatism of the tilted images. Due to large depth of focus and large field of view, the SEM 3D anaglyphs provide a unique tool in studying the morphology of animals, plants and many other biological samples.

5.2.3 Ultra-high Resolution Microscopy

New types of columns have emerged for obtaining ultra-high resolution at low acceleration voltages (i.e. less than 2 nm at 1 keV) using immersion lens technology, either in the single

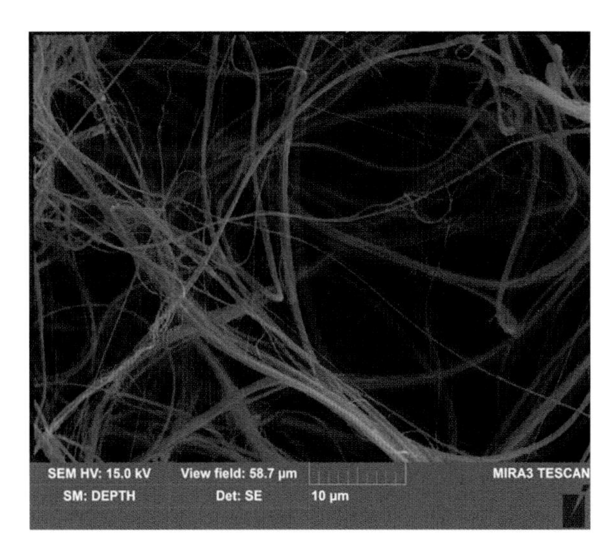

Figure 5.5 A 3D anaglyph of spider web fibers.

pole-piece lens or in a so-called radial-gap or snorkel lens. Here the sample is immersed in a relatively strong magnetic field that facilitates the extraction of secondary electrons again into the magnetic objective lens. An immersion single pole-piece lens is used in the MAIA microscope, reaching the resolution of 1.4 nm at 1 keV (Jiruše *et al.* 2013), that is double resolving power compared to conventional optics. Additional improvement of resolution is possible with beam deceleration principles, see below.

5.2.4 Beam Deceleration Mode

For low energy operation it is beneficial to decelerate the beam just before it arrives at the sample instead of using the low energy of the beam throughout the whole column (Jiruše *et al.* 2014b). A negative voltage of –5 kV is applied to the specimen creating an electrostatic field, the so-called beam deceleration mode (BDM) or also a cathode lens. The primary electrons leave the objective lens with high energy, for example 6 keV, and are decelerated by this field just before landing on the specimen, for example to 1 keV. The electrostatic field not only affects the primary beam, it also influences the electrons emerging from the specimen surface. In the BDM, secondary electrons (SEs) are accelerated and focused inside the electron column (see Figure 5.6). Thus they cannot be detected by a conventional SE detector, but instead the detector placed inside the electron column is optimal for recording a high SE signal. Backscattered electrons can be detected by standard means between the column and the sample; see Section 5.2.6 about the detection system.

BDM improves resolution of conventional MIRA optics from 3.5 nm to 1.8 nm at 1 keV. In the case of MAIA immersion optics, the improvement is from 1.4 nm to 1.0 nm (see Figure 5.7, left). This makes ultra-high resolution imaging at low energies possible. The main advantage of MAIA immersion optics compared to conventional optics is that ultra-high resolution at low energies is achieved even without BDM. This is beneficial for non-conducting or highly topographical samples on which it is not suitable to apply BDM bias voltage. An example is given in Figure 5.8, where uncoated salmonella bacterium is imaged at 500 eV.

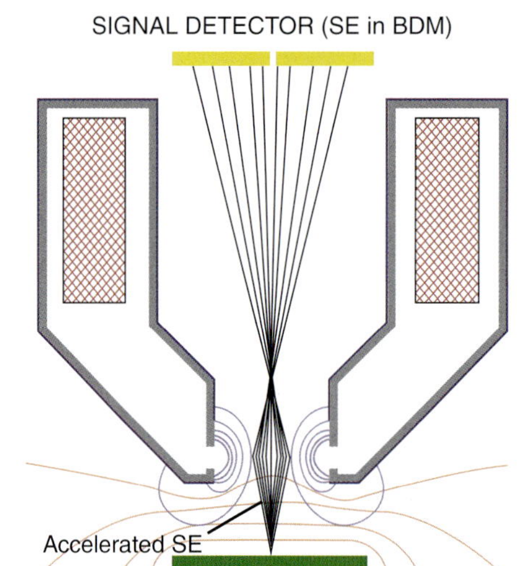

Figure 5.6 Schematic diagram of the beam deceleration mode used, for example, in the MIRA microscope: the beam is focused by a magnetic lens (flux lines shown in blue) and the electrostatic decelerating field between the sample and the objective lens (equipotentials schematically shown in red). Secondary electrons are accelerated into the objective lens where they are detected.

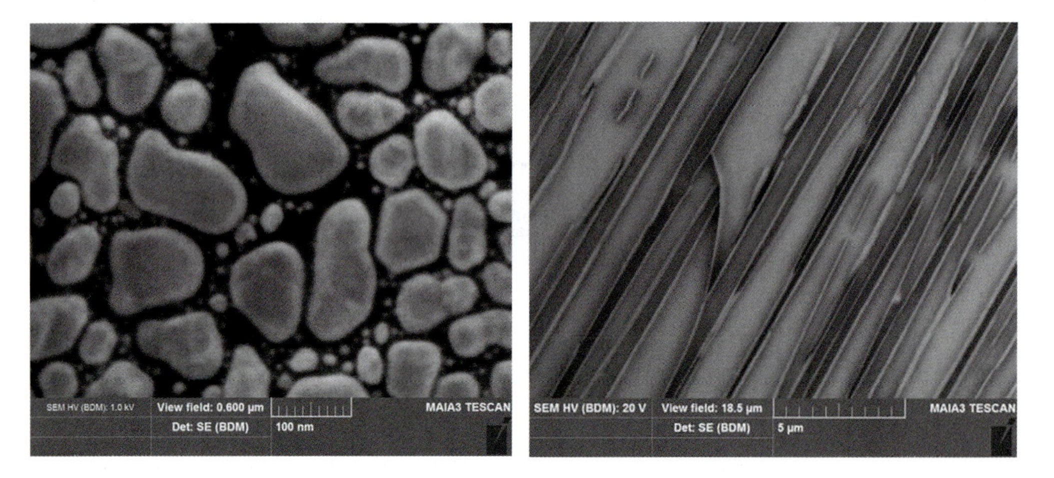

Figure 5.7 Left: high-resolution image of gold on carbon at 1 keV with immersion optics. Right: very low energy of 20 eV imaging with the beam deceleration mode, sample of organic heterostructures of para-hexaphenyl and α-septithiophene grown on muscovite mica.

Another advantage is that with BDM it is possible to slow down the beam to 50 eV in automatic mode or down to 0 eV in manual mode (Jiruše *et al.* 2013). Normally, electron optical columns without deceleration provide incident beam energies from 200 eV because of stray fields and other technical difficulties. BDM gives access to very low energies that are difficult or impossible to achieve by other methods (see Figure 5.7, right. This is especially important for samples sensitive to radiation damage.

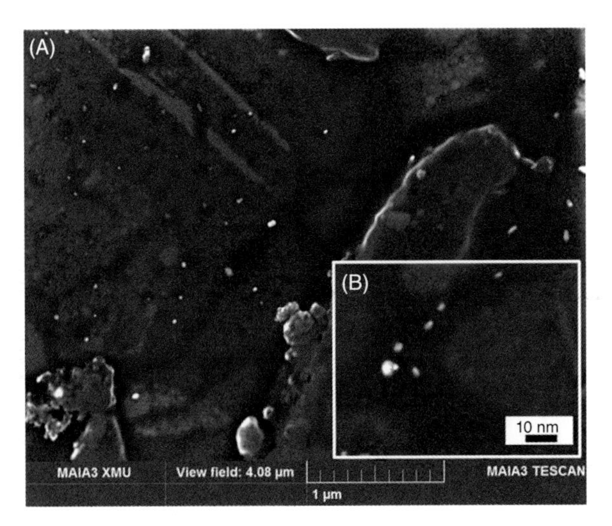

Figure 5.8 Salmonella bacterium imaged uncoated at 500 eV. An overview of bacteria body (A) is supplemented by a detail of its flagellum (B).

5.2.5 In-Flight Beam Tracing

For accurate control of the SEM, that is, obtaining an optimum spot size for each display mode, beam current and working distance, the lens currents are adjusted with the help of an accurate and general method, so called In-Flight Beam Tracing™, which calculates and optimizes all the optical parameters from first principles in real time using the field distribution of the lenses.

This is extremely useful for special modes, for example BDM. The optical power of the electrostatic lens in BDM modifies parameters such as focal length or magnification of the field of view, the impact angle of landing electrons, etc. Using In-Flight Beam Tracing™ we can control optical elements for any combination of beam energy, bias, probe current, SEM magnification or working distance with high accuracy.

5.2.6 Detection System

Conventional signal detection in the vacuum chamber comprises an Everhart–Thornley type detector of secondary electrons (SE), concentric detector of back-scattered electrons (BSE) located between the column and the sample and a detector of transmitted electrons (TEs) below the sample. A dedicated SE detector for low vacuum (LVSTD) is described below in Section 5.4.2 about low vacuum operation. Each detected signal provides different information about the sample and signals from various detectors can be mixed (see Figure 5.9). Another TESCAN special feature is a color and/or panchromatic detector of cathodoluminescence that is fully integrated in SEM software; thus there is no need for external scanners or external image processing as in case of third-party add-ons.

A more advanced set of detectors is included inside the SEM column, the so-called in-beam detectors. They form typical extras of FE-SEMs compared to tungsten SEMs, which are usually equipped only with in-chamber detectors. The in-beam detectors enable sample observation at shorter working distances, and thus at higher resolution, and in the case of BSE clear the space above the sample, which is invaluable for simultaneous integration

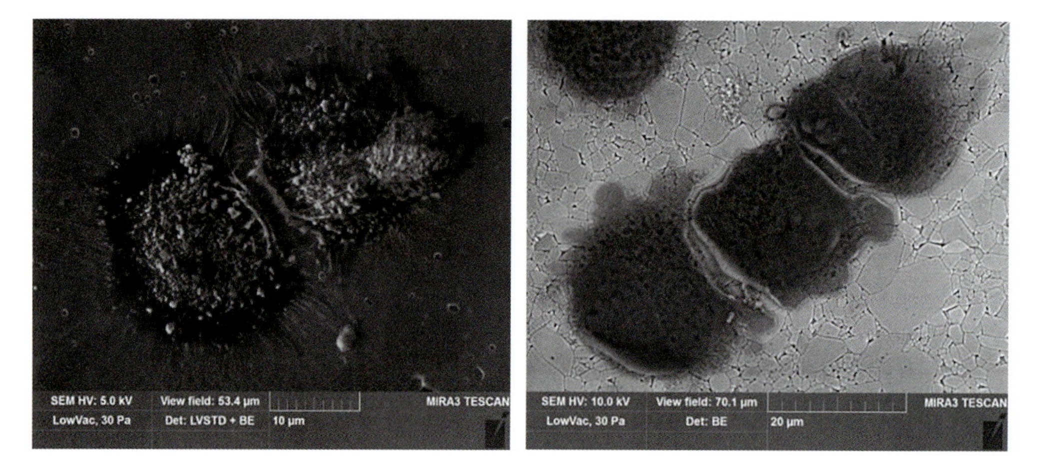

Figure 5.9 Spreading of the osteoblasts on the zirconia ceramics. Left: signal from a BSE detector. Right: mixed signal from LVSTD and BSE detectors.

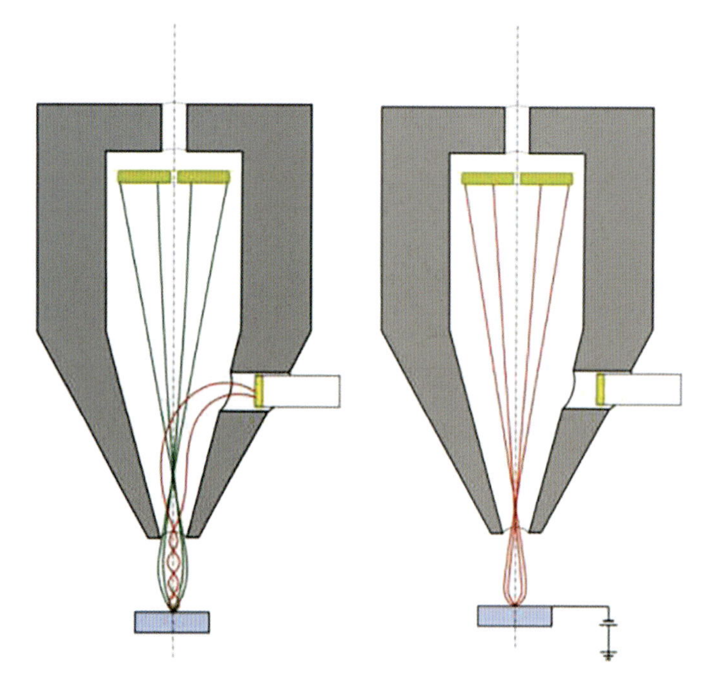

Figure 5.10 A schematic of the detection system inside the FE-SEM column. Left: in-beam SE and in-beam BSE (sample grounded). Right: detection of SE in BDM (sample biased). Paths of secondary electrons are in red and backscattered electrons in green.

of more analyzers. In the standard mode when the sample is grounded, in-beam SE attracts and detects electrons from the side while in-beam BSE is concentric around the optical axis (see Figure 5.10, left). In the beam deceleration mode the signal electrons are accelerated into the column and focused closer to the optical axis. A traditional approach is to use the in-chamber BSE detector below the column, but then only BSE are detected while SE are lost. The detection in the column (see Figure 5.10, right) is able to capture true secondary electrons (Jiruše *et al.* 2014b). Note that the in-chamber BSE detector can still be used with the new system (not depicted in Figure 5.10), and thus enables detection of both BSE and SE signals in BDM.

5.3 METHODS FOR IMAGING BIOLOGICAL SAMPLES

As mentioned earlier in the chapter, working with biological samples requires careful selection of sample preparation and imaging conditions. The main critical issues include low contrast, low conductivity, and beam and vacuum induced damage to samples. All the above mentioned artifacts become more prominent with increasing acceleration voltage of the primary electron beam, and they also lower the signal-to-noise ratio in the observed images. Figure 5.11 shows a typical workflow for preparing a soft biological sample for imaging.

As can be seen from Figure 5.11, the samples can be either (i) fixed, dried, sputter coated and observed in high vacuum or (ii) dried and imaged using a low energy electron beam without any coating in high vacuum, (iii) dried, observed non-coated in the elevated pressure or (iv) imaged directly in a humid environment of water vapor pressure.

A regular workflow for imaging biological samples in a high vacuum environment involves several steps. First, biological structures and cellular activity are fixed and stabilized. This can be achieved by either freezing the samples or chemically fixing them.

In cryo-SEM, the sample is frozen directly in liquid nitrogen slush or by high pressure freezing, carbon or metal sputter-coated and then observed in the frozen state on a cryo-stage under high vacuum conditions. TESCAN routinely incorporates in the SEMs a variety of cryo-holders from different manufacturers, but more details about the technology are beyond the scope of this chapter and are reviewed elsewhere in this book.

Chemical fixation involves treating the sample with formaldehyde or glutaraldehyde, which crosslinks proteins and peptides and stabilizes the tissue. This primary fixation can be followed by a secondary treatment with osmium tetroxide and/or uranyl acetate. These agents can also help in improving sample contrast and reducing charge accumulation. The

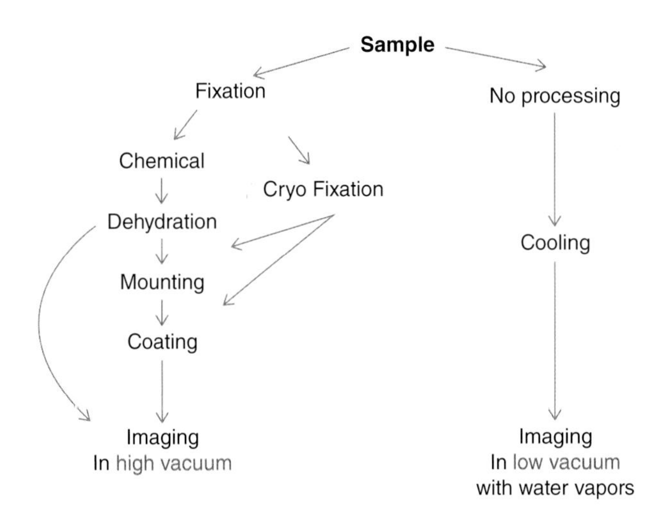

Figure 5.11 A typical workflow for imaging of a biological sample for SEM.

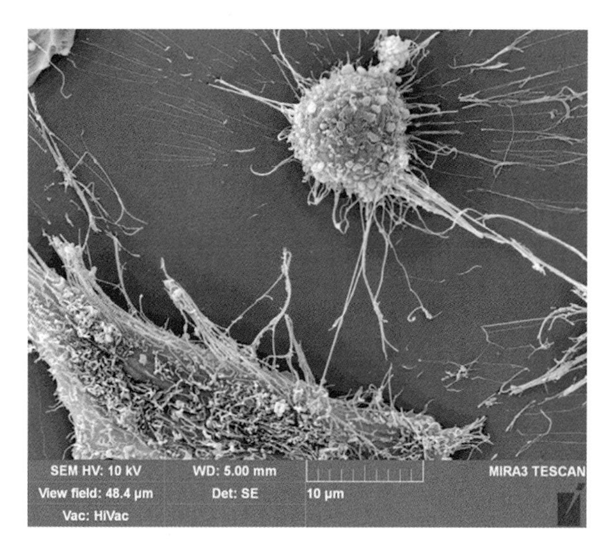

Figure 5.12 Image of fibroblasts. The cells were chemically fixed, critical-point dried, sputter-coated and observed in high vacuum.

next step is dehydration of the sample, which is typically done by treating the tissue with a series of water/alcohol mixtures (e.g. ethanol) with the aim to fully replace water by an organic solvent with minimal changes in the sample volume. The final step is usually critical-point drying, where the organic solvent is replaced with supercritical carbon dioxide and then evaporated out of the sample. Finally, the samples are sputter-coated and then imaged in high vacuum at acceleration voltages typically between 10 and 20 keV. Figure 5.12 shows fibroblasts prepared by this method and imaged in high vacuum.

5.4 IMAGING OF SENSITIVE SAMPLES

Most biological samples are sensitive to radiation damage or to drying. This leads to using low energies of primary electron beam, low vacuum and freezing the samples. These methods are described in the following sections.

5.4.1 Imaging at Low Energies

A commonly observed problem is beam induced damage to the samples (Kirk *et al.* 2008). Many uncoated samples sustain a great deal of damage when exposed to the electron beam with energy higher than 5 keV (Schatten and Pawley 2008; Schatten 2011). Also the large interaction volume can cause a loss of resolution in imaging the fine structures of the sample surface. Low energies are useful also for non-conductive specimens since there exists a balance where the number of emitted electrons is equal to the number of incident ones, thus suppressing charging effects. For these reasons, there has been a continuous trend in lowering the primary beam energy while sustaining high resolution and good contrast of the produced images. On the other hand, at lower energies all aberration effects inside the electron column are increased and result in lower resolution. In particular, the chromatic aberration, being inversely proportional to the electron's energy, becomes more significant. Our

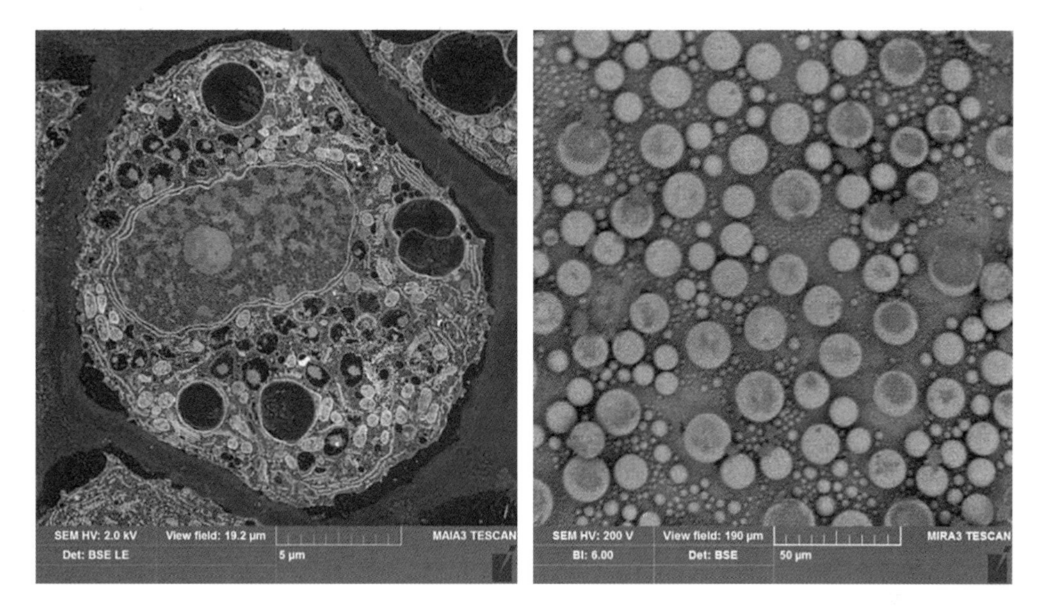

Figure 5.13 Capability of a low-energy BSE detector (with no sample bias). Left: barley root imaged at 2 keV with a good contrast. Right: tin spheres on carbon show the detection limit at 200 eV primary beam energy and a relatively low current of 70 pA.

approach to solve the problem includes the beam deceleration mode and ultra-high resolution immersion optics, which are described above. These approaches can even be combined.

Besides resolving power, signal detection is also important. Common state-of-the-art BSE detectors without BDM (for example, in case of a non-conducting sample) and without any beam acceleration/deceleration inside the column do not enable the use of electron energies below 2–3 keV due to a diminishing signal. We introduced a new type of low-energy BSE detector based on a scintillator with a special surface treatment (Kološová *et al.* 2014) that has a detection limit of 200 eV (see Figure 5.13). Such a detector may be used even in instruments with a conventional optics and can operate in the whole range of acceleration voltages and chamber pressures.

5.4.2 Low Vacuum Operation

Although using low voltage imaging can successfully prevent beam damage to the sample, charging may still be an important issue, for example, if the sample consists of a mixture of materials with different electron emission behaviors. One of the promising techniques is variable pressure (VP-SEM) or environmental scanning electron microscopy (Stokes 2008). This adaptation minimizes the need for the sample preparation steps discussed above.

In general, the instrument can keep higher pressures of tens to two thousand pascals inside the chamber while keeping the low pressure in the column and electron source. This is achieved by dividing the electron optical system into several differentially pumped regions using pressure limiting apertures. The presence of the gas molecules inside the chamber has several consequences on the sample and the observations. The most commonly used gases are N_2 and H_2O (water vapor).

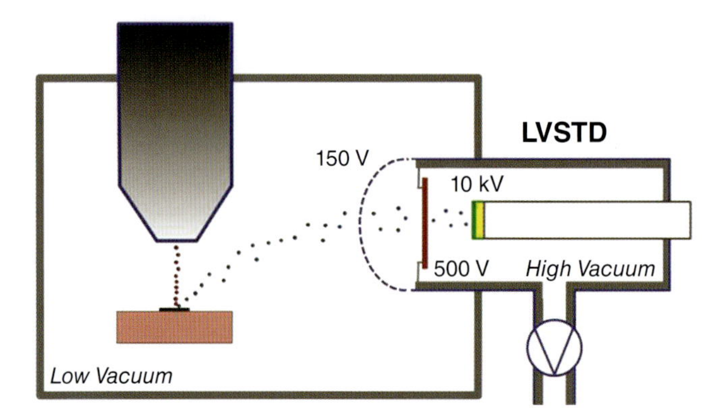

Figure 5.14 Schematic arrangement of low vacuum secondary electron detector (LVSTD). Microlens differential barrier in red separates the detector chamber with lower pressure from main chamber with higher pressure. See the text for an explanation.

The standard Everhart–Thornley detector cannot be used in the VP mode due to discharging between the scintillator and the grounded parts of the microscope if the pressure is higher than approximately 5 Pa. The detector uses a scintillator, usually a single crystal of Y-Al garnet (YAG) doped with Ce ions, held at a voltage between 5 and 10 kV. The patented low vacuum secondary-electron detector (LVSTD) works up to 1000 Pa and optionally to 2000 Pa. It is suitable for the investigation of non-conductive samples. The LVSTD (see Figure 5.14) consists of a standard Everhart–Thornley detector situated in a separate detector chamber pumped by a small turbo molecular pump. The low vacuum of the microscope specimen chamber and the vacuum of the detector chamber are separated by a microlens differential barrier, a thin foil with microscopic openings. It serves as a "wall" for differential pumping and at the same time it creates an array of microlenses focusing the secondary electrons collected by the hemispherical grid into the holes in the disc. The electrostatic field generated by the high potential on the scintillator layer penetrates through the holes of the barrier and, together with the 500 V potential on the barrier, works as lenses focusing electrons into the holes. In the improved vacuum inside the detector chamber no discharges between the detector and the grounded walls of the chamber occur (Jacka *et al.* 2003). The secondary electrons arising in the chamber are collected by the grid, approach the microlens differential barrier, and are subsequently accelerated and focused by the field in the vicinity of the openings, and in the detector chamber they are detected in a standard way. The main advantages of the LVSTD detector are fast response times as well as detection of the true secondary electron signal (see Figure 5.15).

5.4.3 Observation of Biological Samples Without any Preparation

In general, hydrated samples (especially biological samples) should be observed as close as possible to their natural state. Water evaporation is the most important phenomenon that has to be considered when working with biological samples. Low pressure in the SEM chamber causes rapid water evaporation from the sample, which results in sample shrinkage and collapse of the soft structures. Although this can be prevented to some extent by sample preparation, those processes can introduce other artifacts. In the worst case scenario, the shrinkage can be up to 60% in critical-point dried samples and up to 10–15% in cryo-fixed

Figure 5.15 Image of *Gigaspora rosea* obtained in a low vacuum with an LVSTD detector.

samples (Gusnard and Kirschner 1977). One of the commonly used solutions is to keep a humid environment inside the chamber. This is achieved by setting temperature and pressure conditions by either evaporation or sublimation close to the triple point of water, that is, 0 °C and 612 Pa. The evaporation is reduced at lower temperatures, but at temperatures below the freezing point water evaporation is slower but the formation of ice crystals can cause massive damage to the sample. Therefore it is recommended to use temperatures from 1 to 5 °C and pressures of 400–700 Pa. Moreover, all available commercial detectors lose the image quality above 700–1000 Pa and, except for a few experimental systems, they do not give images better than those observed in the light microscope. A comparison of imaging of the same sample in high vacuum and in the presence of water vapor can be seen in Figure 5.16.

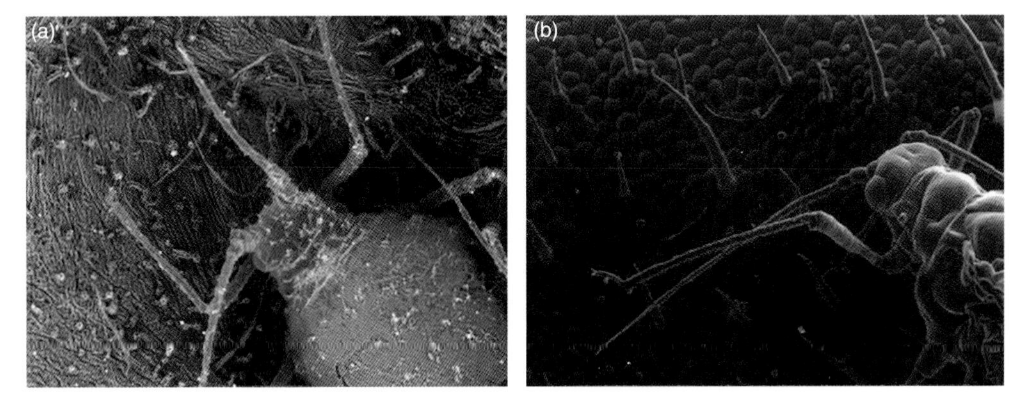

Figure 5.16 An aphid on a leaf observed in (a) high vacuum and (b) low vacuum at –20 °C and 25 Pa of water vapor using SE and LVSTD detectors, respectively. Extensive damage to the leaf surface and shrinkage caused by drying can be seen in panel A. The field of view is 2 mm.

Several factors need to be taken into consideration. Biological samples are not composed of just pure water, but contain dissolved salts and other compounds. These "impurities" lower the partial pressure of water and cause deviations from the dependence given by the phase diagram of pure water. In the case of biological samples, the cytoplasm contains a variety of dissolved polysaccharides, proteins and minerals that lower the equilibrium vapor pressure of the system by up to 20–25% in comparison with pure water. Investigation of the hydrated samples is therefore better carried out at conditions below 100% humidity relative to water. Otherwise, condensation of water molecules on the sample can hide surface structures (usually at higher than 90% relative humidity). Finally, the conditions will largely depend on the specimen and its stability against water loss. Many samples are equipped with mechanisms (e.g., cell membrane) that protect live organisms against unsuitable environments and can withstand slightly dehydrating conditions (Stokes 2003). Therefore, additional lowering of the chamber pressure (another 20–25%) can be tolerated and help during sample investigation. In our experience, pressures around 450 Pa at 2 °C (this corresponds to about 70% relative humidity inside the chamber) usually cause only mild dehydrating conditions of the samples. This is supported by observations of Muscariello *et al.* (2005), who found that the best conditions for observing biomedical samples are approximately 60% relative humidity and about 70% for cell monolayers.

To summarize, working conditions of 450–550 Pa and 2–3 °C are recommended for most practical situations although it is always necessary to take into account the structure and properties of each hydrated sample.

5.5 ADVANCED FE-SEM TECHNIQUES IN BIOLOGY

Standard SEM microscopy analyzes surface features with down to nanometer resolution by monitoring elastic and inelastic interactions of electrons with the sample surface. Intensity of secondary electron emission reflects surface topography, while backscattered electrons provide additional information about atomic contrast. However, more complex applications make it necessary to gather more information from the same region of interest. In this section we will focus on two main trends in advanced SEM instrumentation – correlative microscopy and tomography with focused ion beam systems. In life sciences, correlative microscopy is a rapidly growing field, driven by the search for relationships between structure and function (Jahn *et al.* 2012) and using the focused ion beam (FIB) instruments also allows use of these techniques for investigation of intracellular events.

5.5.1 Correlative Microscopy

In general, correlative microscopy involves characterization of the same region of interest by two or more microscopic or analytical techniques. The main idea of correlating information from more methods is to increase knowledge about the particular sample. In biological science, most applications are done by correlating data from light microscopic (LM) methods with electron microscopy. There are two main setups as to how the correlative microscopy can be arranged depending on the level of integration of the LM method into the SEM.

The first approach involves direct integration of LM into an SEM. The same spot can be observed or analyzed by both techniques without the need to transfer the sample. A fluorescence microscope has been integrated directly into the SEM (Zonneyville *et al.* 2013). There is a large number of selective fluorescence probes such as chemical dyes, fluorescently labeled antibodies and genetic probes, which can target almost any molecule type within the cells or tissues. The presence and changes in fluorescence intensity of the dyes observed in the fluorescence microscope reflects the location and activity of the target molecules. The limitation of fluorescence imaging is its low resolution. With the exception of new super-resolution techniques (such as STORM, PALM, etc. Schermelleh *et al.* 2010), the resolution falls in the range of a few tens of nanometers. When combined with SEM, one can visualize an ultra-fine structure of targeted features with nanometer resolution.

Direct integration of more techniques into one instrument has great benefits for correlative microscopy since all analysis and imaging is guaranteed at the same region of interest as the sample itself, which keeps its state. Transferring the sample among various instruments with changes of pressures and humidity is often disadvantageous. Moreover, some analytical techniques are destructive so simultaneous analyses are needed. Extensive integration of many techniques by TESCAN is described in the next sections.

In the case of separate instruments, offline correlation is required. In this approach, the target areas of interest are usually visualized in the light microscope first. Afterwards, the samples are processed by one of the methods (e.g., fixed and dried, frozen down, variable pressure) for SEM. The sample is placed into an SEM microscope and aligned with the image obtained from the LM using dedicated software. In case the LM has a motorized stage, the coordinate systems of the LM and SEM are easily aligned and then the original coordinates can be used to navigate the stage in the SEM. In case the LM does not have a motorized stage, the coordinate systems of LM and SEM have to be aligned using fiducial marks found in the original LM image and in the SEM scanning window. This step should ideally be very simple for the user. In the case of TESCAN systems, the image alignment takes advantage of the low magnification imaging in dedicated modes of Wide Field Optics, where a large overview of the whole coverslip can be used for SEM alignment. After the alignment, the SEM is navigated according to the image obtained in the LM. Finally, the images are overlaid to correlate them. The major benefit of this approach is that the input data can be obtained from almost any light microscope.

5.5.2 Integration of the Confocal Raman Microscope in SEM

The combined confocal Raman (CRM) and scanning electron microscope is a product of the joint cooperation of TESCAN and WITec companies. CRM is based on a light microscope that is adapted to work inside the SEM chamber and it is therefore capable of imaging in vacuum. A laser beam is introduced through the light microscope objective lens. The two columns are either tilted one with respect to the other and directed at a common intersection point on the sample surface, or both are parallel and the sample is moved automatically between two observation positions. In both cases an identical working point and subsequent correlative microscopy is ensured (Jiruše *et al.* 2014a).

CRM is a major innovation compared to conventional state-of-the-art Raman systems for operation inside a SEM chamber, which are mostly based on focusing the introduced laser beam on the sample with a parabolic mirror located below the SEM column. These systems do not allow scanning the beam on the sample and they are limited either to obtaining only simple Raman spectra or to slow sample scanning. When just a single Raman spectrum is acquired, one can never be sure whether the Raman spectrum is really obtained from a certain position or whether the position calibration is off. Moreover, the resolution of the parabolic mirror based systems is typically only about 2–5 μm, which cannot be compared to the standard stand-alone Raman microscopes (in the air). There are three basic

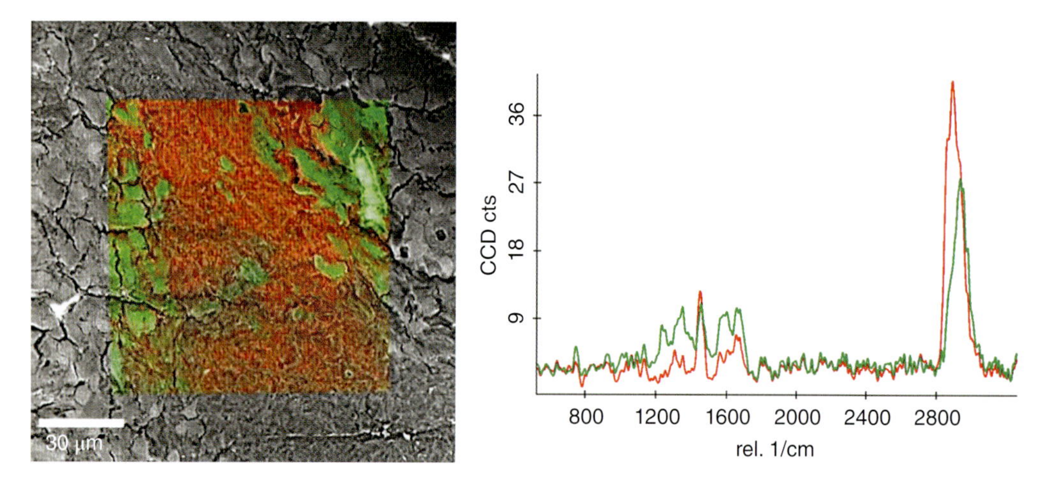

Figure 5.17 BSE image of a hamster brain with overlaid Raman image (left) and spectra (right); the white and gray matter are coded red and green, respectively.

advantages with the light-microscope-based CRM inside SEM: first, it is possible to obtain complete Raman mapping instead of only point spectra, second, to increase the resolution considerably to about 360 nm and third, to overlay the Raman and SEM images for comparison. Moreover, due to its confocal nature, CRM also allows 3D non-destructive Raman tomography spectral data-cubes to be created for transparent samples.

An example in Figure 5.17 shows a part of a thin section of the hamster brain, where one can see white and gray brain matter. Both brain constituents can be distinguished by their Raman spectra and they also show differences in SEM contrast. The SEM image was taken in low vacuum 20 Pa at 10 keV using a BSE detector. The Raman image, taken at the same position, is overlaid with the SEM image. The white and gray brain matter is shown in the Raman image as red and green, respectively. The spectra obtained in the different areas are shown as well.

5.5.3 FIB-SEM Instrumentation

Focused ion beam-scanning electron microscopy (FIB-SEM) combines a standard FE-SEM column with an additional ion column that coincides at the same observation point. The Ga FIB column COBRA from Orsay Physics with a minimum spot size of 2.5 nm was seamlessly integrated with TESCAN SEM using a unified software system (called LYRA for conventional electron optics and GAIA for immersion optics). It gives the FE-SEM microscopes the ability to locally mill materials or to deposit new materials on to the sample surface with electron or ion beam assisted deposition from metalorganic precursor gases. The rapid development of this technique has been driven by the needs of uncovering subsurface features in materials science research and the semiconductor industry. The ability to perform cross-sectioning and 3D visualizations of analyzed samples has drawn significant attention from researchers in the biomedical engineering and the life sciences. Ga FIB can be used to prepare lamellas for TEM, expose selected parts of the object for observation, even to perform 3D tomography in the SEM (see below) and, in conjunction with the gas injection system, to deposit nano-size objects both with the electron or ion beams (Giannuzzi and Stevie 2005).

In place of Ga FIB, an Xe plasma source FIB-SEM (called FERA for conventional electron optics and XEIA for immersion optics) with up to 50 times higher sputtering efficiency is finding many applications where large volume milling is needed. An example is given in Figure 5.18, where 3D tomography of a dental tissue using FERA plasma FIB-SEM is shown (Shahidi *et al.* 2015). Such a large-volume milling of 400 × 1000 × 400 μm^3 would

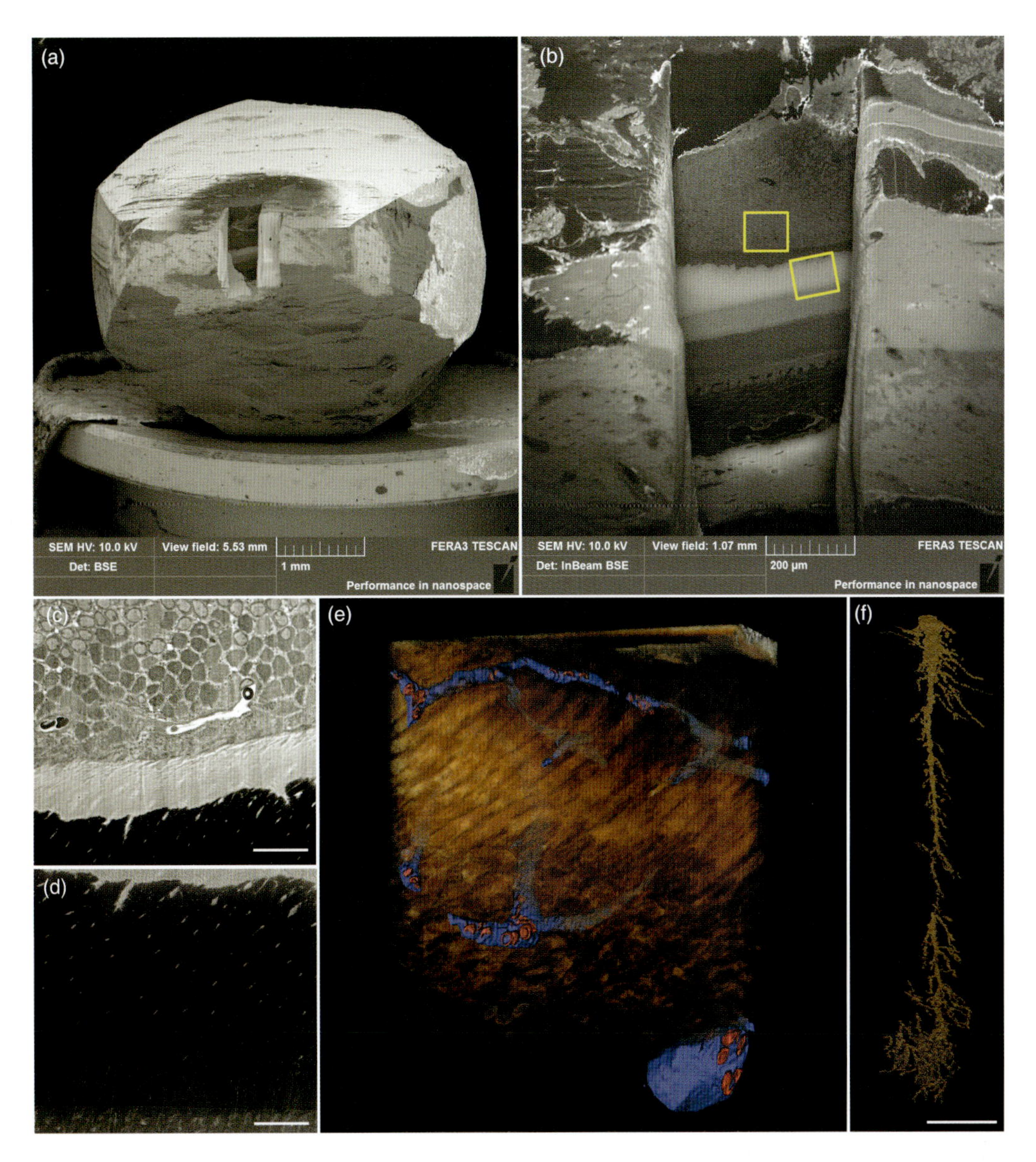

Figure 5.18 3D tomography of a dental tissue using FERA plasma FIB-SEM. (a) Mouse embryo tooth embedded in a resin. Large scale image taken by the wide field mode. The removal of 400 × 1000 × 400 μm^3 of bulk resin was done using a 2 μA ion current. (b) Localization of tissue layers and selection of regions of interests for serial slicing. (c) BSE image of veins in the odontoblast layer. (d) BSE image of channels in the dentin layer (the scale bar is 20 μm). (e) 3D reconstruction of the vein system (veins are shown in blue, red blood cells in red and odontoblasts in brown). (f) 3D visualization of the dentin channels architecture (the scale bar is 20 μm).

be unthinkable with a Ga ion beam. This application is not suitable for serial block-face imaging either because of the heterogeneous nature of the sample. Dental layers strongly differ in hardness, which may result in slicing artifacts or knife damage. Plasma FIB is ideally suited here.

In all types of FIB-SEM a time-of-flight secondary ion mass spectrometry (TOF-SIMS) can be added, which allows distinguishing isotopes and very light elements with lateral resolution down to 50 nm (Whitby *et al.* 2012). Future applications of this technique in biology are to be expected.

Unprecedented levels of integration are achieved by combining FIB-SEM with TOF-SIMS, confocal Raman microscope (CRM), traditional analytical techniques of X-ray spectroscopy (EDX/WDX), electron backscattered diffraction (EBSD) and even scanning probe microscope (SPM) in a single instrument (Jiruše *et al.* 2012b; Zadražil *et al.* 2012) (see Figure 5.19). This novel multipurpose nanoanalytical tool meets growing needs to visualize, manipulate, modify and analyze nano-objects in a single tool. SPM instruments with lateral resolution of 1 nm and depth resolution of 0.1 nm in combination with SEM yield precise information about the topography of the sample, while elemental and compositional analysis is delivered by TOF-SIMS, EDX and CRM. The last technique also allows conventional light microscopy imaging. While EDX gives elemental composition, light microscope-based confocal CRM can analyze chemical bonds with 5–10 times better resolution than a conventional Raman spectrometer in the SEM using a parabolic mirror. Structural characterization of the sample is done by EBSD. The user can then choose an arbitrary combination of the above-mentioned techniques in addition to the standard set of SEM detectors of secondary and backscattered electrons, both in chamber and in column, transmitted electrons, panchromatic or color cathodoluminescence, etc. Moreover, all of this is in a single analytical instrument, eliminating the usual problems of correlative microscopy of finding the same region of interest while transferring a sample among several stand-alone techniques.

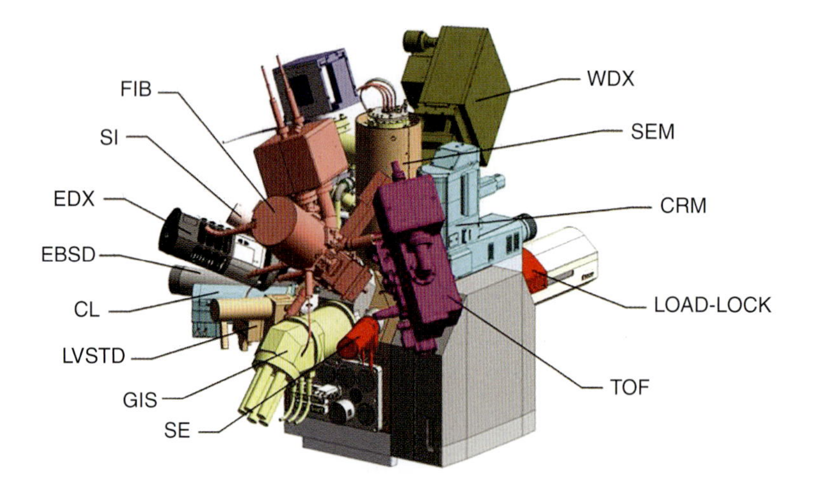

Figure 5.19 Integration of FIB-SEM with secondary ion mass spectrometry (TOF-SIMS), scanning probe (SPM) and confocal Raman microscope (CRM). Standard analyzers also include X-ray spectroscopy (EDX, WDX), electron backscattered diffraction (EBSD), gas injection system (GIS), detectors of secondary electrons (SE), secondary ions (SI), cathodoluminescence (CL) and secondary electrons in low vacuum (LVSTD) and a load-lock system. This enables sample imaging, manipulation, modification and analysis – all in one multifunctional tool.

Some techniques can provide 3D analysis on their own. TOF-SIMS can create 3D mapping during the process of destructive analysis. CRM, due to its confocal nature, can also perform non-destructive 3D Raman tomography of transparent samples. Other techniques like EDX, EBSD, EBIC, BE or TE enable 3D tomography in connection with FIB in automatic repeating of FIB slicing followed by analysis, as described in the following section. The multifunctional tool in Figure 5.19 enables correlation between alternative 3D methods.

5.5.4 Preparation of Cross-sections and 3D Tomography in FIB-SEM

3D reconstruction of biological tissues has mostly been a domain of TEM. It is commonly performed either by sequential slicing the sample using the ultra-microtome followed by imaging of each section, quite a tedious and time consuming procedure. Another approach, 3D tomography, uses a thicker slice and performs projections by imaging under different tilt angles, from which the 3D model is reconstructed with a resolution that can reach down to 3 nm (Barcena and Koster 2009). The cross-sections are particularly useful in studies of cell material interactions, site specific subcellular visualization and revealing the 3D organization of subcellular structures within cells and tissue.

In FIB-SEM the tissue or cell samples can be prepared in similar fashion as those for TEM, which includes sample fixation, contrasting with heavy metals and embedding into a resin (Leser *et al.* 2008). The 3D analysis includes milling a trench into the sample surface, followed by imaging or analyzing the cross-sections created in different depths, or producing a lamella as a sample for TEM observations. 3D tomography of biological samples with Ga FIB-SEM is capable of resolution down to 5–10 nm, given by the thickness of the removed slice, but typically it is in the range of 20–50 nm. This is much thinner than the standard 80–100 nm slices produced by a diamond knife in an ultra-microtome. In this way even large volumes of thousands of cubic micrometers can be successfully reconstructed. The quality of the reconstruction strongly depends on the resin type, and since the volumes are visualized according to the gray levels, the right selection of contrast agents is crucial. Figure 5.20 shows a 3D reconstruction of cerebellum tissue using the FIB-SEM system with a low energy BSE detector.

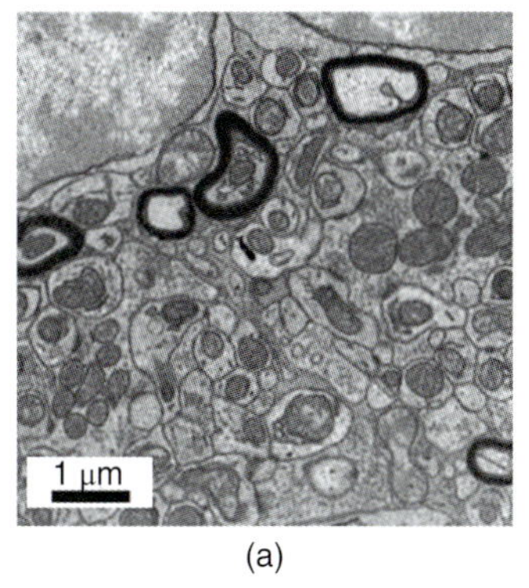

(a)

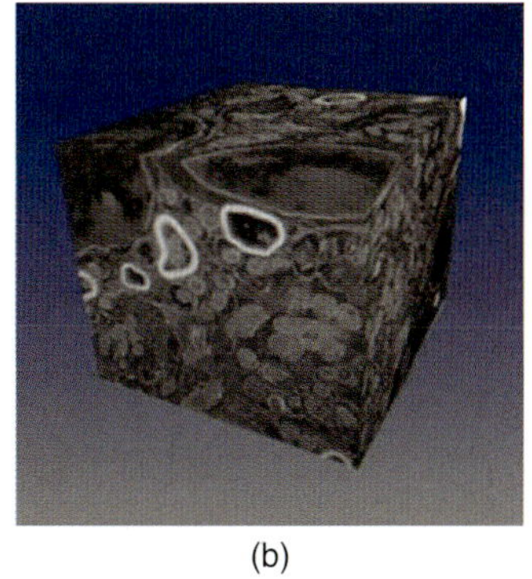

(b)

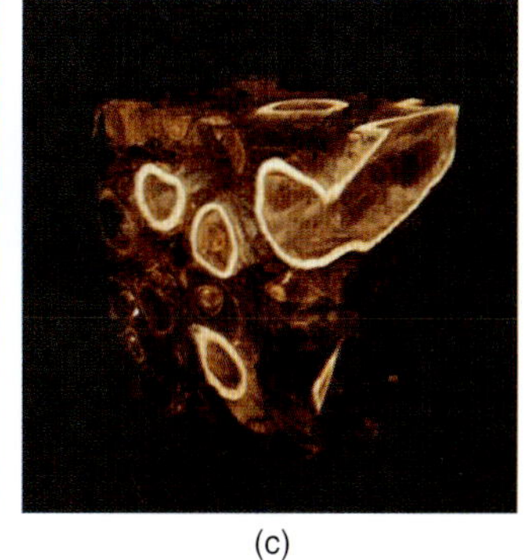

(c)

Figure 5.20 3D reconstruction of mouse cerebellum tissue embedded in resin. FIB slicing was done with a 200 pA beam at 30 kV and SEM images acquired at 2 keV using a low energy BSE detector. (a) A BSE image of one slice, (b) a 3D image stack containing 560 slices, each 5 nm thick, and (c) its 3D visualization using an ORS Visual, http://www.theobjects.com. software package.

CONCLUSIONS

SEM is a technique with a great potential in the life sciences. In particular the FE-SEM enables the operator to obtain high resolution data that may be further enhanced by the application of advanced imaging techniques based on the beam deceleration method and immersion optics. The display modes used in TESCAN microscopes allow imaging with enhanced depth of focus or enhanced field of view that is especially useful for navigation. The ability to investigate surfaces with very high resolution can provide valuable data about cell functions, membrane processes, adhesion mechanisms or host–pathogen interactions. Although a lot of research has been done to establish this technology in various fields in the life sciences, there is a lot of work ahead in order to transform it into a routine tool for cell and molecular biologists. It is especially desirable to investigate different sample preparation techniques, FIB resistant resins, fixing procedures and optimizing FIB conditions, including, for example, acceleration voltages and milling currents, or extend the possibilities of correlative microscopy. The combination with other analytical technologies available in TESCAN microscopes, such as TOF-SIMS, Raman spectroscopy, SPM or correlative microscopy, has reached unprecedented levels of integration that broaden the analytical possibilities of FE-SEMs.

ACKNOWLEDGEMENTS

The authors thank Kristýna Rosíková, Filip Lopour and Miloslav Havelka from TESCAN and Olaf Hollricher from WITec for their support.

REFERENCES

Barcena, M. and Koster, A.J. Electron Tomography in Life Science. *Sem. Cell Dev. Biol.*, 20, 920–930 (2009).

Giannuzzi, L.A. and Stevie, F.A. *Introduction to Focused Ion Beams. Instrumentation, Theory, Techniques and Practice*, Springer (2005).

Gusnard, D. and Kirschner, R. Cell and Organelle Shrinkage During Preparation for Scanning Electron Microscopy: Effects of Fixation, Dehydration and Critical Point Drying. *Journal of Microscopy*, 110, 51–57 (1977).

Hrnčíř, T., Lopour, F., Zadražil, M., Delobbe, A., Salord, O. and Sudraud, P. Novel Plasma FIB/SEM for High Speed Analysis and Real Time Imaging of Large Material Removal, in *ISTFA: Conference Proceedings from the 38th International Symposium for Testing and Failure Analysis*, pp. 26–29 (2012).

Jacka, M., Zadražil, M. and Lopour, F. A Differentially Pumped Secondary Electron Detector for Low-Vacuum Scanning Electron Microscopy. *Scanning*, 25, 243–246 (2003).

Jahn, K.A. *et al.*, Correlative Microscopy: Providing New Understanding in the Biomedical and Plant Sciences. *Micron*, 43, 565–582 (2012).

Jiruše, J., Haničinec, M., Havelka, M., Hollricher, O., Ibach, W. and Spizig, P. Integrating Focused Ion Beam – Scanning Electron Microscope with Confocal Raman Microscope into a Single Instrument. *Journal of Vacuum Science and Technology*, B32, 06FC03 (2014a).

Jiruše, J., Havelka, M. and Lopour, F. Novel Field Emission SEM Column with Beam Deceleration Technology. *Ultramicroscopy*, 146, 27–32 (2014b).

Jiruše, J., Havelka, M., Haničinec, M., Polster, J. and Hrnčíř, T. New High-Resolution Low-Voltage and High Performance Analytical FIB-SEM System. *Microsc. Microanal.*, 20 (Suppl. 3), 1104 (2014c).

Jiruše, J., Havelka, M., Polster, J. and Lopour, F. New Ultra-High Resolution SEM for Imaging by Low Energy Electrons. *Microsc. Microanal.*, 19 (Suppl. 2), 1302–1303 (2013).
Jiruše, J., Havelka, M., Polster, J. and Hrnčíř, T. Xe Plasma FIB-SEM with Improved Resolution of Both Ion and Electron Columns. *Microsc. Microanal.*, 21 (Suppl. 3), 1995 (2015).
Jiruše, J., Hrnčíř, T., Lopour, F., Delobbe, A. and Salord, O. Combined Plasma FIB-SEM. *Microsc. Microanal.*, 18 (Suppl. 2), 652–653 (2012a).
Jiruše, J., Sedláček, L., Rudolf, M., Friedli, V., Östlund, F. and Whitby, J. Combined SEM-FIBTOF-EDX-EBSD as a Multifunctional Tool. *Microsc. Microanal.*, 18 (Suppl. 2), 638–639 (2012b).
Kirk, S.E., Skepper, J. N. and Donald, A.M. Application of Environmental Scanning Electron Microscopy to Determine Biological Surface Structure. *J. Microsc.*, 233, 205–224 (2008).
Kološová, J., Beránek, J., Jiruše, J. and Horodyský, P. New scintillation low-energy BSE detector, in *18th IMC Proceedings*, p. 442 (2014).
Leser, V. *et al.*, Comparison of Different Preparation Methods of Biological Samples for FIB Milling and SEM Investigation. *J. Microsc.*, 233, 309–319 (2008).
Muscariello, L., Ross, F. *et al.*, A Critical Overview of ESEM Applications in the Biological Field. *J. Cell Physiol.*, 205, 328–334 (2005).
ORS Visual, http://www.theobjects.com.
Schatten, H. and Pawley, J. *Biological Low-Voltage Scanning Electron Microscopy*, Springer (2008).
Schatten, H. Low Voltage High-Resolution SEM (LVHRSEM) for Biological Structural and Molecular Analysis, *Micron*, 42, 175–185 (2011).
Schermelleh, L., Heintzmann, R. and Leonhardt, H. A Guide to Super-Resolution Fluorescence Microscopy. *J. Cell Biol.*, 190 (2), 165–175 (2010).
Shahidi, M. *et al.*, Three Dimensional Imaging Reveals New Compartments and Structural Adaptations in Odontoblasts. *J. Dental Research*, 94 (7), 1–10 (2015); doi: 10.1177/0022034515580796.
Stokes, D.J. Recent Advances in Electron Imaging, Image Interpretation and Applications: Environmental Scanning Electron Microscopy. *Phil. Trans. R. Soc. A*, 361, 2771–2787 (2003).
Stokes, D.J. *Principles and Practice of Variable Pressure/Environmental Scanning Electron Microscopy (VP-ESEM)*, John Wiley & Sons, Ltd (2008).
Whitby, J.A. *et al.*, High Spatial Resolution Time-of-Flight Ion Mass Spectrometry for the Masses: A Novel Orthogonal ToF FIB-SIMS Instrument with in situ AFM. *Adv. Mater. Sci. Engng*, 180437, 13 (2012).
Zadražil, M., Jiruše, J., Lencová, B., Dluhoš, J., Hrnčíř, T., Rudolf, M., Sedláček, L. and Šamořil, T. The Step Towards an Ultimate Multifunctional Tool for Nanotechnology, in *Proceedings of the 2012 International Conference on Manipulation, Manufacturing and Measurement on the Nanoscale (3M-NANO)*, pp. 175–179 (2012).
Zonneyville, A.C. *et al.*, Integration of High-NA Light Microscope in a Scanning Electron Microscope. *J. Microsc.*, 252, 50–70 (2013).

6

FEG-SEM for Large Volume 3D Structural Analysis in Life Sciences

Ben Lich, Faysal Boughorbel, Pavel Potocek and Emine Korkmaz
Thermo Fisher Scientific, Eindhoven, The Netherlands

6.1 INTRODUCTION

Recent advances in the design of SEM using field emission guns (FEGs) permit rapid acquisition and reconstruction of three-dimensional (3D) structures with a resolution sufficient to trace neuronal networks and other cellular and subcellular features through relatively large volumes. Here we review general considerations relevant to the optimization of a workflow using SEM for biological applications and then look specifically at the requirements for high speed 3D applications.

Biological systems are massively complex, containing structural details down to the atomic scale that may be functionally relevant over distances approaching the millimeter or even larger scales. Neuronal networks are an example where functional analysis of network connections requires the ability to follow small cellular processes and identify small organelles such as synaptic vesicles or microtubules. At the same time these processes can span over distances of hundreds of micrometers or larger. These features, and many other cellular structures and subcellular organelles involved in physiologically important processes, are too small to be resolved in the context of the cellular architecture by light microscopy. Electron microscopes have sufficient resolution but have been challenged by the sample contrast and susceptibility to beam damage of biological materials. Recent advances in two aspects of SEM design, namely the efficient design of modern FEG as well as the ability to image with very low electron beam energy (beam deceleration) capabilities, have dramatically improved imaging performance with (of) biological materials. These new technologies at low accelerating voltages have enhanced both spatial and depth resolutions by reducing beam penetration. At the same time advances in automation and analytical

Biological Field Emission Scanning Electron Microscopy, First Edition.
Edited by Roland A. Fleck and Bruno M. Humbel.

and computational methods enabled the rapid acquisition and reconstruction of structures spanning hundreds of micrometers with isotropic resolution approaching the nanometer scale.

6.2 HIGH RESOLUTION SEM IMAGING AT LOW ACCELERATING VOLTAGE

An SEM creates an image by scanning a finely focused beam of electrons over the sample surface and collecting various signals that the beam electrons create as they interact with atoms in the sample. These signals are spatially correlated with the instantaneous position of the scanning beam and are usually presented as an image in which the gray scale value at any point represents the measured signal strength at the corresponding location in the scanned area. The resolution of the image is determined by the size of the region within the sample from which the signal emanates at any instant in time, which, in turn, is determined by the diameter of the beam when it hits the sample surface and the direction and distance the beam electrons penetrate into the sample as they give up energy and generate secondary events, resulting in different imaging signals. There must be sufficient current in the beam to yield a detectable signal and hence build contrast and suppress noise in the image. The beam should not damage the sample significantly over the period of time the image data are acquired. Additionally, preparation procedures required to make the sample compatible with the vacuum environment of the microscope, or to enhance imaging contrast, should alter the observed structures as little as possible. Finally, after the images are recorded, typically these need to be processed to form a reconstruction in 3D where segmentation of the relevant cellular organelles or processes allow the researcher to analyze the sample and find an answer to the biological questions that started the process.

6.3 SPOT SIZE

Conventional wisdom in electron optics has always encouraged the use of higher accelerating voltages to achieve higher resolution. This is based on the effects of chromatic aberration, which describes the dependence of the power of a lens on the energy of the electrons. An electron beam is composed of individual electrons having a range of energies. Chromatic aberration causes each electron to be affected differently by the lens field and thus limits the ability of the objective lens to form a small spot on the sample surface. Since the magnitude of the chromatic aberration is determined by the ratio of the range of electron energies (energy spread), which can be considered constant, to the accelerating voltage, its degrading effect becomes worse at lower voltages. The energy spread of the beam can be reduced by a monochromator, resulting in a beam with improved coherence and hence resulting in higher resolution imaging. By reducing the effects of chromatic aberration it permits subnanometer beam diameters at accelerating voltages down to 1 keV. Particularly for workflows typically using low voltage, low current applications like Cryo-SEM, the monochromator adds a lot of value for looking at surface details using secondary electron detectors. In Figure 6.1 cryo-fractured and metal-coated yeast is used to demonstrate the high quality imaging that can be achieved with a modern, monochromated microscope. Here not only the transmembrane particles can be visualized, but even the central channels of these particles can be observed in some cases.

The spherical aberration is another important factor in determining the point resolution of a microscope. It is caused by the lens field acting inhomogeneously on the off-axis electrons.

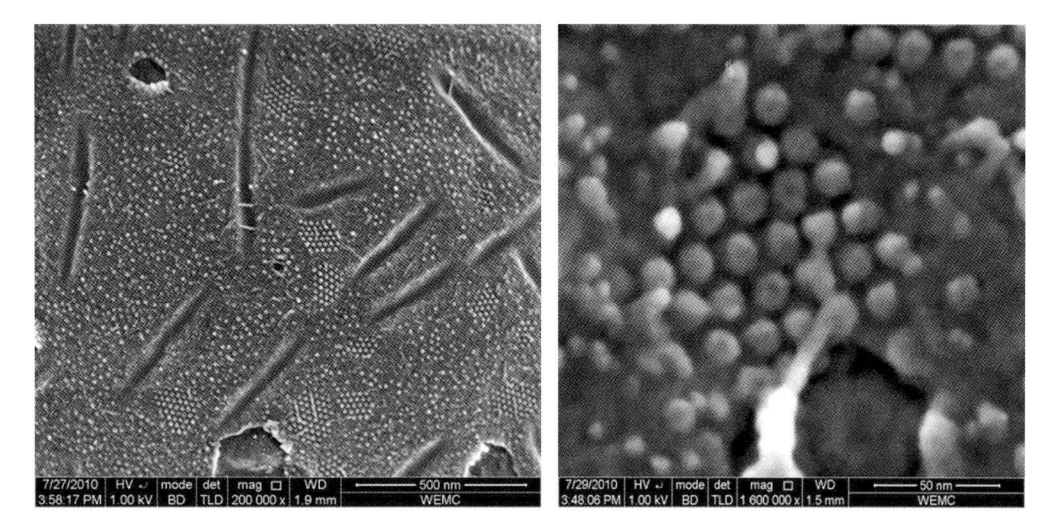

Figure 6.1 Freeze-fractured, metal-coated yeast plasma membrane. The sample was imaged with a monochromated 1 keV landing energy (5 keV primary beam – 4 keV BD). Sample courtesy of Dr A. van Aelst, Wageningen University, The Netherlands; imaging: I. Gestmann, Thermo Fisher Scientific Company; microscope: Magellan; Cryo-system: Leica VCT 100, Leica Microsystems, Vienna, Austria.

In other words, the electrons that are parallel to the optic axis but at different distances from the optic axis fail to converge at the same point. The further off-axis the electron is, the more strongly it is bent back toward the axis. This poses limitations on the ability to magnify details since the features are degraded by the imaging process and therefore do not produce a perfect focal point. By decreasing the effects of spherical aberration, it greatly improves the resolution at low keV imaging. In Figure 6.2, chemically fixed, heavy metal stained, and

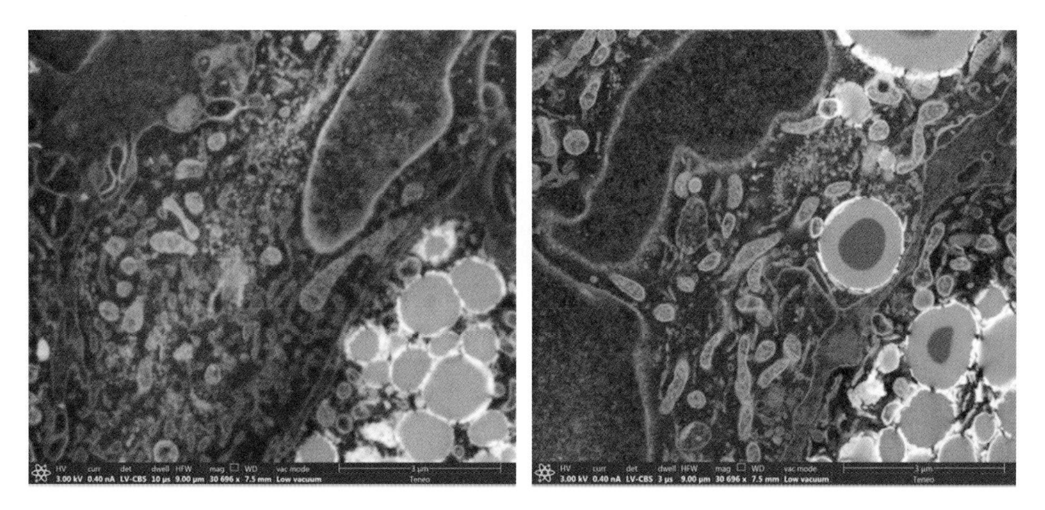

Figure 6.2 Chemically fixed, heavy metal stained, resin embedded biological sample. A comparison of spherical and chromatic aberration coefficients (WD 7.5 mm, landing energy 3 keV) for given use-cases; standard mode versus VolumeScope mode. In the VolumeScope mode, use of beam deceleration for soft landing energy via accelerating tube (AT) increases resolution and detection sensitivity, even with a shorter dwell time of 3 µs compared to the standard mode (without AT) with a 10 µs dwell time.

resin embedded sample is used to reveal the resolution improvements overcoming the limitations posed by the spherical and chromatic aberrations. For challenging, non-conductive biological samples, it is crucial to mitigate charge build-up on the sample via low vacuum (LoVac) mode imaging using water vapour, since it provides excellent resolution on sensitive samples without resolution deterioration from charging artifacts.

6.4 BEAM PENETRATION

The other determinant of resolution, particularly when using backscattered electrons (BSE) for image formation, like serial block face imaging, array tomography, or focused ion beam SEM, is the size of the volume of interaction: the region below the beam spot on the sample surface within which beam electrons interact with sample atoms to generate secondary imaging signals. It is generally known that the depth of interaction is determined by the beam landing energy and the material it interacts with. The shape of this interaction volume,

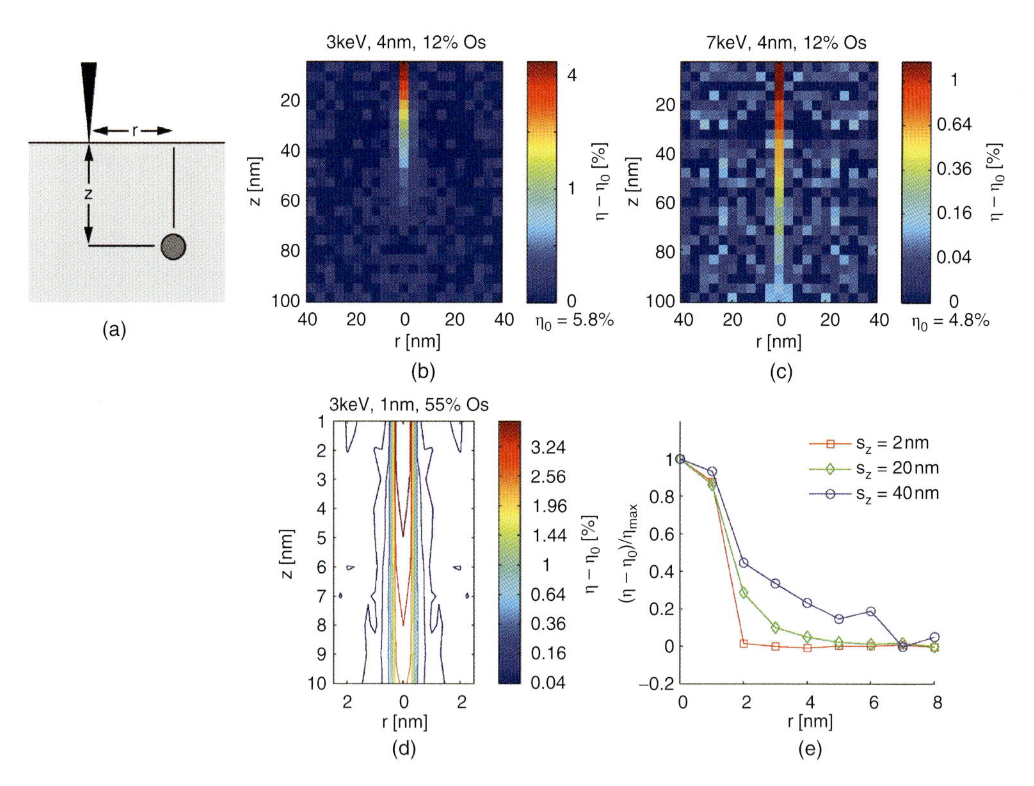

Figure 6.3 Point-spread functions. Principle of PSF mapping: (a) an Os containing sphere (12% Os and 4 nm diameter for panels (b), (c), and (e)) is placed at different depths, z, and radial distances, r, from the BIP. The PSFs for PBEs are 3 (b) and 7 keV (c). The region (d) close to the beam impact point was sampled again at higher resolution, using a smaller sphere (1 nm diameter, 55% Os). Each simulation contains 106 traces. The spacing in the color scale is roughly proportional to the square root of the signal. T0 is 5.82% for 3 and 4 keV, 4.76% for 7 keV. Horizontal (e) cross-sections through the PSF at different depths, normalized to their peaks, sampled at a higher lateral resolution, using 106 traces per data point. Note that absolute values are displayed (due to noise T – T0 can go slightly negative).

however, particularly for life science samples (dehydrated, stained, and resin embedded), has only recently been studied in more detail (Hennig and Denk 2007). This Monte Carlo simulation study showed that although different compositions interact in a nonlinear way during signal generation, heavy metal stained biological samples staining is sufficiently dilute to allow an approximately linear treatment; hence the interaction volume is well confined laterally and in depth for beam energies up to a few keVs. For obtaining high resolution BSE images a small interaction volume is required and hence a low landing voltage is required (see Figure 6.3).

In the above section landing energy was mentioned. Just to be complete in the explanation of the difference between accelerating voltage and the landing energy, the accelerating voltage is the energy to generate the primary beam, whereas the landing energy is the energy of the beam that interacts with the sample. The difference between the two can be set by applying a negative bias voltage to the sample (beam deceleration) (Phifer *et al.* 2009). Beam deceleration slows down the beam towards the sample, reducing the landing energy and hence the beam-penetration depth. In a typical case an accelerating voltage in the column of 4 keV may be opposed by a decelerating voltage of –2 keV on the sample for a net landing energy of 2 keV. In Figure 6.1 we have used beam deceleration to reduce the landing energy to 1 keV. In Figure 6.4 we have used beam deceleration to reduce the beam energy to 2 keV on a mouse brain specimen in an embedded sample block. Here one can start to see (resolve) the bilayer structure of the membranes in the synaptic density region and in the synaptic vesicles. Beam deceleration on flat samples, as in the case with serial block face imaging and array tomography, lack of topology alleviates any generation of aberration due to the electrical field that is created.

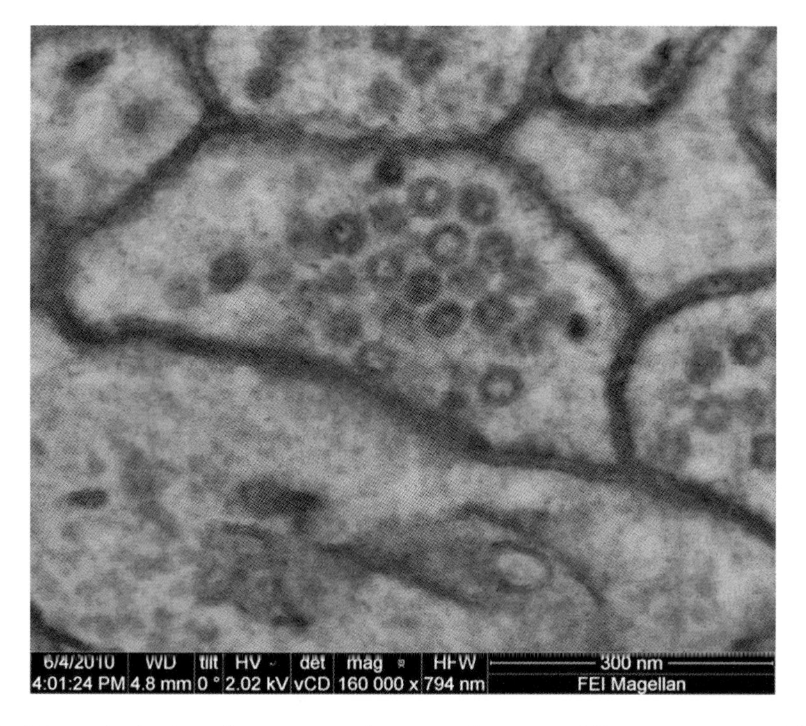

Figure 6.4 Mouse brain. Sample courtesy of Dr M. Ellisman, UCSD; imaging: D. Wall, Thermo Fisher Scientific Company; microscope: Magellan, detector: vCD (high sensitivity low keV BSE detector).

Beam deceleration has a couple of important benefits for SEM imaging:

1. The electrical field between the specimen and the column, or any detectors directly below the column, generates an electrostatic lens (cathode lens) that considerably affects the electron optical properties of the system. The cathode lens can improve image resolution at a low landing energy.
2. All the signal electrons that escape from the negatively biased specimen are accelerated towards the grounded parts in the specimen chamber. The signal electrons have higher energy when reaching the detectors and the signal electron trajectories are modified. This increases the efficiency of signal collection and hence significantly higher SNR.

6.5 CONTRAST AND SIGNAL-TO-NOISE RATIO

Although a small beam diameter is desirable, a small beam is useless if it does not contain sufficient current to generate enough secondary imaging signal needed for the corresponding detectors. Resolution is not the sole determinant of image quality. Contrast and low noise are also important factors and are fundamentally related. The more signal collected, the greater is the number of statistically significant gray levels that form the image. As the signal accumulates, random variations in the signal average out spatially and become less significant at any one point. "Adequate" signal levels might be best specified as a level sufficient to yield the required precision in the data (which translates into contrast and low noise in the image) over an acceptable acquisition time for the next step after acquisition, typically 3D reconstruction and segmentation.

Three factors bear consideration here: the current available in the beam, the nature of the specimen, and the efficiency with which the imaging signal is collected and measured. The amount of current available is determined largely by the type of electron source, in particular by a parameter known as its brightness, which measures its angular intensity. The brighter the source the more current can be focused into a beam of a given diameter. The brightest sources are field emission sources and they come in two varieties: cold field emission (CFE) and Schottky (thermally assisted) field emission. Historically, CFE offered the highest brightness and lowest energy spread and thus the highest performance at low accelerating voltages. However, CFE are less suitable for automated long duration runs as they are subject to variations in current as contamination accumulates at the very sharp tip. The restoration to a well performing tip following contamination is achieved by regular (programmed) "flashing". However, "flashing" is interruptive and the lack of stability introduces uncertainty, particularly in longer duration experiments. For this reason Schottky sources have traditionally been preferred for long acquisition procedures (e.g., quantitative elemental analysis) over the higher resolution CFE sources. However, current Schottky sources provide high, stable currents and high brightness and can be combined with monochromator to reduce the energy spreads to less than CFE without the requirement for flashing.

The other side of the signal level equation comprises the efficiency of the detectors and the amplification infrastructure. Different signals provide different information about the sample. Different signals use different detectors with different efficiencies. However, it is generally true that the higher the detection efficiency the better. This is particularly true with delicate biological samples where, in order to avoid damage to the sample, the goal is often to use the lowest possible beam current dose that still generates adequate signal levels.

In conventional SEM, most imaging is done using the secondary electron signal. Because of their low energies, most secondary electrons that reach the detector originate from the near surface region directly below the beam spot. They show less effect from increased

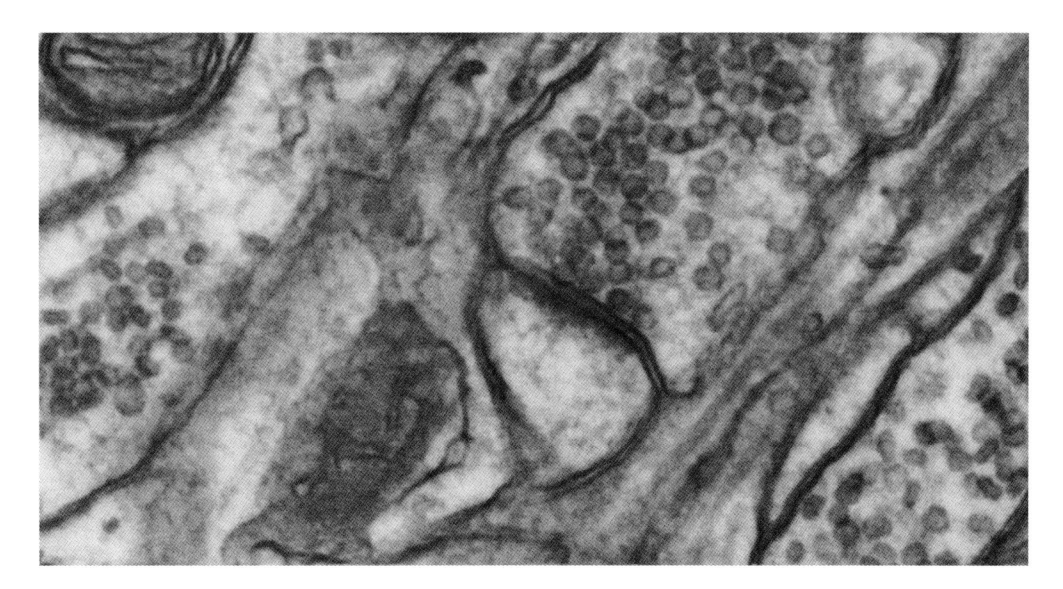

Figure 6.5 Mouse cerebellum, 60 nm section on a TEM grid in STEM mode; pixel size 0.5 nm. Sample courtesy of Dr G. Knott, EPFL; imaging: A. Burgess Thermo Fisher Scientific Company; microscope: Thermo Fisher Scientific Verios using an STEM III detector.

beam penetration at higher accelerating voltages and generate images with high sensitivity to surface topography. Backscattered electron (BSE) imaging is typically reserved for lower resolution applications that take advantage of its atomic number contrast. However, at low voltages BSE resolution improves dramatically because of the reduction in beam spread and penetration, and the directionality of the BSE signal provides an opportunity to explore contrast mechanisms other than atomic number. Current generation BSE detectors provide greatly improved performance at low voltages, with detection efficiencies at 1 keV equivalent to previous generation detectors at 4 keV.

Detectors may also be incorporated to collect electrons transmitted through a thin sample – scanning transmission electron microscopy (STEM). The STEM image of brain tissue in Figure 6.5 was acquired at 30 keV as a tileset of 7×6 frames of 4096×4096 pixel with 10 μs dwell time and at a working distance of 4.5 mm.

6.6 SERIAL BLOCK FACE IMAGING

The SEM, as a surface-imaging technique was first recognized several decades ago by Leighton in 1981, who constructed a microtome for cutting sections inside the microscope chamber (Leighton 1981). However, Winfried Denk and Heinz Horstmann were the first to demonstrate an automated serial block-face imaging (SBFI) in 2004 (Denk and Horstmann 2004). In SBFI the surface of a bulk sample is imaged, then a layer of several tens of nanometers thick is removed from the surface by an automated microtome located inside the sample chamber, and the newly revealed surface is imaged. The process is repeated many times, resulting in a stack of digital images that can then be computationally reconstructed into an accurate representation of the three-dimensional structure of the sectioned volume. Denk used BSE imaging and looked at samples that had been stained with heavy metals using techniques that are routine for transmission electron microscopy. When sample conductivity proved not to be adequate to avoid sample charging, low pressure (water vapor) SEM was used to neutralize the charges. They were able to achieve section thickness of about 50 nm and spatial contrast/resolution that was sufficient to trace membranes and identify synaptic regions.

6.7 CHALLENGES OF SBFI

Although the original SBFI work was groundbreaking and caused an increase of interest in application of SEM techniques in biology, it also identified a number of challenges in SBFI:

- Z-resolution is limited by the thickness of the mechanical slices. Initially, since most users were used to TEM sections, a thicknesses of around 50 nm, the section thickness, was perceived as adequate. However, when focused ion beam methods (Knott *et al.* 2008; de Winter *et al.* 2009) were introduced for serial section SEM data acquisition, thin sections down to 3 nm became achievable. This helped to provide "ground truth" data for testing automated segmentation methods and algorithms for alternative sectioning techniques such as SBFI and Array tomography. The ultra-thin sections that can be achieved in FIB has prompted the desire to cut thinner and thinner sections, both in SBFI and in Array tomography approaches. Although on some samples serial sections of less than 10 nm have been achieved in SBFI (Collinson *et al.* 2016), typically section thicknesses of 25–30 nm are achievable on most samples.
- Data acquisition times can be very long. Given the equipment used in early studies, scanning a volume of 100 μm × 100 μm × 100 μm with a voxel size of 10 nm × 10 nm × 50 nm, the total acquisition time using a scan speed of 100 kHz (voxels/s or dwell time of 10 μs) would take about a month. Clearly, for making the SEM more attractive for whole cell tomography or for brain-connectivity studies higher speeds are desirable, both for throughput and also for stability reasons.
- Accurate registration of images in the vertical stack requires instrument stability over the period of the data acquisition. Inter- and intraframe variability, such as drift, might result from local sample charging effects, and/or temperature variability in the microscope room can degrade the accuracy and align-ability of the resulting image stack. This is particularly the case when extremely large frame images (larger than 12k × 12k pixel) are acquired.
- A dedicated stage for serial sectioning limits the usability of the SEM for applications other than acquiring 3D datasets. For many labs this is not a problem, but for typical service labs it is preferred to have a more flexible solution that allows the SEM to function as a routine SEM for conventional applications (see other chapters in this book) and, when required, quick and easy modification of the SEM for SBFI.

6.8 RECENT ADVANCES

In the ten years since Denk published his original paper, significant advances have been made in SEM technology in general and also in specific areas that are important in the optimization of an instrument for SBFI. In order to overcome the resolution limit in the Z-direction multienergy deconvolution SEM (MED-SEM) techniques have been developed that can resolve details in the vertical direction within an individual (physical) slice. The first complete, integrated system recently became available that combines SBFI with MED-SEM with advanced data acquisition and data processing capabilities using Amira: Thermo Fisher Scientific's Teneo VolumeScope, shown in Figure 6.6.

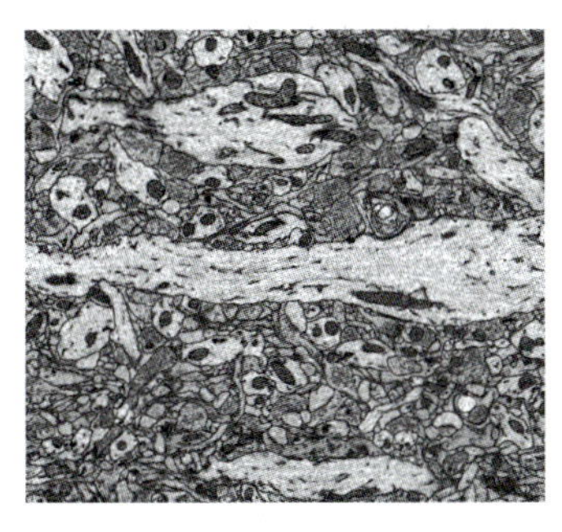 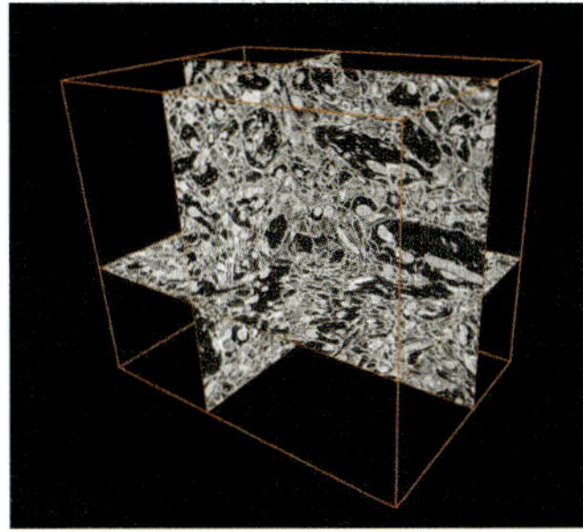 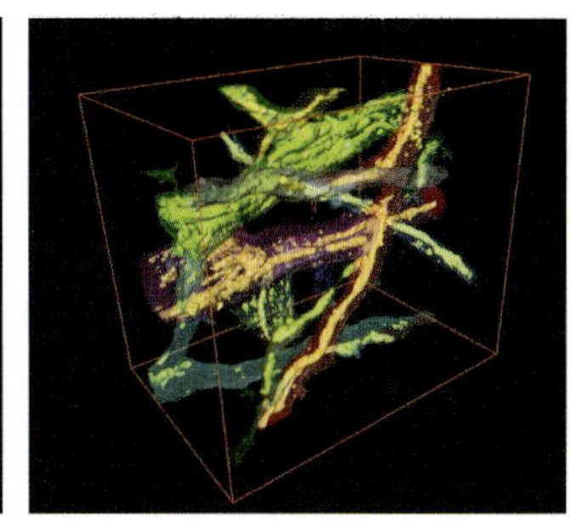

Figure 6.6 Volume reconstruction of a mouse brain acquired with a combination of physical and optical sections at high vacuum mode. After each physical section, the block-face was imaged with a high contrast in-lens backscattered electron (BSE) detector, at increasing accelerating voltages ranging from 1.2 to 3.1 keV. Collected images are automatically registered and deconvolved. 3D data visualization and reconstruction was done with 3D imaging software, Amira. Isotropic resolution of 10 nm × 10 nm × 10 nm pixels (x, y, z), physical section thickness at 50 nm, 1040 sections including physical and optical sections. Reconstructed volume 15.00 μm × 12.9 μm × 10.4 μm (1040 slices).

6.9 MULTIENERGY DECONVOLUTION

A direct result of the understanding that the interaction volume is well confined laterally, Thermo Fisher Scientific Company has developed a method (Boughorbel *et al.* 2012) that uses multiple scans from the same regions with varying interaction depths, yielding a series of images that can be deconvolved to a stack of virtual section images that represent the 3D volume. Multienergy deconvolution allows acquisition of data at high isotropic resolution without the need for a focused ion beam sectioning or the need for very special sample preparation to achieve sub 10 nm cuts (Collinson *et al.* 2016).

The MED-SEM method was evaluated and calibrated using a modified Slice and View approach in the DualBeam. In this early experiment an energy sequence was acquired after every sixth physical section of 8 nm, while recording the serial section images after every FIB cut as well. During a 24 hour experiment 1400 images were acquired and the energy sequence images were deconvolved into 8 nm optical layers, to obtain a dataset with isotropic resolution. Three orthogonal views after aligning the images stacks are represented in Figure 6.7, the top one shows the orthogonal reconstruction after aligning the data from 48 nm sections without deconvolution, the middle one shows the stack when MED-SEM has been applied on the 48 nm sections, and the last one shows a straight FIB stack with 8 nm sections. It is clear that MED-SEM allows a much better interpretation of the cellular structures when looking at the orthogonal sections, but that the FIB dataset still looks better. In order to judge the value of this deconvolution technique further experiments using automated segmentation techniques are desired.

6.10 STABILITY

A stable beam current is essential for long duration experiments. As described above, the latest generation Schottky sources provide high current levels and stability.

Charging can be a troublesome source of instability for biological materials embedded in plastic resin. Currently various groups are looking into further developing resins that have good sample penetration, cutting, and improved conductivity properties. However, to-date

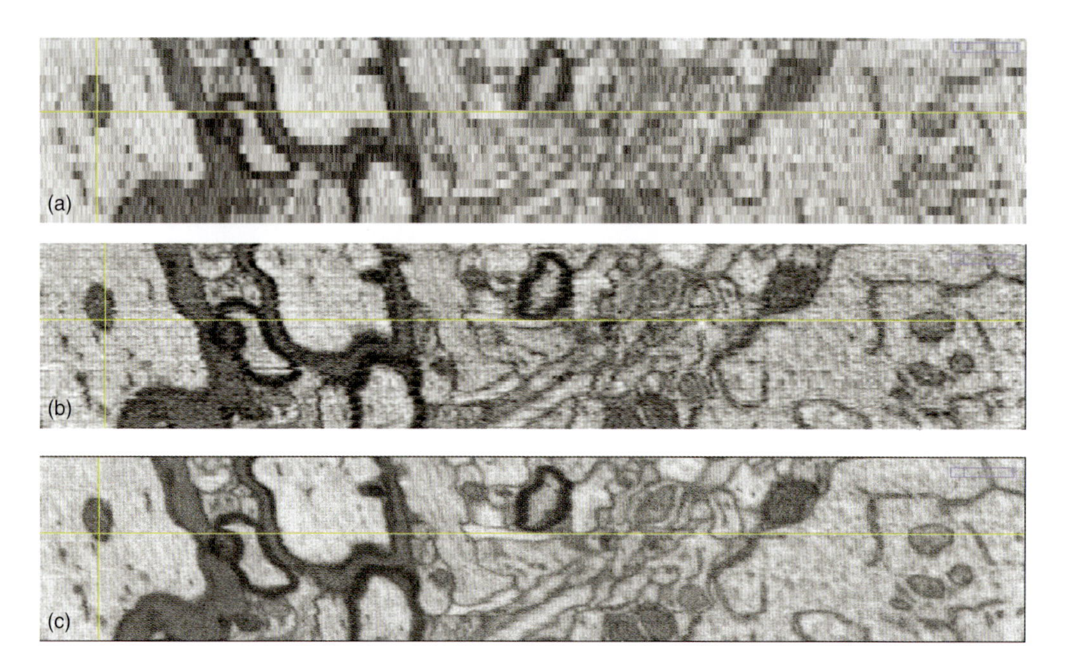

Figure 6.7 Orthogonal view on an aligned stack of mouse brain tissue: (a) 48 nm sections without MED-SEM, (b) 48 nm sections with MED-SEM reconstruction, (c) 8 nm sections DualBeam slice and view. Sample courtesy of Dr G. Knott, EPFL; image acquisition: P. Potocek and I. Gestmann, Thermo Fisher Scientific; MED-SEM deconvolution: Dr F. Boughorbel, Thermo Fisher Scientific; microscope: Helios 650, using through the lens backscatter detector (TLD) with modified Slice and View software.

results obtained with low vacuum techniques based on water vapor have been proven to be good at neutralizing charge, while maintaining resolution to an adequate level for most biological experiments. Additionally specific scanning strategies, such as interlaced scans and multiple line scans, may also be used to avoid charge accumulation.

BSE imaging is generally less susceptible to charging artifacts than secondary electron imaging because BSE have higher energies. However, this immunity is degraded when BSE are imaged at lower accelerating voltages. Segmented BSE detectors provide filtering capability, based on the angular distribution of BSE energies, which can be used to reduce charging artifacts.

Serial process of SEM image acquisition and requirements for resolution and signal-to-noise ratio (SNR) fundamentally determine the time required to acquire an image. Hence, the signal strength and efficiency are important considerations. The strength of signal is determined largely by the amount of beam current the sample can endure before damage occurs; biological samples, even when embedded, can be especially sensitive to beam damage. Consequently, the detection system needs to be optimized to collect low-loss BSE signals with high efficiency. This is best done using an in-lens detection geometry that generates images with strong contrast and high SNR.

During charge accumulation of insulating/nonconductive samples, charging artifacts could occur in SEM images (de Winter *et al.* 2009). Therefore plastic-embedded biological samples can pose difficulties during image acquisition. In routine imaging applications, conductive coatings may be applied to nonconductive samples to provide a path to ground that prevents charge accumulation. Coatings can be utilized in SBFI following a freshly exposed surface using an in situ ultra-microtome; however, this increases time to data considerably. Low vacuum (LoVac) operation delivers a means to dissipate charging that is compatible with SBFI. In LoVac mode, a gas, in this instance water vapor, is introduced into the specimen chamber and sustained at a low enough pressure that allows the beam to reach the sample without excessive scattering. Some of the gas molecules are ionized by the electron beam or by electrons originating from the specimen. These ionized gas molecules are then available to neutralize any charge build-up on the sample. LoVac operation is essential to secure high-quality imaging of difficult samples.

6.11 SPEED

The primary determinants of acquisition speed are the amount of current in the beam spot (beam current density) and the efficiency with which emitted electrons are detected. Schottky field emitters deliver high currents in small spots. An immersion objective lens, in which the sample is immersed in the lens field, provides additional gains in beam current density and sophisticated in-lens detectors channel signal electrons toward an in-lens detector to optimize detection efficiency. Current generation BSE detectors offer significant improvements in detection efficiency at lower energies.

For high speed scanning Thermo Fisher Scientific's high end SEMs are equipped with electrostatic scanning. This allows faster scan speeds because the beam can be positioned accurately without the need to compensate for hysteresis, typical for electromagnetic scanning systems.

Developments in the last decade of detectors, amplifiers, electron optics for high beam current density, and electrostatic scanning systems allow single beam serial block face imaging tools that are able to scan up to 20 MHz.

Obviously when scanning extremely fast, the beam current density is critical to enable image acquisition with an adequate signal-to-noise (S/N) ratio, but also with an adequate resolution. Larger beam currents result in larger beam sizes, hence reducing the resolution. Figure 6.8 shows the result of imaging with 1.6, 3.2, and 6.4 nA, respectively. It is clear that

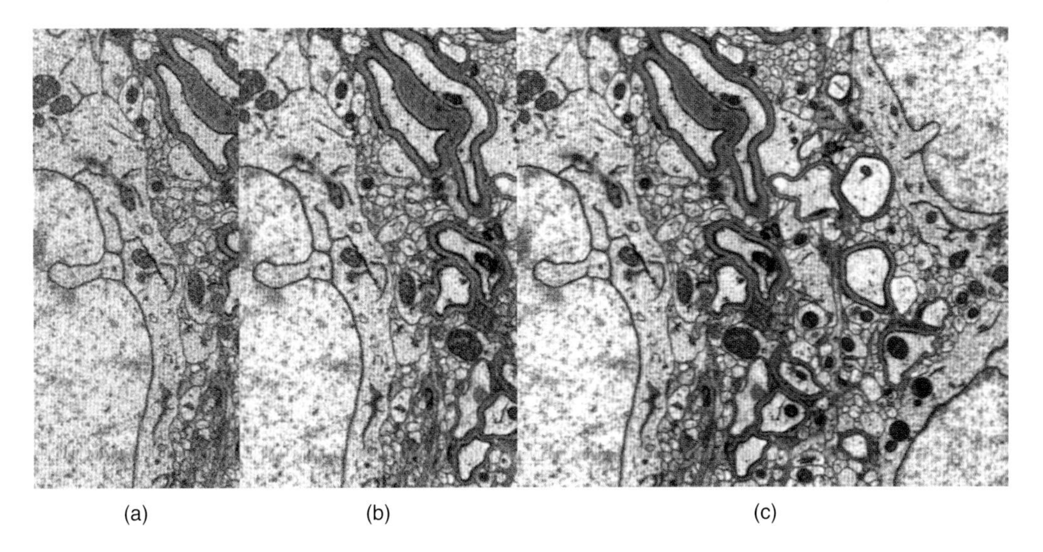

Figure 6.8 Mouse brain, 40 nm sections on Si wafer imaged at 20 MHz, at (a) 1.6 nA, (b) 3.2 nA, and (c) 6.4 nA. Sample courtesy of Prof J. Lichtman, Harvard University; image courtesy of I. Gestmann, Thermo Fisher Scientific; microscope: Verios 460 using DBS detector.

S/N increases with the current but that the resolution decrease is subtle. The resolution at 6.4 nA is still perceived acceptable for many applications.

6.12 AUTOMATED ACQUISITION, RECONSTRUCTION, AND ANALYSIS

Perhaps the advances with the greatest practical value have come in the area of automation and data management. Modern SEMs have evolved from specialized scientific instruments to become highly productive, easy to use laboratory tools. Automation is pervasive, everywhere from instrument setup procedures, such as alignment and calibration, to routine operations, such as changing beam current or accelerating voltage. Operations that once required sophisticated knowledge and training are now accomplished automatically with the click of a mouse.

6.12.1 Maps

Automated routines permit unattended data acquisition, reducing operator fatigue and improving data reproducibility. This is especially important for the long acquisition times required by data intensive procedures such as SBFI. Specialized software (Maps) stitches adjacent fields of view into a single large image and permits the overlay of images from other techniques, such as light microscopy, to facilitate navigation to previously identified features of interest. Maps runs the Thermo Scientific microscopes (SEM, Dual Beam, TEM) but it also allows images to be imported from other manufacturer's light microscopes for correlation. Downstream in the workflow Maps connects automatically with Thermo Scientific Amira to channel acquired and imported data to the visualization, reconstruction, and segmentation capabilities that Amira has been providing to the community for a long time.

6.13 CONCLUSION

Modern SEMs have evolved from specialized scientific instruments operated by highly trained microscopists to high productivity, easy to use tools used by scientists to find answers to questions in their own fields of expertise. This is especially true in the life sciences, where the evolution of microscope technology is enabling dramatic and scientifically important discoveries about the relationship between structure and function at spatial scales ranging from cells and tissues (Briggmann *et al.* 2011) down to molecule dynamics (Fischer *et al.* 2010). Life scientists are actively developing novel techniques like SBFI to pursue these goals and SEM manufacturers are collaborating with leaders in the field to enhance the capabilities of their instruments based on practical workflows and the ability to deliver valuable scientific results. In the case of SBFI this is reflected in the development of specific new capabilities, such as multienergy deconvolution, that substantially improve the quality of the results, and in more practical general developments that improve the speed and ease of use of the instruments. As a result of these collaborative developments, new techniques like SBFI have the opportunity to become primary tools in life science research.

REFERENCES

Boughorbel, F. *et al.*, SEM imaging method, Patent US 8232523 B2 (2012).

Briggmann, K. *et al.*, Wiring specificity in the direction-selectivity circuit of the retina. *Nature*, 471, 183–188 (2011).

Collinson, L.M. *et al.*, 3D correlative light and electron microscopy of cultured cells using serial block-face scanning electron microscopy. *J. Cell Sci.*, 188433, July 21, 2016, ii (2016).

Denk, W. and Horstmann, H. Serial block-face scanning electron microscopy to reconstruct three-dimensional tissue nanostructure. *PLoS Biology* (2004).

Fischer, N. *et al.*, Ribosome dynamics, a tRNA movement by time-resolved electron cryomicroscopy. *Nature*, 466, 329–333 (2010).

Hennig, P. and Denk, W. Point spread functions for backscattered imaging in the scanning electron microscope. *Journal of Applied Physics*, 102, 123101 (2007).

Knott, G. *et al.*, Serial section scanning electron microscopy of adult brain tissue using focused ion beam milling. *Journal of Neuroscience*, 2959–2964 (2008).

Leighton, S.B. SEM images of block faces, cut by a miniature microtome within the SEM – a technical note. *Scan Electron Microsc.*, 1981 (Pt 2), 73–76 (1981).

Phifer, D. *et al.*, Improving SEM imaging performance using beam deceleration, FEI Company, 5350 NE Dawson Creek Drive, Hillsboro, OR 97124, HYPERLINK "mailto:Daniel.Phifer@fei.com" Daniel.Phifer@fei.com (2009).

de Winter, M. *et al.*, Tomography of insulating biological and geological materials using focused ion beam (FIB) sectioning and low-kV BSE imaging. *Journal ofMicroscopy*, 233, 372–383 (2009).

7

ZEISS Scanning Electron Microscopes for Biological Applications

Isabel Angert, Christian Böker, Martin Edelman, Stephan Hiller, Arno Merkle and Dirk Zeitler

Carl Zeiss Microscopy GmbH, Oberkochen, Germany

For centuries men have built instruments to get a closer view of living objects (Gest, 2004). This, in the first place, was done to yield a much desired insight into the mysteries of life. However, soon enough the goal was extended to understanding the processes of life and – closely interlinked – the processes of disease. Ever since its foundation by Carl Zeiss in 1846, starting from a small workshop building components for microscopes dedicated to the investigation of biological objects, ZEISS as a company has contributed substantially to the development of innovative products in all fields of optical technology. While the technologies certainly have evolved and been extended from those early days, the challenge remains unaltered at its core, namely to understand the interaction in living systems on a functional and structural level in three or even four dimensions.

Pioneering the development of light microscopes since the nineteenth century, ZEISS has introduced many technical innovations over more than 150 years and transferred them to successful and highly accepted tools (Figure 7.1). After advancing the field of transmission electron microscopes, for example with the first in-column energy-filter microscope in 1984, in 1985 ZEISS introduced the first digital scanning electron microscope, the DSM 950 with computer controlled scan generation and signal processing gaining considerably in automation and ease of use. The introduction of the ZEISS GEMINI® technology for the field emission scanning electron microscopes (FESEMs). In 1992 was another major

Biological Field Emission Scanning Electron Microscopy, First Edition.
Edited by Roland A. Fleck and Bruno M. Humbel.

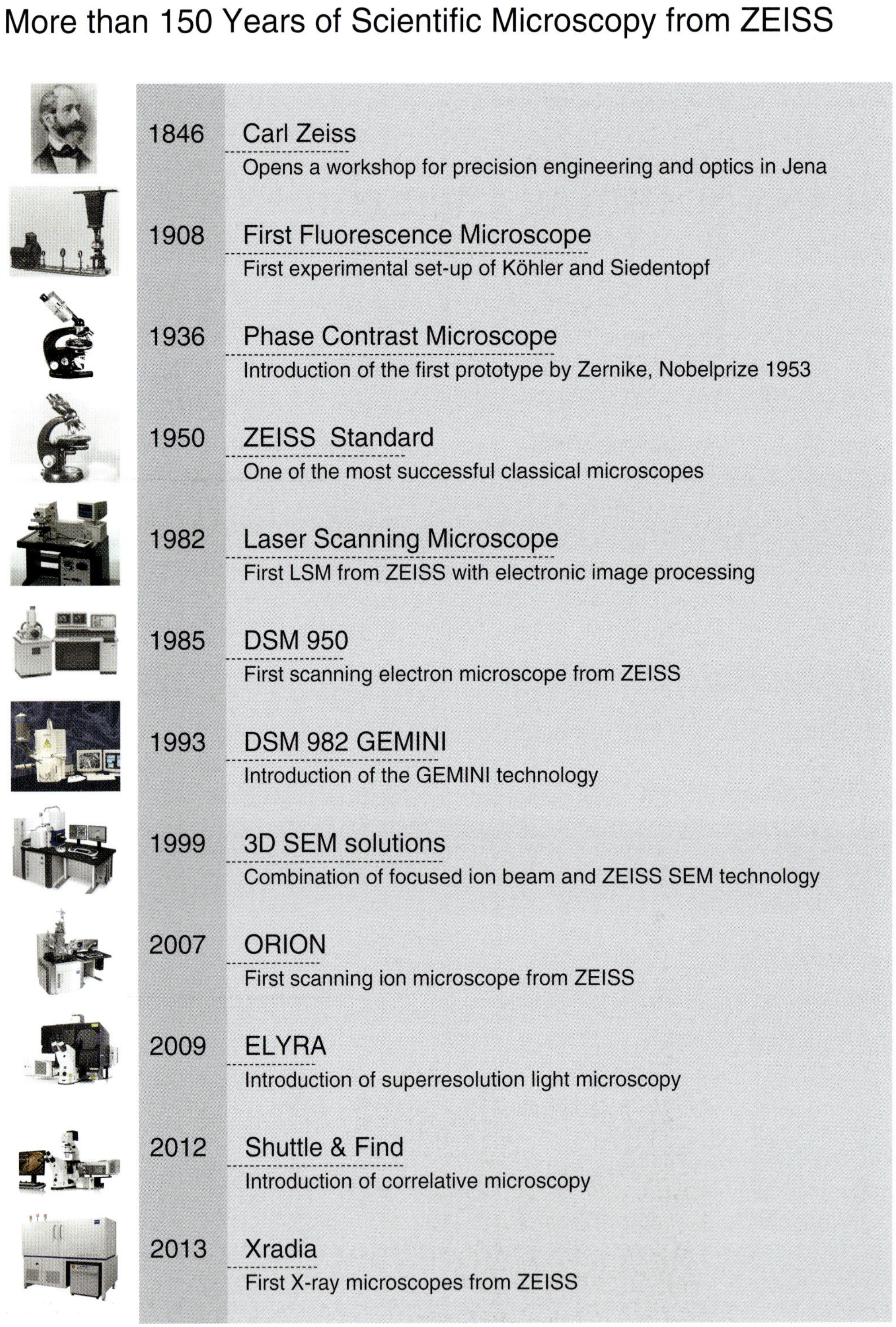

More than 150 Years of Scientific Microscopy from ZEISS

1846	Carl Zeiss Opens a workshop for precision engineering and optics in Jena
1908	First Fluorescence Microscope First experimental set-up of Köhler and Siedentopf
1936	Phase Contrast Microscope Introduction of the first prototype by Zernike, Nobelprize 1953
1950	ZEISS Standard One of the most successful classical microscopes
1982	Laser Scanning Microscope First LSM from ZEISS with electronic image processing
1985	DSM 950 First scanning electron microscope from ZEISS
1993	DSM 982 GEMINI Introduction of the GEMINI technology
1999	3D SEM solutions Combination of focused ion beam and ZEISS SEM technology
2007	ORION First scanning ion microscope from ZEISS
2009	ELYRA Introduction of superresolution light microscopy
2012	Shuttle & Find Introduction of correlative microscopy
2013	Xradia First X-ray microscopes from ZEISS

Figure 7.1 Milestones spanning 150 years of innovative microscope solutions by Carl Zeiss Microscopy.

milestone: Launching the focused ion beam (FIB) tool in combination with a ZEISS FESEM in 1999 – the Crossbeam series – for the first time enabled real 3D imaging on a nanometer scale.

Over a long period of time FESEM techniques were predominantly considered to be material science tools and their use in biosciences was only claimed by a specialized community of users. Only in recent years has the scientific community started to see the huge potential in using field emission SEM techniques for studying the structural properties of biological objects. The main reason for this change in perception is that classic transmission electron microscopy (TEM) methods are inherently limited in their ability to acquire 3D data sets of large volumes like whole cells, while FESEM and, in particular, the associated automated imaging routines still have not reached their full potential. Here, ZEISS Microscopy with its developments in automated imaging (Atlas software) and high speed acquisition technology (multiSEM technology, multibeam SEM), is supporting and in fact enabling cutting edge research in today's biological research. Beside well established approaches new technologies such as ion beam microscopy and X-ray microscopy are also being pioneered.

In addition, a comprehensive understanding of biological processes requires structural and functional information of exactly the same location in the identical sample. Here, correlative imaging techniques are of major importance. Being a leading supplier to biosciences researchers in the area of light microscopy, it was only consequently that ZEISS introduced the first commercial correlative microscopy solution in 2012 – Shuttle & Find.

7.1 BIOLOGICAL IMAGING USING ZEISS TECHNOLOGY

For over 50 years the procedures to prepare biological samples for electron microscopic imaging have been under constant development and are now well established for most biological systems (Schatten, 2008; McDonald and Webb, 2011; Cavalier, Spehner, and Humbel, 2008). However, imaging technologies have also developed with increasing pace. In this chapter we introduce unique ZEISS contributions to this field of biological imaging such as the GEMINI® lens, ion beam microscopy, and the multibeam SEM.

7.1.1 ZEISS GEMINI® Technology

Within the FESEM imaging workflow, key performance aspects to consider are optimum resolution and contrast, surface sensitivity, control of sample charging and the need of speeding up the data acquisition for high-throughput imaging.

In the following discussion we will illustrate how the ZEISS electron-optical design and, more specifically, the GEMINI® lens technology (Figure 7.2a) addresses these aspects perfectly.

7.1.1.1 Optimum Resolution, Contrast, and Surface Sensitivity

One of the most impressive design features of the ZEISS GEMINI® column (Frosien, Plies, and Anger, 1989; Martin *et al.*, 1994; Jaksch and Martin, 1995) is the fact that, unlike conventional lens design, the resolution limiting aberrations are actually decreasing with decreasing beam energy (Figure 7.2b). This is enabled by accelerating the electrons to a high potential that is kept throughout the column and only decelerated by the beam booster voltage to the target landing energy within the tip of the objective lens. Furthermore, the

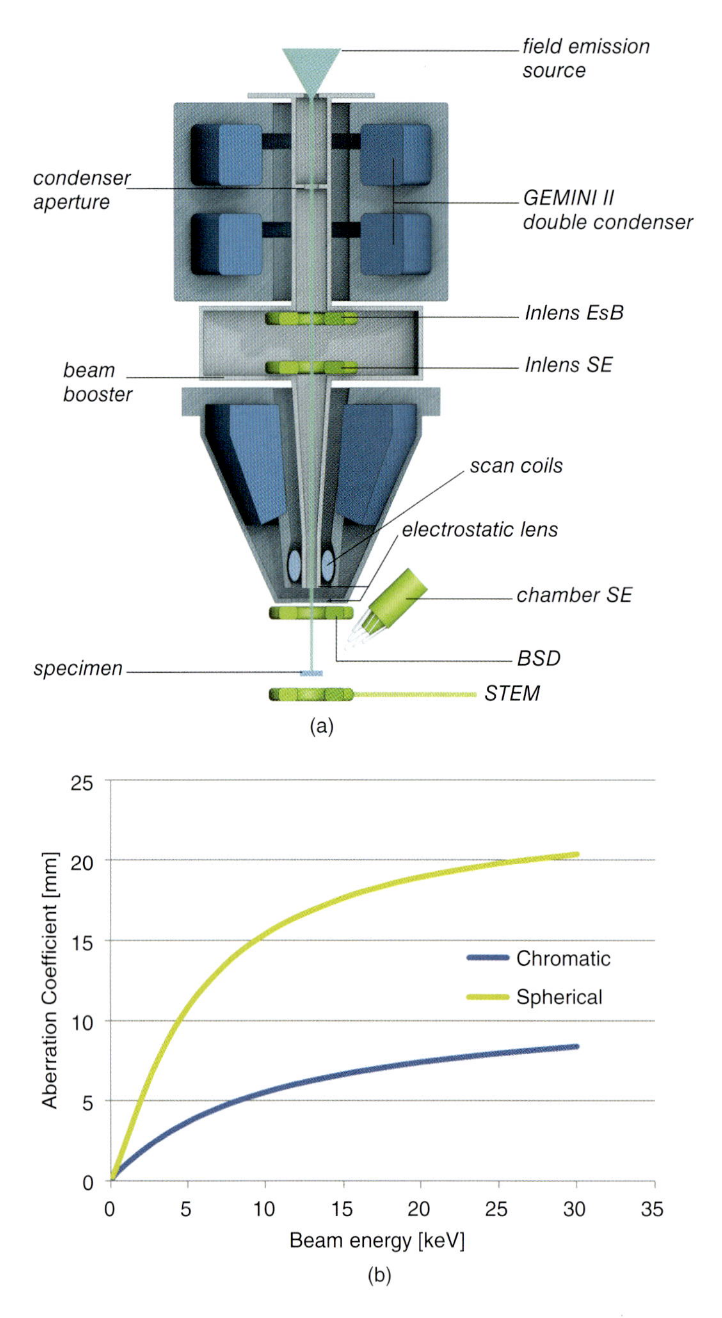

Figure 7.2 (a) Cross-section of the GEMINI® column in ZEISS FESEMs. The primary electron beam passes a combined magnetic and electrostatic field before it is focused on the specimen. The beam booster decelerates the electrons before they hit the sample and accelerate them again back to the in-lens and EsB (energy-selective backscatter) detector. (b) Spherical and chromatic aberrations in the GEMINI® lens at different beam energies. Owing to the beam booster the chromatic and spherical aberration coefficients are negligibly at very low acceleration voltages.

magnetic–electrostatic lens combination (hence the name GEMINI for twin lenses) increases the incident beam aperture angle at the specimen, which improves both the signal-to-noise ratio (SNR) and the resolution. The GEMINI® lens design therefore is tuned to deliver excellent low-voltage performance, allowing ZEISS FESEMs to achieve a resolution in the 1 nm range across the entire voltage range without limiting the overall beam current by a filter system. Likewise for high-throughput imaging, required in most 3D data acquisition jobs, limiting the total beam current compromises the SNR and thus the imaging speed. For imaging flat and conductive samples like serial sections on wafers at energies between 1 and 5 keV the use of sample biasing methods can enhance contrast and speed up data collection significantly.

For low energies the beam booster concept of the GEMINI® lens not only ensures low primary beam aberrations at low landing energies but also enhances secondary (SE) and backscatter electron (BSE) yield. The positive booster voltage acts as an accelerator for SE and BSE electrons and effectively sucks them back into the objective lens. This allows for an extremely high in-lens detection efficiency of SE and BSE electrons (Jacksch *et al.*, 2003). Furthermore, applying a negative voltage to a grid in front of the in-lens BSE detector (ZEISS proprietary EsB™ detector) enables efficient discrimination of SE and BSE signals and thus simultaneous observation of surface, voltage, and material contrasts without disturbing the primary electron beam (Steigerwald *et al.*, 2004). While low landing energies are essential for limiting the signal depth of the primary electrons, which is beneficial especially for 3D applications such as FIB or *in situ* ultramicrotomy (see Section 7.2.1.2), the EsB detector's energy filtering capability allows even better and genuinely live signal depth control. This is again accomplished by using the filter grid to exclude all backscattered electrons with high energy loss from being detected. Thus only the so-called "low loss" backscatter electrons are detected, enabling an effective z resolution in the nm range.

7.1.1.2 Advanced STEM Imaging

Using a transmission imaging mode in the scanning electron microscope (STEM imaging) is a widely applied method for imaging transmissible samples, often resulting in higher resolution than classical secondary or backscatter imaging mode. For this method a STEM detector is placed underneath ultrathin sections (e.g., cut by an ultramicrotome) in the FESEM. Electrons scattered at various angles can be detected and used for image formation. With its annular design consisting of a central element and three additional concentric rings the ZEISS annular STEM detector (Salzer *et al.*, 2011) is perfectly adapted to the scattering behavior of electrons with a primary energy between 2 and 30 keV. This allows to differentiate classical bright-field (BF), dark-field (DF), oriented dark-field, and high-angle annular dark-field (HAADF) signals for image formation. Its improved sensitivity lets the detector maintain excellent SNR, even at fast scan speeds. The detector electronics offers the unique possibility of reading out up to four signal configurations, including mixing of different detector segments, simultaneously. This results in an advantage over more conventional designs where investigation with different STEM detection modes needs to be done in a serial manner, which leads to higher sample exposure and increased beam damage. The ZEISS STEM detector provides excellent contrast and dynamic range. Images of ultrathin sections clearly show the lipid bilayer ultrastructure of membranes (Figure 7.3). This is close to the limit in the amount of detail that can be expected from conventionally prepared, chemically fixed, and resin embedded samples.

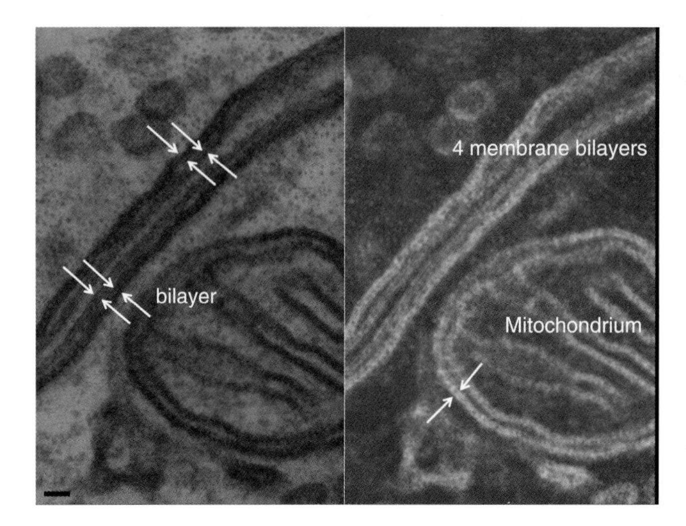

Figure 7.3 Bright-field (left) and dark-field (right) STEM images of an ultrathin section from a mouse brain mounted on a grid acquired with the new ZEISS annular STEM detector. The double membrane of mitochondria is resolved down to the lipid bilayer structure. Scale bar: 10 nm. Sample: courtesy of Graham Knott, EPFL Lausanne, Switzerland.

7.1.1.3 Concepts of Imaging Non-conductive Samples

As mentioned earlier, one of the challenges in imaging biological samples is charging under the electron beam. FESEMs offer different strategies to overcome this, namely low keV imaging, the variable pressure mode (VP), and local charge compensation (CC).

Whether an insulating specimen is charged positively or negatively depends on the total electron yield of SE and BSE electrons with respect to the incoming primary electrons. It is well known that two defined energies for the primary electron beam lead to a neutral point where no charge is accumulated on the sample surface (Cazaux, 2004). Depending on the material, the lower neutral point is around a hundred to a few hundred electron volts whereas the upper critical energy for most biological specimens is around 1.5 keV. One strategy to avoid charging is to adjust the energy of the primary electrons to this point. Operating in this energy regime offers the additional advantages of minimized specimen damage and increased surface sensitivity (Figure 7.4a), especially in imaging block faces or thicker sections.

Given their low energy, SE electrons and hence SE imaging is much more affected by charging effects as compared to BSE imaging. Hence one simple approach is to use the BSE electrons (Figure 7.4b) to make charging practically invisible. While this is a means to overcoming the effects of charging on image formation, it does not remove charging itself.

In FESEMs equipped with the VP mode (Gnauck, Drexel, and Greiser, 2001) the pressure in the specimen chamber can be adjusted from 1 to about 100 Pa. Using several pressure stage apertures within the electron column and an adapted vacuum system allows one to keep the gun area pressure range of 10^{-8} Pa required for safe operation of the field emission filament. In VP mode low-energy secondary electrons collide with gas molecules, producing positively charged ions, which are attracted to negatively charged sample areas and neutralize them, hence removing the charging. Here, imaging is accomplished by detecting photons emitted by ion relaxation using a special VP detector (the so-called VPSE

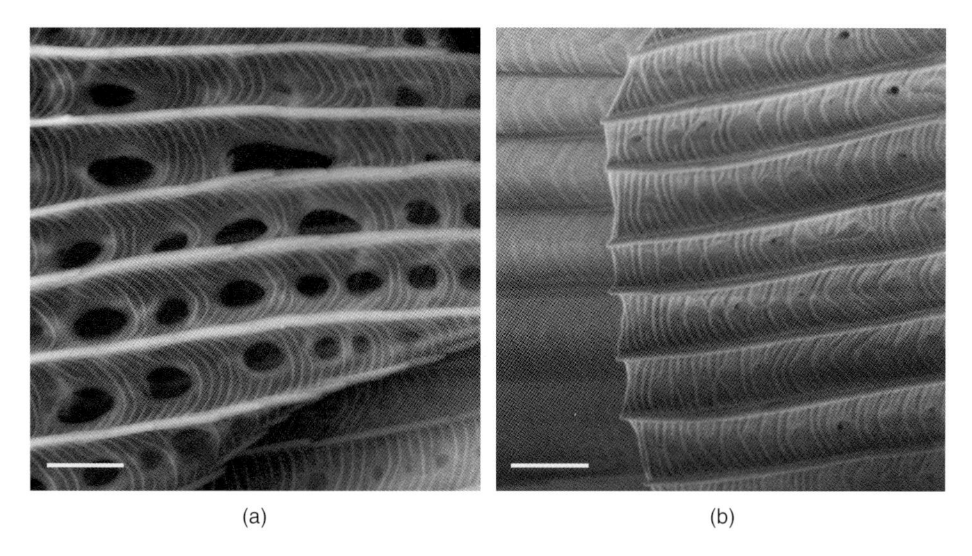

Figure 7.4 (a) Image of a moth wing detected with an in-lens detector at 100 eV primary energy of the electrons. (b) Increasing the primary energy of the electrons to 5 kV makes subsurface structures visible without charging artifacts using a backscatter detector. Scale bar: 1 μm.

detector, Figure 7.5a). This signal essentially generates an indirect secondary electron image. Backscatter imaging is still possible using standard detectors as a significant fraction of the high energy backscattered electrons will penetrate the gas.

High-resolution imaging at elevated pressure is also possible when a differential pumping aperture is additionally inserted below the objective lens in the "NanoVP" mode. Here the pressure in the specimen chamber can be adjusted up to 500 Pa. The shielding of the aperture substantially reduces the path length of the incident beam in the gas atmosphere, thus reducing beam broadening and enabling high-resolution imaging with all SE and BSE detectors.

The charge neutralization process using the local charge compensator (CC) is similar to the one in VP mode and is a unique option of the ZEISS FESEMs. Technically the CC is based on a needle mounted to a pneumatic retraction mechanism. After insertion of the needle the local gas jet of dry nitrogen (Figure 7.5b) is applied to the sample surface near the imaging area. This allows toggling between the charge compensation mode and high-vacuum operation within seconds, just by stopping or allowing gas injection. Because of the fairly small increase in total chamber pressure, all standard in-lens as well as chamber detectors can be used in a regular way. In addition, by applying small amounts of oxygen through a local CC it can be converted into a "cleaning" device by ionizing the oxygen with the electron beam for local decontamination of the sample.

7.1.1.4 High-Throughput and Ease-of-Use Imaging

Imaging large sample areas at high resolution is extremely time-consuming. Using large imaging frame-stores help to minimize acquisition time by reducing the need for image tiling and the computational effort in post-acquisition stitching. For a given field of view (FOV), the effective deflection angle in the extreme corners of the scan field are smaller when using GEMINI® lens technology compared to more classic designs (e.g., a single pole piece) due

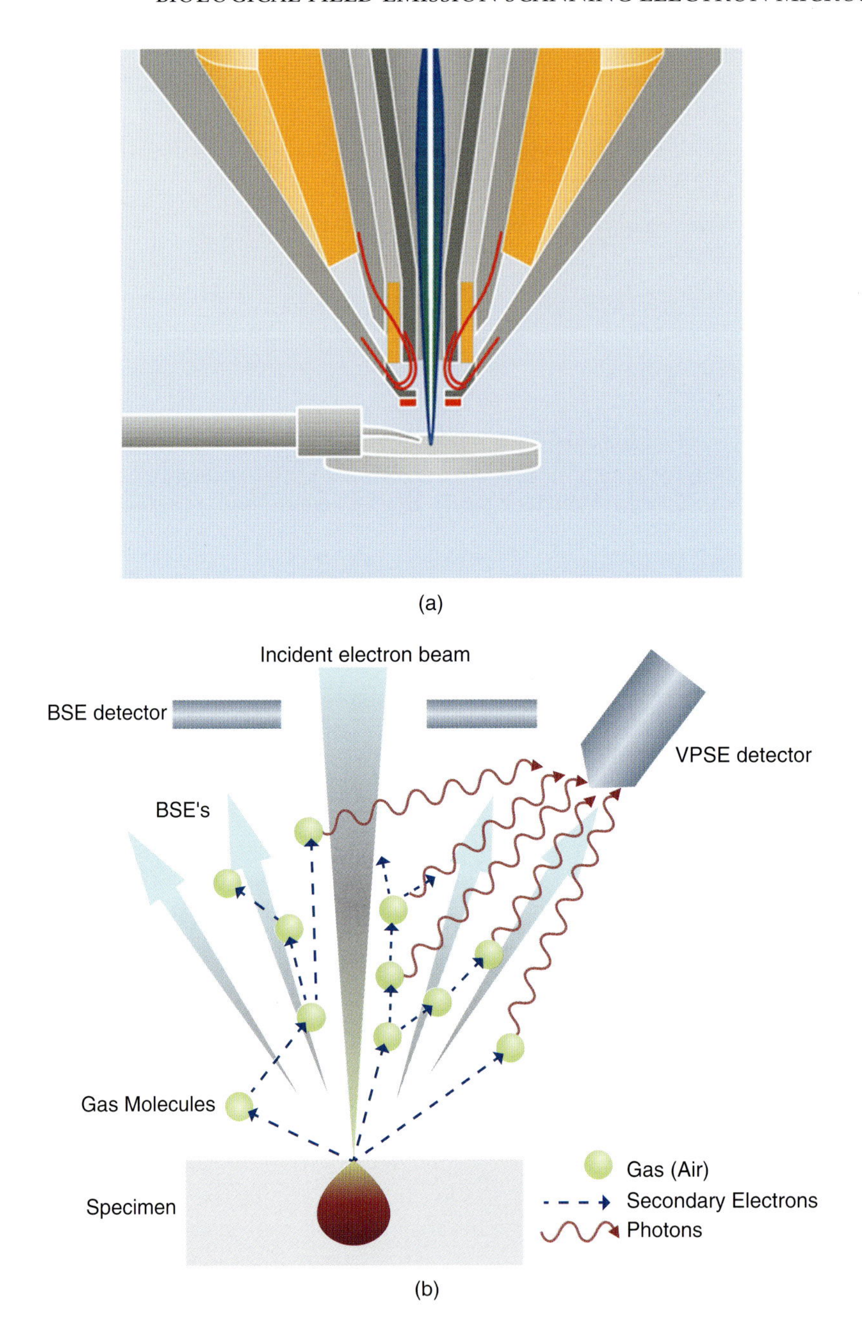

Figure 7.5 (a) On gas injection into the column SE electrons emerging from the sample ionize gas molecules and these positively ionized gas molecules compensate local charging on the sample surface. Due to multiple interaction with the gas in the variable pressure (VP) mode the SEs produce photons emitted on gas–electron interaction, which are then detected by a specialized VPSE detector. BSE imaging also remains possible. The BSE can still penetrate the gas because of their higher energy. (b) Local gas injection by the CC needle. The small amount of gas, which is applied only locally, allows unlimited use of all chamber and in-lens detectors in the GEMINI® column.

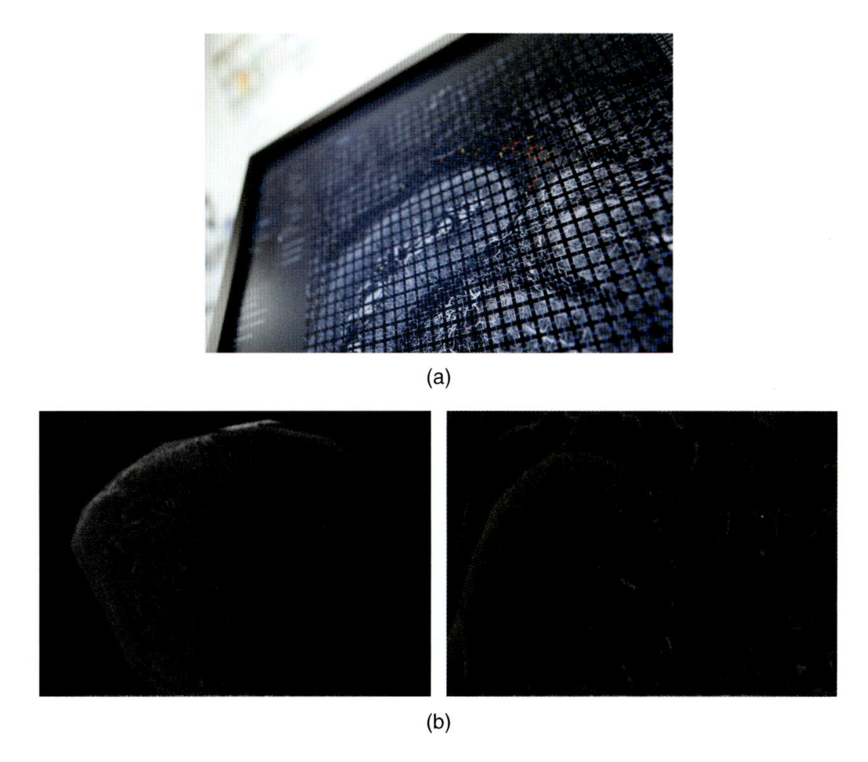

Figure 7.6 (a) The Atlas graphical user interface for the automated tiling and stitching tool is shown. ZEISS SEMs can image data in a frame-store up to 32 k × 24 k. In addition the ZEISS Atlas solution allows automated multiresolution acquisition of frame-store up to 32 k × 32 k and FOVs covering several cm^2 with a pixel resolution of a few nm corresponding to tens of thousands of typical SEM images by automated acquiring and stitching. (b) Left: large-scale SE image (FOV ~60 mm) of a monkey brain's vasculature. Preparation of brain blood vessels using the corrosion cast technique, where 3259 images from a mosaic (each image is 4096 × 4096 pixel, image pixel size 150 nm) were automatically acquired and stitched with ZEISS Atlas software solution. Right: enlarged area out of left image (FOV ~700 μm). Sample: courtesy of Anna Lena Keller, Carl Zeiss Microscopy, before Max Planck Institute of Biological Cybernetics, Tübingen, Germany.

to the long focal distance of the twin lens design. As distortions are a function of deflection angle the resulting beam blurring is comparatively small. This enables the image acquisition of an FOV with several hundred micrometers at less than 10 nm pixel resolution with a frame-store resolution of at least 24 k × 32 k. In addition, using ZEISS Atlas software for automated image acquisition enables a user to easily image FOVs of several square cm at pixel resolutions down to the resolution limit in a fraction of time compared to frame-store resolutions that are typically used (Figure 7.6).

However, large scan fields are only one aspect of high-throughput imaging. The other important factor is scan speed. The special electron optics of the GEMINI® lens keeps the probe size in perfect shape at all conditions, even at high probe currents for enhanced signal and brightness. Extremely fast scan speeds can be used in combination with highly efficient in-lens detection. The overall image acquisition time is a linear function of the dwell time of the electron beam and depends only on the desired image quality demanded by the application (SNR, contrast levels, and pixel resolution).

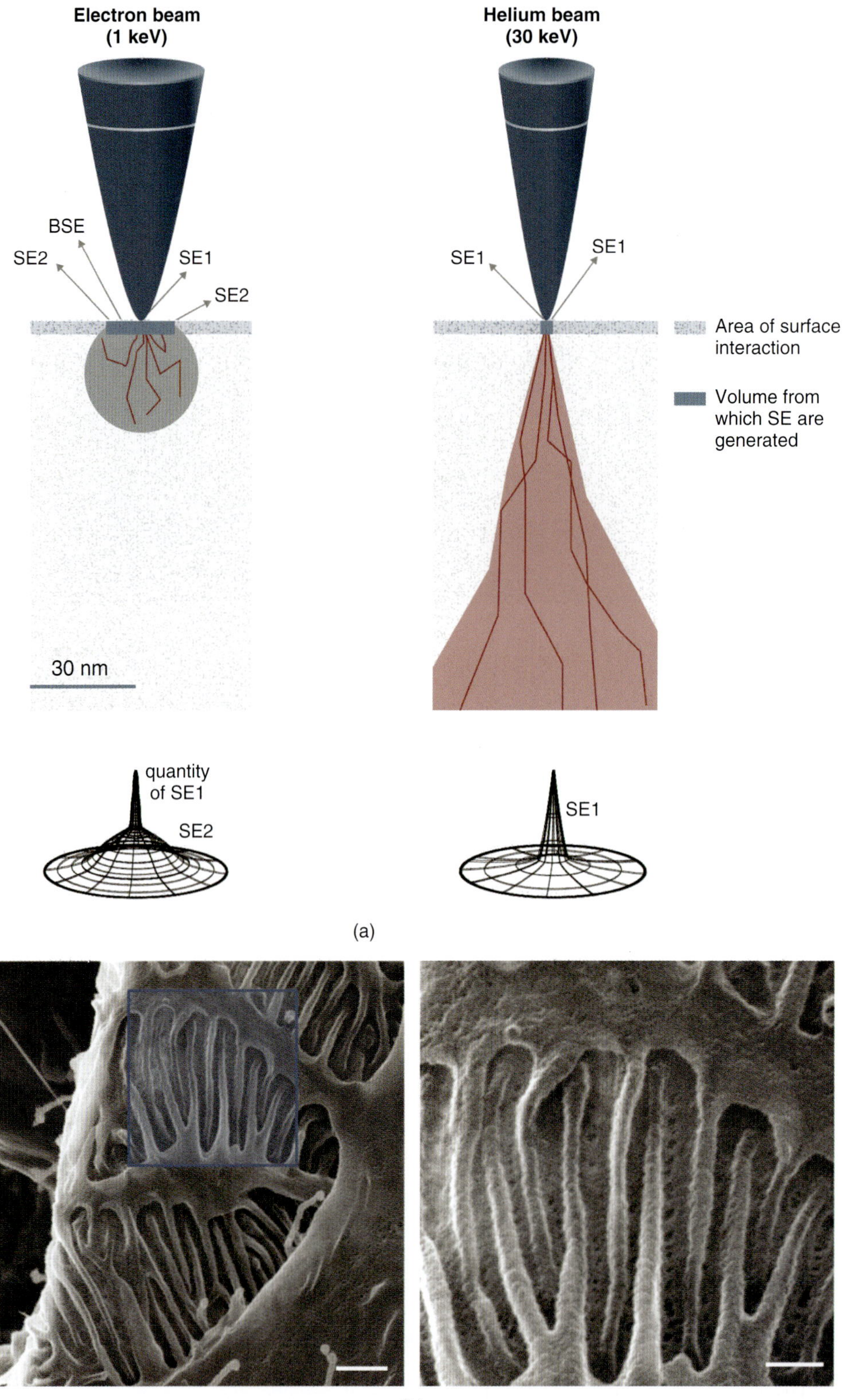
Electron beam
(1 keV)
Helium beam
(30 keV)
BSE
SE2
SE1
SE2
SE1
SE1
Area of surface interaction
Volume from which SE are generated
30 nm
quantity of SE1
SE2
SE1

(a)

(b)

7.1.2 Helium Ion Microscopy in Biology

Besides electron microscopy (EM) scanning, ion microscopy has been established in recent years as a new technique for a broad application landscape from sample nano-patterning to imaging. Especially for imaging biological samples it retains all the advantages of a SEM– for example, well-known sample preparation procedures and straightforward interpretation of images – but brings the additional advantages of superior resolution and higher depth-of-field (Figure 7.7) (Joens *et al.*, 2013; Boden *et al.*, 2013; Rice *et al.*, 2013).

The ZEISS scanning ion microscope (ORION NanoFab) is equipped with a gas field ion source generating a helium or a neon ion beam with high brightness and a very low ion energy spread (<1 eV). Helium ions are about 7000 times heavier than electrons and neon ions exceed the weight of an electron by a factor of 40 000. Therefore the diffraction effect at apertures and edges can be neglected, which results in a minimized probe size. The ion-optical system of the microscope consists of electrostatic lenses and can generate an ion beam with a probe size down to 0.5 nm, scanning the sample surface with ion energies of 10 to 35 keV.

The ion beam hitting the sample surface generates secondary electrons by energy transfer to the sample. The first-order secondary electrons escaping the specimen for detection are thereby confined to the landing area, which is limited by the probe size. The emission of higher-order secondary electrons – limiting the imaging resolution in an SEM – is suppressed by the characteristic shape of the ion–matter interaction volume of helium and neon ions (Notte, 2012). This leads directly to higher image resolution and higher surface sensitivity compared to classic SEM. For signal detection the ORION NanoFab is equipped with a standard detector dedicated to the collection of the SEs.

As discussed in Section 7.1.1.3, imaging non-conductive samples is prone to charging. Because of the positively charged ion beam and the emission of negatively charged SEs out of the sample, a positive net charge is accumulated at the sample surface. Using an electron flood gun integrated into the microscope chamber, a negatively charged beam of low-energy electrons is applied directly to the area being imaged. This approach guarantees effective charge neutralization, enabling high-contrast imaging of non-conductive samples in the ion microscope (Bazou *et al.*, 2011).

Figure 7.7 (a) A schematical comparison of 1 keV electrons and 30 keV helium ions shows the principle differences in secondary electron contribution to the image formation in both systems. Heavy ions may travel very large distances before finally giving up all their residual energy and coming to rest. The ion beam hitting the sample surface generates secondary confined only to the landing area, which is limited by the probe size. The emission of higher-order secondary electrons – limiting the imaging resolution in a SEM – is suppressed. (b) Ion microscopic imaging of glomerulus of mouse kidney after critical point drying, with no additional coating. Left: primary podocyte foot processes arising from the podocyte soma (lower right corner). Scale bar: 500 nm. Right: zoom in showing pores of approximately 40 nm diameter between secondary podocyte foot processes reflecting part of the glomerular filter. Scale bar: 200 nm. Sample: courtesy of Stefan Eimer, Department for Structural Cell Biology, BIOSS Center for Biological Signaling Studies Center for Systems Biology Analysis (ZBSA), Albert-Ludwig University, Freiburg, Germany, and Oliver Kretz, Huber lab – Clinical Research Center, Renal Division – Department of Medicine, University Hospital Freiburg, Germany.

7.1.3 Multibeam SEM

The needs of high throughput and large volume imaging in modern biological EM were discussed extensively in Section 7.1.1. Here we show how the unique MultiSEM technology meets these needs.

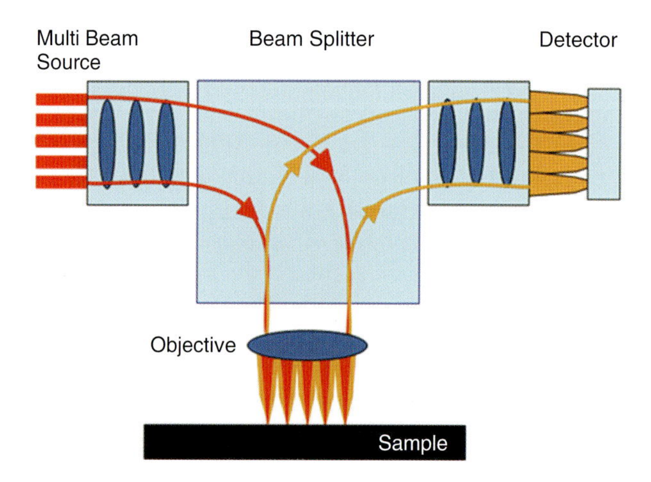

Figure 7.8 MultiSEM technology uses many beams in parallel to image a sample area of 100 μm × 100 μm. Primary electrons (red, left) are focused on to the sample and separated by a beam splitter from the secondary electrons (yellow, right), which are detected simultaneously. All electron beams form many individual images, which are then merged into a single, large area micrograph.

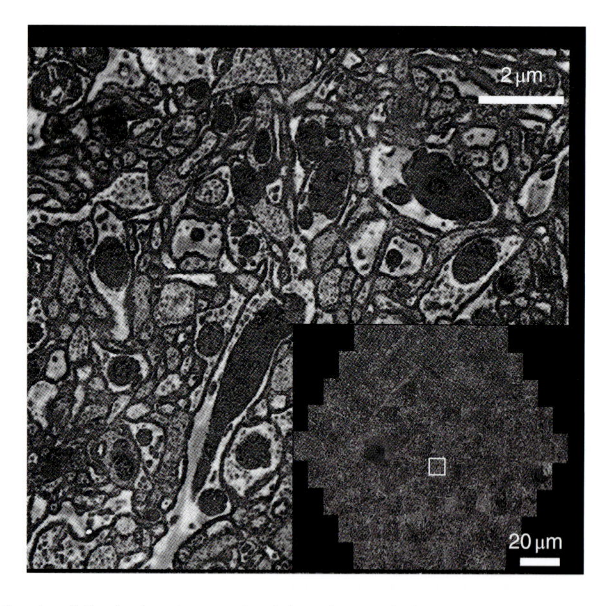

Figure 7.9 Mouse brain (block face) acquired by the multibeam SEM at 0.43 Gpixel/s, scale bar: 2 μm. Inset lower right: large area of 100 μm × 100 μm imaged by 61 beams in parallel, right, scale bar: 20 μm. Area of main image is marked by a white box. Data: courtesy of S. Mikula, W. Denk, MPI für Medizinische Forschung, Heidelberg, Germany.

Both mapping neuronal circuits (Lichtman and Denk, 2011; Helmstaedter *et al.*, 2013) and the visual analysis of cellular structures and networks (Holcomb *et al.*, 2013) require imaging large volumes of tissue in great detail. Only EM enables imaging at the resolutions needed for these experiments. Until now, the long acquisition time of conventional EM limits sample size that is accessible within a reasonable time. By using multiple electron beams in parallel, the MultiSEM technology offers a dramatically higher imaging speed compared to conventional EM techniques.

The multibeam SEM uses 61 beams in a regular array in a single electron-optical column that are scanned in parallel over the sample. Secondary electrons emitted at each primary spot are projected on to a multidetector that records all beams simultaneously. A magnetic beam splitter separates primary and secondary electron beams (Figure 7.8). Each secondary electron yield is recorded at each scan position, forming 61 individual images in parallel, which are merged into a final single image. The final image contains up to 600 million pixels, is close to 30 000 pixels wide, and covers an area of approximately 100 μm × 100 μm to 250 μm × 250 μm. Imaging speed can be as high as 1.22 Gpixel/s (Figure 7.9). Landing energies range from 1 to 3 keV for optimum image contrast on resin embedded samples.

The multibeam SEM is designed for studies of macroscopic areas and objects (mm to cm) at EM resolution (Marx, 2013). Looking ahead, the MultiSEM technology will enable wholly new fields of research that require sample sizes much larger than those accessible with currently available EM throughput.

7.2 3D IMAGING – LIVE HAPPENS IN 3D

Optimum imaging parameters for 2D imaging as described in section 7.1 are the prerequisite for collecting high-resolution structural information on a level where cellular interaction and tissue structure become visible in an SEM. However, to fully understand the functionality of biological systems one must first understand the morphological structure in a three-dimensional context. Several concepts have been developed in the last few years to meet these needs by using SEM imaging technology to generate 3D data sets. Starting at the point where a scientist has a suitable preparation of the structure of interest – mainly this will be resin embedded samples – there are two principle paths for retrieving 3D data information from such samples:

- Using a separate microtome to cut the sample mechanically into serial sections, collecting them on a sample carrier, and imaging one after the other: array tomography (AT) (Micheva and Smith, 2007; Horstmann *et al.*, 2012; Wacker and Schröder, 2013).
- Alternate imaging of the surface of a sample block and removing a slice of a certain thickness, using mechanical devices (block-face imaging, SBF-SEM tomography) (Leighton, 1981; Denk and Horstmann, 2004) or a focused ion beam (FIB-SEM tomography) (Drobne *et al.*, 2008; Knott *et al.*, 2008).

7.2.1 GEMINI®: One Technology for Several 3D Approaches

Optimum speed and resolution are the key issues for all 3D imaging workflows. How fast can a 3D data set of the volume of interest be generated? What is the minimum resolution or voxel size in this data set that reveals the structures of interest? GEMINI® technology allows for large fields of view, thus reducing the time needed for stitching many smaller images (see also Figure 7.6a). With the GEMINI II column the beam current can be adjusted continuously. That gives the user the opportunity to find the optimal current/resolution conditions specific for the application. Compared to aperture-based current settings this accelerates the acquisition by a factor of up to 10, depending on the application.

In the following discussion we will give a short overview of how ZEISS technology meets the demands of the different 3D approaches. We will also introduce the different Atlas solution packages – namely Atlas, Atlas 3D, and Atlas 5 Array Tomography – which make the difference in workflow oriented imaging.

7.2.1.1 *Imaging of Serial Sections – Array Tomography*

Using the serial section approach, ribbons of consecutive serial sections are laid on a solid substrate such as a silicon wafer or indium tin oxide (ITO) coverslips, and are imaged afterwards. The resulting images are then aligned and can be reconstructed into a three-dimensional data set representing the object (Micheva and Smith, 2007; Horstmann *et al.*, 2012: Wacker and Schröder, 2013). Alternatively, sections can be mounted on a grid and imaged in a TEM or in an SEM in the STEM mode (Porter and Blum, 1953; White *et al.*, 1988; Kuwajima *et al.*, 2013).

To do all this in an automated way, ZEISS SEMs can be provided with the Atlas 5 Array Tomography software package. Atlas 5 has a comfortable graphical user interface (GUI) as the front-end of a workflow- and protocol-oriented software. One important timesaver for high-throughput data acquisition is the possibility of reducing the area that is being scanned with highest resolution (Figure 7.10). Very often the structure of interest covers only a small part of the whole sample. Using the ZEISS Atlas software package the user can define arbitrarily shaped regions of interests (ROI) around these structures and assign different imaging protocols. The reduction of the scanned area by user-defined ROIs avoids imaging of void volumes. On serial sections these ROIs can be cloned from one section to all others, which remarkably reduces the time needed for upfront definition of ROIs. The protocols are chosen for the best speed-resolution balance. When the region of interest exceeds the maximum scan field an automated tiling routine for generating larger images is enabled. In addition auto-focus and auto-stigmation algorithms are implemented to offer maximum ease of use and provide a robustness essential for long-term image acquisition sessions.

The obvious advantage of imaging serial sections is the conservation of the original sample. Depending on preparative constraints, it is possible to combine diverse imaging modalities or to take a truly correlative imaging approach. Since the sample is imaged in a classic FESEM, all detector modes can be used for imaging.

7.2.1.2 *Block-Face Imaging*

Block-face imaging using an in-situ ultramicrotome mounted in the SEM chamber (e.g., a Gatan 3View® system) has become well established since its realization by Denk in 2004 (Leighton, 1981; Denk and Horstmann, 2004). Here the sample is repeatedly cut and imaged, resulting in a 3D data set after computational alignment. The *z*-resolution is limited by the slice thickness and thus can go down at best to 15–20 nm (Hughes *et al.*, 2014). GEMINI® technology provides two major advantages in this approach. Block-face imaging requires low acceleration voltages to avoid charging and to get the signal only from the top surface of the sample block. The low beam penetration at landing energies

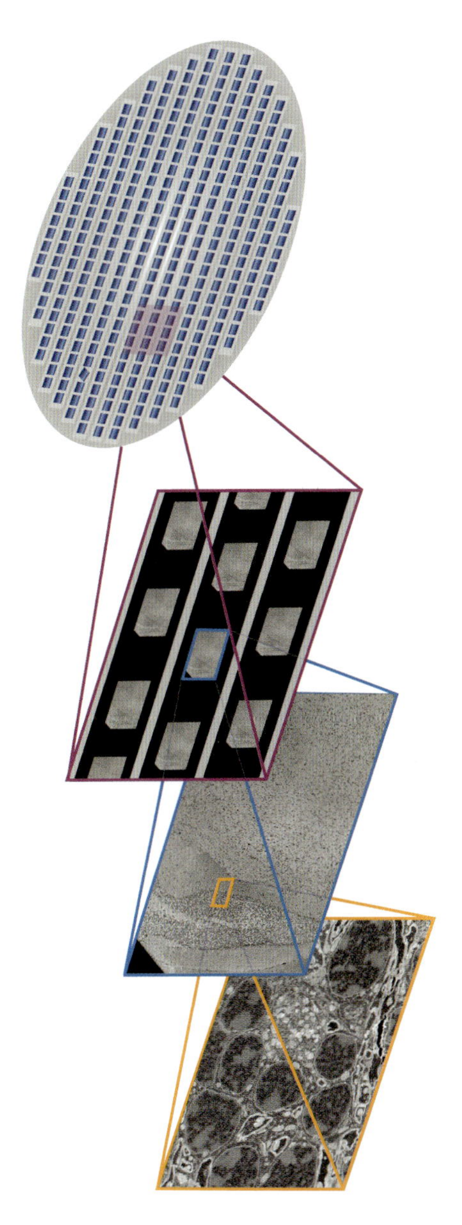

Figure 7.10 Workflow for automated array tomography imaging with the ZEISS Atlas 5 Array Tomography software solution. Consecutive imaging of serial sections on a wafer, automated recognition of sections, zoom-in images, definition of arbitrarily shaped ROIs followed by high-resolution imaging of the final structure. For each step a different imaging protocol can be assigned.

of about 1 keV is matched optimally to the slice thickness. As described in the previous section, acceleration voltages down to 100 eV can be realized in the GEMINI® column by in-column beam deceleration without any disadvantage in beam quality. In addition a very large field of view can be imaged without aberrations, regardless of the image coordinates. An example is shown in Figure 7.11. Block-face imaging using in-situ ultramicrotomy in combination with GEMINI® technology is the fastest method of getting large volume data

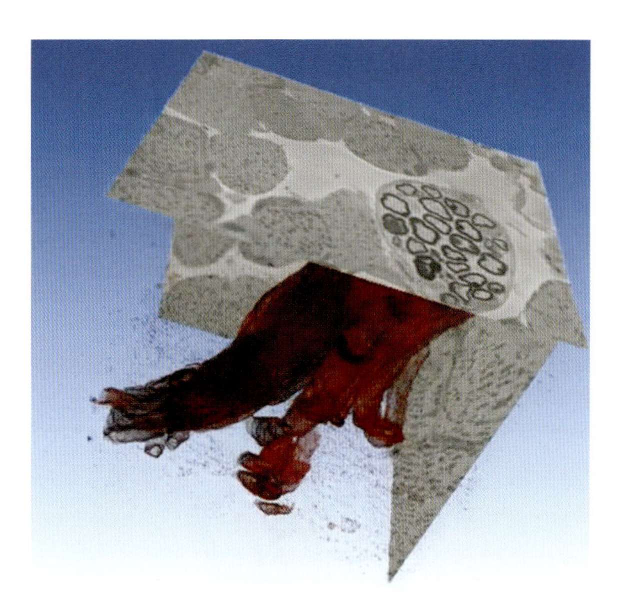

Figure 7.11 3D data reconstruction of a resin embedded block-face sample of mouse extra-ocular muscle. The reconstructed peripheral nerves are shown in red. Data were acquired with a ZEISS Sigma equipped with a Gatan 3View®. A 100 μm × 100 μm × 100 μm 3D data set was reconstructed out of 1000 slices. Data: courtesy of Peter Munro, University College London, UK.

sets, for example, 100 slices in a few hours or 1000 slices overnight. Without doubt this is the preferred solution for 3D imaging of rather large samples where the *z*-resolution of 15 nm is acceptable and where it is not necessary to preserve the sample.

7.2.1.3 FIB-SEM Tomography

Instead of a microtome, an FIB column is attached to the electron microscope in the ZEISS Crossbeam series (Schneider *et al.*, 2011), being the perfect tools for high-resolution FIB-SEM tomography. Here the ion beam takes over the role of the ultramicrotome's knife, which allows material to be removed repeatedly from the sample by subsequent imaging of the surface. Crossbeam is the section tool for many different samples including compound materials of varying hardness. This approach also destroys the sample but, compared to the ultramicrotome, the ion beam is able to ablate even thinner layers – if need be, below 5 nm (Figure 7.12) (Wei *et al.*, 2012). The biggest advantage of the FIB-SEM approach for 3D data acquisition is the possibility of generating high resolution 3D data sets with a homogeneous voxel size (Kreshuk *et al.*, 2011). In addition, the region of interest can be selected with extremely high precision, making Crossbeam a tool for target preparations (Hekking *et al.*, 2009; Narayan *et al.*, 2014). However, there is a difficulty with block-face applications: the user cannot in any case foresee which part of the sample will become interesting over time. For ZEISS Crossbeam systems, Atlas 3D uses the key frame concept to overcome this obstacle. Here ROIs on a selected number of consecutive sections are

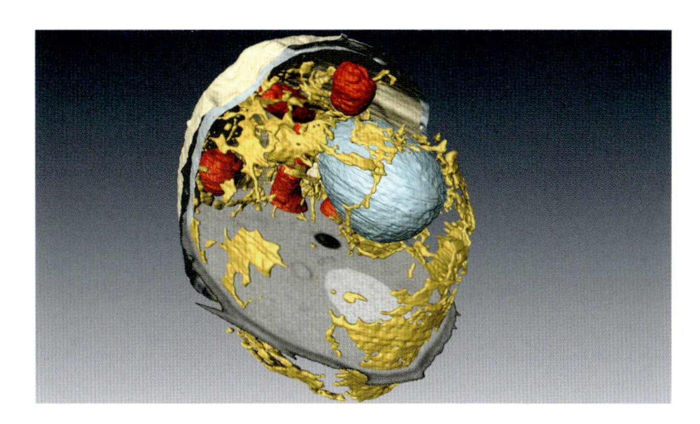

Figure 7.12 3D reconstruction of a yeast cell. Image data taken with a ZEISS Crossbeam. 3D segmentation of ER (yellow), mitochondria (red), and nucleus (blue) of a whole yeast cell at 5 nm isotropic voxels (Wei *et al.* 2012). Cell diameter: 3.5 μm. Sample: courtesy of Jeff Caplan, Delaware Biotechnology Insitute, University of Delaware, Newark, NJ, USA.

imaged and from time to time a larger part of the surface is scanned with an overview imaging protocol, allowing the user to change ROIs or even add more of them during the acquisition process. This reduces the acquisition time to a minimum and simultaneously offers the user the possibility of interactively following up the interesting features of the sample, making Crossbeam a perfect tool for high-resolution structural imaging in biological applications.

7.3 CORRELATIVE SOLUTIONS – FOR A DEEPER INSIGHT

In recent times, more and more researchers have been wanting to combine complementary information using correlative microscopy to gain new insights into the interdependency of function and structure (Müller-Reichert and Verkade, 2012). This can involve the combination of any microscopic methods, but usually the term refers to the most widespread techniques: light microscopy (LM) and EM. LM and EM have been successful imaging technologies for many decades yet mainly they have been used independently. LM – and especially fluorescence light microscopy (FLM) – allow the investigation of dynamic events in living specimens. Introducing fluorescent proteins and thus the possibility of specifically tagging proteins of interest in living cells and organisms with fluorescent signals led to an enormous rise in the use of LM in life sciences research. Because of the diffraction limit the resolution of classic LM is limited to approximately 200 nm. With new LM methods such as PALM (photo-activated localization microscopy) and STORM (stochastic optical reconstruction microscopy) (Betzig *et al.*, 2006; Rust, Bates, and Zhuang, 2006) – often referred to as "super-resolution LM" – the calculated resolution was improved down to approximately 20 nm. However, even with this enhanced resolution, while being highly specific and sensitive enough to identify single molecules, FLM does not reveal the structural context of where the signals come from. This is the domain of EM: visualization of morphological details with single-digit nanometer resolution. Modern SEMs make it easier and more comfortable than ever before to access this level of ultrastructural details. Combining the advantages of multi-imaging modes by bridging the gap between LM and EM gives new insight to many biological and medical questions (Takizawa, Suzuki, and Robinson, 1998; Polishchuk *et al.*, 2000; Schwarz and Humbel, 2014; Jahna *et al.*, 2012; Kopek *et al.*, 2012, 2013; Lucas *et al.*, 2012; Murphy *et al.*, 2011).

7.3.1 Shuttle & Find – The Interface for Easy Correlation

A prerequisite for a successful correlative approach is to reposition the specimen precisely in LM and EM in order to find the same ROIs within the sample in both imaging systems. The different image contents can make this a tedious, time-consuming, nearly impossible task. With Shuttle & Find, ZEISS offers a simple straightforward workflow solution that allows easy calibration of the sample carrier and transfer of the sample together with the corresponding coordinates between LM and SEM, both running ZEISS Efficient Navigation (ZEN) software. As the microscope systems remain unchanged, the full performance and flexibility to adapt to the exact requirements of the sample and to choose the most suitable imaging modes are always guaranteed. This is very important as the example in Figure 7.13 nicely illustrates. Different experimental setups with varying imaging modes

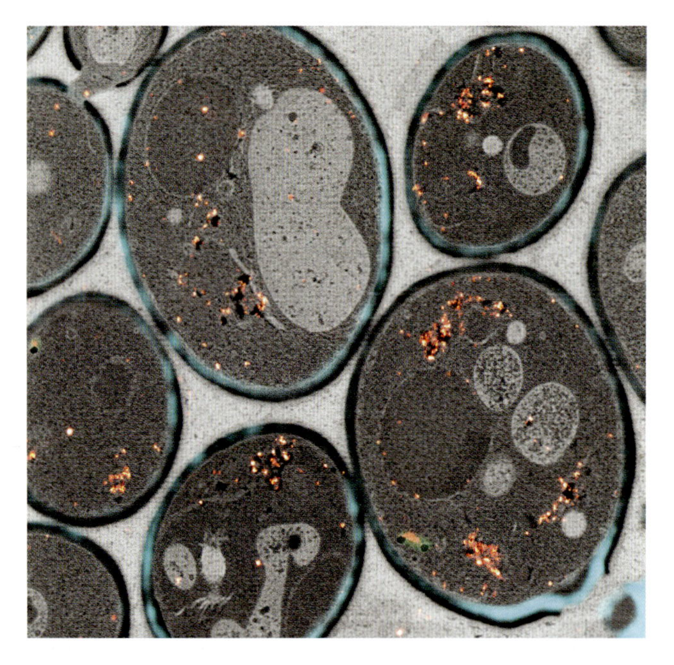

Figure 7.13 High magnification correlative super-resolution microscopy of yeast ultrathin sections. Cell walls are shown in blue (imaged by structured illumination microscopy SIM), Alexa Fluor® 647 labeled hA1aR-Cerulean imaged via dSTORM (red) overlaid with an FE-SEM BSE image. Locations imaged by super-resolution microscopy were automatically recovered in the SEM using Shuttle & Find. FOV in image = 8 μm. Data: courtesy of Jeffrey L. Caplan and Kirk Czymmek, University of Delaware, Newark, NJ, USA.

will need various adaptations. What remains the same is that the combination of LM and SEM data from exactly the same sample areas neatly demonstrate the benefits of correlative microscopy: while LM allows one to specifically identify the proteins of interest and super-resolution (in this case STORM) helps to further resolve the signals, SEM immediately reveals their subcellular localization.

7.3.2 Correlative Microscopy Going 3D

7.3.2.1 Correlative Array Tomography

A widely spread method for uncovering 3D information from resin embedded samples is array tomography (AT) (see Section 7.2.1.1) or, as will be described here, correlative array tomography (CAT). After alignment of the image series from LM and SEM into *z*-stacks, the corresponding areas and volumes can be registered. For correlative microscopy this method offers a major advantage: the *z*-resolution for LM imaging is now limited to the thickness of the ultrathin sections, typically in the range between 50 and 100 nm, if not specific on-section labeling reduces this even more. This is significantly better than what can be achieved with confocal or other optical sectioning approaches. The example in Figure 7.14 shows the detection of acetylcholine receptors in fluorescence and its localization within the neuromuscular junction. Imaging of consecutive sections and reconstruction of the fluorescent data reveals the net-like protein distribution in 3D.

7.3.2.2 Correlative Volume Imaging

Probably the most demanding approach is the correlation of 3D LM data from living samples with the corresponding 3D data from EM. Many steps of sample preparation and processing have to be taken between LM and EM imaging. This makes it extraordinarily difficult to keep position and *xyz* orientation of the structures of interest throughout the whole workflow. Figure 7.15 shows a typical workflow of correlative imaging of mouse neuronal dendrites *in vivo* using two-photon and FIB Crossbeam imaging within research projects on neurodegenerative diseases in the Lab of Prof. Jochen Herms, German Center for Neurodegenerative Diseases, LMU (Blazquez-Llorca *et al.*, 2015). To determine the expression and distribution of certain GFP-tagged proteins, a defined area in the brain of a living mouse is imaged over time. At a certain point in time the animal is fixed by perfusion fixation and the now fixed brain area is once more imaged with multiphoton laser scanning microscopy, resulting in data sets of the neurons with spines that are later reconstructed in 3D. As the following heavy metal contrasting and resin embedding for EM destroys all fluorescent signals, laser marks are burned into the tissue to define the regions of interest (Bishop *et al.*, 2011) that need to be relocalized in the SEM. For 3D FESEM imaging a Crossbeam system was used. After segmentation of the neuron of interest within the individual planes of the *z*-stack, it was possible to directly correlate the information from LM and FESEM, giving an insight into the fine structural changes that occurred in the spines after protein expression. Given that this process is time-dependent it is essential to follow the development in the living tissue to find the right moment, and also to image exactly the same structure in the SEM to document the morphological changes at this point of development.

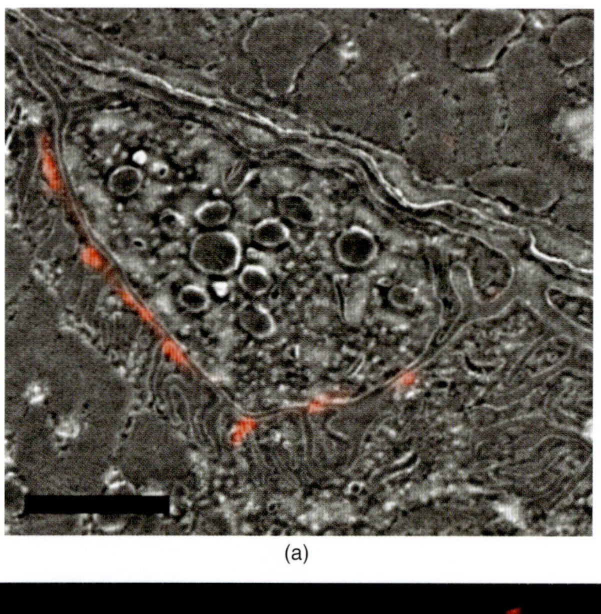

(a)

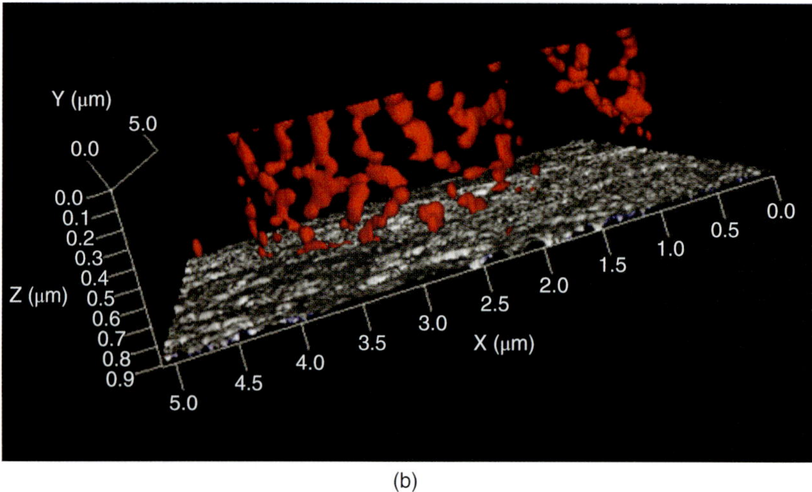

(b)

Figure 7.14 Resin embedded section of a neuromuscular junction imaged by correlative wide-field LM and SEM (a) combining the structural SEM information with functional LM information. The acetylcholine receptors of the postsynaptic membrane of a neuromuscular junction are fluorescently labeled. An increased density of receptors was found in the yellow regions. (b) 3D reconstruction of 40 LM images projected on to the corresponding area in an SEM image. Scale bar: 1 µm. Image data: courtesy of Jochen Fuchs and Christian Dietrich, Corporate Research and Technology, Carl Zeiss AG, Germany. Sample: courtesy of Ira Röder, Bioquant, University of Heidelberg. Data compiled as part of a project financially supported by the German Federal Ministry for Education and Research – “NanoCombine” FKZ 13N11401/13N11402.

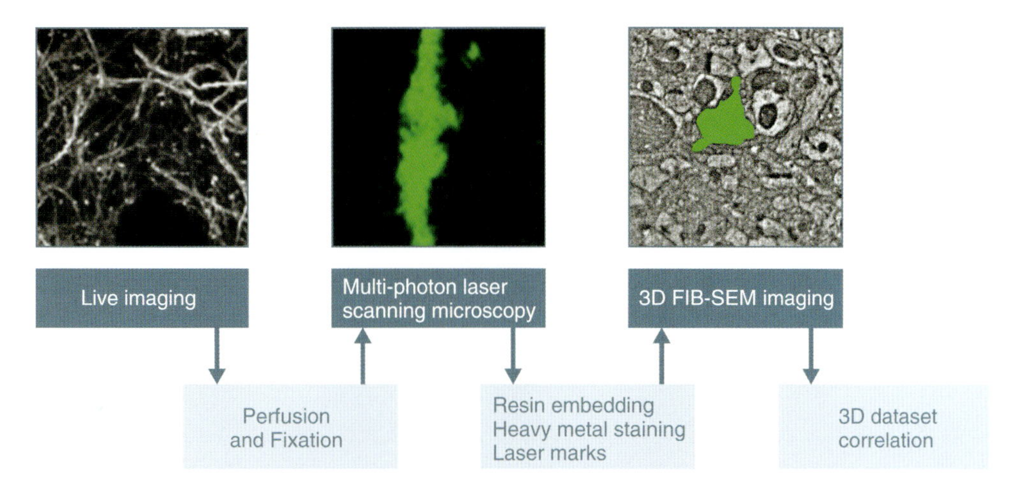

Figure 7.15 Showing a correlative workflow from functional live cell imaging to 3D data acquisition in an FIB-SEM. At a certain point in time live imaging of the GFP tagged mouse brain area is stopped by perfusion and fixation. Then high-resolution multiphoton laser scanning microscopy generates a 3D data set of the chosen area. After fixation and heavy metal staining, laser marks are imprinted in the sample to refind the region of interest in the FIB-SEM where a 3D data set of the morphological structure is imaged. Finally, a correlative 3D data set is generated. Data: courtesy of Jochen Herms and Lidia Blazquez-Llorca, Department of Translational Brain Research, German Center for Neurodegenerative Diseases (DZNE), LMU München, Germany, and Eric Hummel and Hans Zimmermann, Carl Zeiss Microscopy GmbH, Germany.

7.3.3 New Possibilities in Correlative Microscopy

Though the idea of correlative microscopy is very old, it is only recently that it has become trendy. Now the field is developing very dynamically and new results and approaches are frequently published.

7.3.3.1 Cryo-FIB

One very promising option for correlative microscopy is just emerging: 3D cryo-FIB-SEM imaging of unstained vitreous samples. The possibility to work with near native samples without any artifacts induced by chemical fixation, embedding, heavy metal contrasting, etc., will be a major improvement. Especially for correlative approaches, the preservation of fluorescence that usually is destroyed by EM sample preparation protocols seems very promising. In the case of vitrified samples the uncovering of the third dimension via FIB technology is significantly easier than by mechanical serial sectioning. A recent study (Schertel *et al.*, 2013) shows that at least certain samples can provide sufficient contrast for FIB-SEM imaging under cryo conditions without any additional treatment. Data sets of 127 serial images were recorded by repeated FIB milling and block face FESEM imaging under cryo conditions covering a volume of $X = 7.72\,\mu m$, $Y = 5.79\,\mu m$, and $Z = 3.81\,\mu m$, lateral image pixel size of 7.5 nm, and an FIB slice thickness of 30 nm. As Figure 7.16 documents, not only myelin sheaths but a variety of cellular structures including mitochondria, nuclear pores, Golgi compartments, etc., are visible.

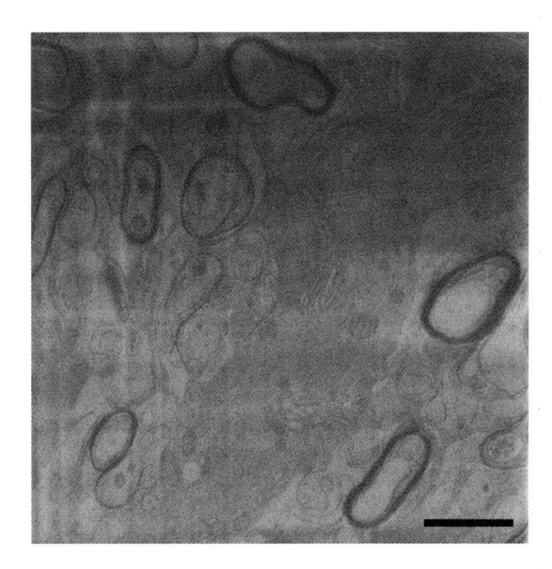

Figure 7.16 Exemplary image of the oligodendrocyte of the optical nerve of the mouse. The sample was high-pressure frozen. Structural details like myelinated axons, mitochondria with their cristae, Golgi complex, and vesicles are clearly visible. Scale bar: 1 μm. Data: courtesy of Andreas Schertel Carl Zeiss Microscopy GmbH, Germany, and Wiebke Möbius, Max-Planck Institute of Experimental Medicine, Göttingen, Germany. Data compiled as part of a project financially supported by the German Federal Ministry for Education and Research – "NanoCombine" FKZ 13N11403.

7.3.3.2 X-Ray Microscopy

Another increasingly important technology in the correlative workflow, bridging length scales between light and electron microscopy, is X-ray microscopy (XRM). Laboratory sources and optics have made it possible to image 3D samples non-destructively that are otherwise opaque to light. Depending on the choice of source and architecture (ZEISS Versa or ZEISS Ultra) resolution from 50 nm to 700 nm can be realized (Merkle and Gelb, 2013). Primary contrast mechanisms in XRM are absorption, phase contrast, and diffraction contrast. Using the difference in absorption of soft X-ray in the water window region at wavelengths between 2.34 and 4.4 nm by carbon atoms and oxygen atoms has a primary focus on imaging frozen hydrated samples (cells) with resolution in the 30–50 nm range. Chemical imaging utilizing tunable energy sources is a unique strength of synchrotron beamlines, but some functionalities have been enabled on laboratory systems that make it more straightforward to combine and align dual-energy acquisitions, which can assist in the discrimination of discrete phases of material. In doing so, laboratory XRM has evolved past conventional microcomputed tomography (microCT) methods, offering levels of contrast and resolution only rivaled by synchrotron sources, and often surpassing them (Maire and Withers, 2014). In the field of neuroscience research there is great interest in creating complete neural network maps of the brain and the need for information across multiple length scales in 3D has spawned the development of high-throughput 3D electron microscopy (3DEM) techniques (as discussed in Sections 7.1 and 7.2). However, even if long acquisition times, sensitivity to sample preparation, and quality can be overcome by intelligent software solutions and more sophisticated preparation methods the challenges in locating regions of interest (buried subsurface features) still have to be averted.

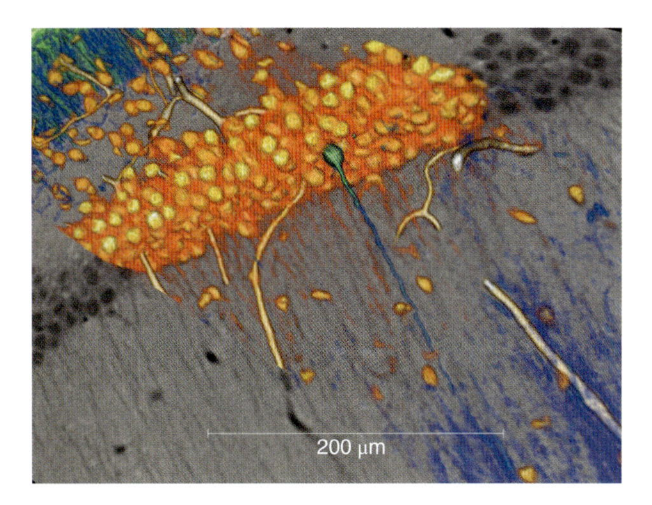

Figure 7.17 ZEISS Xradia Versa XRM acquisition of stained mouse brain tissue, used as a navigational aid for destructive serial block-face EM techniques. 3D reconstructed volume rendering (color) is overlaid on one 2D XRM virtual slice (grayscale). A single diaminobenzidine-labeled neuron can be identified (green). This example is from work being carried out by the National Center for Microscopy and Imaging Research (NCMIR) at the University of California, San Diego, together with Carl Zeiss X-ray Microscopy (Miekle *et al.*, 2013; Tapia *et al.*, 2013).

Figure 7.17 shows an impressive example where 3D XRM is being used to image an embedded sample, prepared for SEM imaging with cellular resolution resulting in a 3D roadmap for efficient cutting, trimming, or orientation via block-face imaging or serial section techniques. The ZEISS XRM technique is the perfect solution for acting as a bridge between light and electron microscopy, and as an efficiency multiplier in the field of intelligent volume imaging.

Here we conclude our short overview of how ZEISS contributes to the field of biological and medical research with a broad spectrum of tools and solutions, spanning from high-end FESEMs over 3D SEM imaging tools for classical imaging and correlative applications to novel and unique techniques like ion microscopy, multiSEM technology, and X-ray imaging.

ACKNOWLEDGEMENT

The authors wish to thank all their cooperation partners who have contributed to this chapter with helpful discussions and image material.

REFERENCES

Bazou, D., Behan, G., Reid, C., Boland, J.J., and Zhang, H.Z. (2011) Imaging of human colon cancer cells using He-Ion scanning microscopy. *Journal of Microscopy*, 242 (3), 290–294.

Betzig, E., Patterson, G.H., Sougrat, R., Lindwasser, O.W., Olenych, S., and Bonifacino, J.S. (2006) Imaging intracellular fluorescent proteins at nanometer resolution. *Science*, 313 (5793), 1642–1645.

Bishop, D., Nikić, I., Brinkoetter, M., Knecht, S., Potz, S., Kerschensteiner, M., and Misgeld, T. (2011) Near-infrared branding efficiently correlates light and electron microscopy. *Nature Methods*, 8 (7), 568–570.

Blazquez-Llorca, L., Hummel, E., Zimmermann, H., Zou, C., Burgold, S., Rietdorf, J. and Herms, J. (2015) Correlation of two-photon *in vivo* imaging and FIB/SEM microscopy. *Journal of Microsocpy*, 259 (2), 129–136.

Boden, S.A., Asadollahbaik, A., Rutt, N.A., Bagnal, D. (2013) Helium ion microscopy of Lepidoptera scales, *SCANNING*, 34, 107–120.

Cavalier, A., Spehner, D., and Humbel, B.M. (eds) (2008) *Handbook for Cryo-Preparation Methods for Electron Microscopy*, CRC Press, Inc., Boca Raton, FL, USA.

Cazaux, J. (2004) Charging in scanning electron microscopy "from inside and outside". *SCANNING*, 26,181–203.

Denk, W. and Horstmann, H. (2004) Serial block-face scanning electron microscopy to reconstruct three-dimensional tissue nanostructure. *PLoS One Biology*, 2 (11), e329.

Drobne, D., Milani, M., Leser, V., Tatti, F., Zrimec, A., Znidarsic, N., Kostanjsek, R., and Strus, J. (2008) Imaging of intracellular spherical lamellar structures and tissue gross morphology by a focused ion beam/scanning electron microscope (FIB/SEM). *Ultramicroscopy*, 108 (7), 663–670.

Frosien, J., Plies, E., and Anger, K. (1989) Compound magnetic and electrostatic lenses for low-voltage applications. *J. Vac. Sci. Technol.*, B7, 1874–1877.

Gest, H. (2004) The discovery of microorganisms by Robert Hook and Antoni von Leuwenhoek, Fellows of the Royal Society. DOI: 10.1098/rsnr.2004.0055.

Gnauck, P., Drexel, V., and Greiser, J. (2001). A new high resolution field emission SEM with variable pressure capabilities. *Microsc. Microanal.*, 7 (Suppl. 2), 880–881.

Hekking, L.H.P., Lebbink, M.N., De Winter, D.A.M., Schneijdenberg, C.T.W.M., Brand, C.M., Humbel, B.M., Verkleij, A.J., and Post, J.A. (2009) Focused ion beam-scanning electron microscope: exploring large volumes of atherosclerotic tissue. *J. Microsc.*, 235, 336–347.

Helmstaedter, M., Briggman, K.L., Turaga, S.C., Jain, V., Seung, H.S., and Denk, W. (2013) Connectomic reconstruction of the inner plexiform layer in the mouse retina. *Nature*, 500, 168–174.

Holcomb, P.S., Hoffpauir, B.K., Hoyson, M.C., Jackson, D.R., Deerinck, T.J., Marrs, G.S., Dehoff, M., Wu, J., Ellisman, M.H., and Spirou, G.A. (2013) Synaptic inputs compete during rapid formation of the calyx of Held: A new model system for neural development. *The Journal of Neuroscience*, 33, 1295412969.

Horstmann, H., Körber, C., Sätzler, K., Aydin, D., and Kuner, T. (2012). Serial section scanning electron microscopy (S3EM) on silicon wafers for ultra-structural volume imaging of cells and tissues. *PLoS One*, 7, e35172.

Hughes, L., Hawes, C., Monteith, S., and Vaughan, S. (2014) Serial block face scanning electron microscopy – the future of cell ultrastructural imaging. *Protoplasma*, 251, 395–401.

Jahna, K.A., Barton, D.A., Kobayashia, K., Ratinaca, K.R., Overallb, R.L., and Braeta, F. (2012) Correlative microscopy: Providing new understanding in the biomedical and plant sciences. *Micron*, 43, 565–582.

Jaksch, H. and Martin, J.P. (1995) High-resolution, low-voltage SEM for true surface imaging and analysis. *Fresenius J. Anal. Chem.*, 353, 378–382.

Jaksch, H., Steigerwald, M., Drexel, V., and Bihr, H. (2003) New detection principles on the GEMINI SUPRA FE-SEM. *Microsc. Microanal.*, 9 (Suppl. 3), 106–107.

Joens, M.S., Huynh, C., Kasuboski, J.M., Ferranti, D., Sigal, Y.J., Zeitvogel, F., Obst, M., Burkhardt, C.L., Curran, K.P., Chalasani, S.H., Stern, L.A., Goetze, B., and Fitzpatrick, J.A.J. (2013) Helium ion microscopy (HIM) for the imaging of biological samples at sub-nanometer resolution. *Scientific Reports* 2013, 3: Art. No. 3514; DOI: 10.1038/srep03514.

Knott, G., Marchman, H., Wall, D., and Lich, B. (2008) Serial section scanning electron microscopy of adult brain tissue using focused ion beam milling. *J. Neurosci.*, 28 (12), 2959–2964.

Kopek, B.G., Shtengel, G., Xu, C.S., Clayton, D.A., and Hess, H.F. (2012) Correlative 3D superresolution fluorescence and electron microscopy reveal the relationship of mitochondrial nucleoids to membranes. *PNAS*, 109 (16), 6136–6141.

Kopek, B.G., Shtengel, G., Grimm, J.B., Clayton, D.A., and Hess, H.F. (2013) Correlative photoactivated localization and scanning electron microscopy. *PLoS One*, 8 (10), e77209.

Kreshuk, A., Straehle, C.N., Sommer, C., Koethe, U., Cantoni, M., Knott, G., and Hamprecht, F.A. (2011) Automated detection and segmentation of synaptic contacts in nearly isotropic serial electron microscopy images. *PLoS One*, 6 (10), e24899.

Kuwajima, M., Mendenhall, J.M., Lindsey, L.F., and Harris, K.M. (2013) Automated Transmission-mode scanning electron microscopy (tSEM) for large volume analysis at nanoscale resolution. *PLoS One*, 8 (3), e59573.

Leighton, S.B. (1981) SEM images of block faces, cure by a miniature microstome within the SEM – A technical note. *Scan Electron Microsc.*, 2, 73–76.

Lichtman, J.W. and Denk, W. (2011) The big and the small: Challenges of imaging the brain's circuits. *Science*, 334, 618–623.

Lucas, M.S., Guenthert, M., Gasser, P., Lucas, F., and Wepf, R. (2012) Bridging microscopes: 3D correlative light and scanning electron microscopy of complex biological structures. *Method. Cell Biol.*, 111, 325–356.

Maire, E. and Withers, P.J. (2014) Quantitative X-ray tomography. *International Materials Review*, 59, 1–43.

Martin, J.P., Weimer, E., Frosien, J., and Lanio, S. (1994) Ultra-high resolution SEM – a new approach. *Microscopy and Analysis*, 28, 43.

Marx, V. (2013) Neurobiology: Brain mapping in high resolution. *Nature*, 503, 147–152.

McDonald, K.L. and Webb, R.I. (2011) Freeze substitution in 3 hours or less. *Journal of Microscopy*, 243, 227–233.

Merkle, A. and Gelb, J. (2013) The ascent of 3D X-ray microscopy in the laboratory. *Microscopy Today*, 21, 10–15.

Micheva, K.D. and Smith, S.J. (2007) Array tomography: A new tool for imaging the molecular architecture and ultrastructure of neural circuits. *Neuron*, 55, 25–36.

Mielke, R.E., Priester, J.H., Werlin, R.A., Gelb, J., Horst, A.M., Orias, E., and Holden, P.A, (2013) Differential growth and nanoscale TiO2 accumulation in *Tetrahymena thermophila* by direct feeding versus cactivory of *Pseudomonas aeruginosa*. *Applied and Environmental Microbiology*, 79 (18), 5616–5624.

Müller-Reichert, T. and Verkade, P. (eds) (2012) *Correlative Light and Electron Microscopy*, Associated Press, Elsevier, Amsterdam.

Murphy, G.E., Narayan, K., Lowekamp, B.C., Hartnell, L.M., Heymann, J.A.W., Fu, J., and Subramaniam, S. (2011) Correlative 3D imaging of whole mammalian cells with light and electron microscopy original. *Journal of Structural Biology*, 176 (3), 268–278.

Narayan, K., Danielson, C.M., Lagarec, K., Lowekamp, B.C., Coffman, P., Laquerre, A., Phaneuf, M.W., Hope, T.J., and Subramaniam, S. (2014) Multi-resolution correlative focused ion beam scanning electron microscopy: Applications to cell biology. *Journal of Structural Biology*, 185 (3), 278–284.

Notte, J.A. (2012) Charged particle microscopy: Why mass matters. *Microscopy Today*, 20, 16–22. DOI: 10.1017/S1551929512000715.

Polishchuk, R.S., Elena, V., Polishchuk, P.M., Saverio, A.R., Buccione, A.L., and Mironov, A.A. (2000) Correlative light-electron microscopy reveals the tubular-saccular ultrastructure of carriers operating between Golgi apparatus and plasma membrane. *Journal of Cell Biology*, 148, 45–58. DOI: 10.1083/jcb.148.1.45.

Porter, K.R. and Blum, J. (1953) A study in microtomy for electron microscopy. *Anat. Rec.*, 117, 685–709.

Rice, W.L., Hoek, A.N., Paunescu, T.G., Huynh, C., Goetze, B., Singh, B., Scipioni, L., Stern, L.A., and Brown, D. (2013). High resolution helium ion scanning microscopy of the rat kidney. *PLoS One*, 8 (3), e57051.

Rust, M.J., Bates, M., and Zhuang, X. (2006) Sub-diffraction-limit imaging by stochastic optical reconstruction microscopy (STORM). *Nature Methods*, 3 (10), 793–795.

Salzer, R., Ackermann, J., Arnold, R., Meyer, S., and Kübler, C. (2013) New low voltage scanning transmission electron microscope detector for fastest image acquisition in BF, DF and HAADF. *Microsc. Microanal.*, 19 (Suppl. 2), 1182–1183.

Schatten, H. (2008) High resolution, low voltage, field-emission scanning electron microscopy applications for cell biology and specimen preparation protocols, in *Biological Low-Voltage Scanning Electron Microscopy* (eds H. Schatten and J. Pawlea), Springer Science+Business Media, New York, pp. 145–170.

Schertel, A., Snaidero, N., Han, H.M., Ruhwedel, T., Laue, M., Grabenbauer, M., and Möbius, W. (2013) Cryo FIB-SEM: Volume imaging of cellular ultrastructure in native frozen specimens. *Journal of Structural Biology*, 184 (2), 355–360.

Schneider, P., Meier, M., Wepf, R., and Müller, R. (2011). Serial FIB/SEM imaging for quantitative 3D assessment of the osteocyte lacuno-canalicular network. *Bone*, 49, 304–311.

Schwarz, H. and Humbel, B.M. (2014) Correlative light and electron microscopy using immunolabeled resin sections. *Methods Molecular Biology*, 1117, 559–592.

Steigerwald, M., Arnold, R., Bihr, J., Drexel, V., Jaksch, H., Preikzas, D., and Vermeulen, J.P. (2004) New detection system for GEMINI. *Microsc. Microanal.*, 10 (Suppl. 2), 1372–1373.

Takizawa, T., Suzuki, K., and Robinson, J.M. (1998). Correlative micros-copy using FluoroNanogold on ultrathin cryosections: Proof of principle. *J. Histochem. Cytochem.*, 46, 1097–1102.

Tapia, J.C., Kasthuri, N., Hayworth, K., Schalek, R., Lichtman, J.W., Smith, S.J., Buchanan, J.A. (2013). High contrast en bloc staining of neuronal tissue for field emission scanning electron microscopy. *Nat. Protoc.*, 7 (2), 193–206.

Wacker, I. and Schröder, R.R. (2013) Array tomography. *Journal of Microscopy*, 2013. DOI: 10.1111/jmi.12087.

Wei, D., Jacobs, S., Modla, S., Zhang, S., Young, C.L., Cirino, R., Caplan, J., Czymmek, K. (2012) High-resolution three-dimensional reconstruction of a whole yeast cell using focused-ion beam scanning electron microscopy. *BioTechniques*, 53, 41–48.

White, J.G., Southgate, E., Thomson, J.N., and Brenner, S. (1986) The structure of the nervous system of the nematode *Caenorhabditis elegans*. *Phil. Trans. R. Soc. Lond. A*, 1, 314–340.

8

SEM Cryo-Stages and Preparation Chambers

Robert Morrison
Quorum Technologies Ltd, Laughton, UK

8.1 OVERVIEW

Preparation instruments for SEM samples, such as critical point dryers and freeze dryers, are designed to controllably remove water from biological and similar liquid-based specimens. This partially overcomes the adverse effects of air drying (namely, severe collapse of cellular structure as the drying front passes through the specimen).

Prior to drying, most biological specimens require chemical stabilization ('fixation') in order to preserve them and allow them to better withstand subsequent dehydration and sputter coating processes.

Many modern scanning electron microscopes (SEMs) can operate without exposing the specimen to high vacuum. Often referred to as environmental/low-vacuum/high-pressure/variable-pressure SEMs, these instruments use higher pressures to minimise out-gassing from volatile specimens. However, variable pressure techniques are generally limited to certain specimen types and have a number of disadvantages in terms of specimen stability and the information that can be obtained (see Chapter 20 in Volume II, Reimer, Eggert and Hohenberg).

For water-based, liquid or semi-liquid and beam-sensitive materials, cryo-SEM overcomes many of the problems associated with drying protocols and variable pressure techniques. Cryo-SEM of biological material removes the need for conventional preparation methods, such as chemical fixation and critical point drying, and allows observation of the specimen in its 'natural' hydrated state. Figure 8.1 shows two examples of preservation of delicate structures. The left hand image is of pollen stuck to a spider's web and the right hand image is wax on a leaf surface.

Biological Field Emission Scanning Electron Microscopy, First Edition.
Edited by Roland A. Fleck and Bruno M. Humbel.

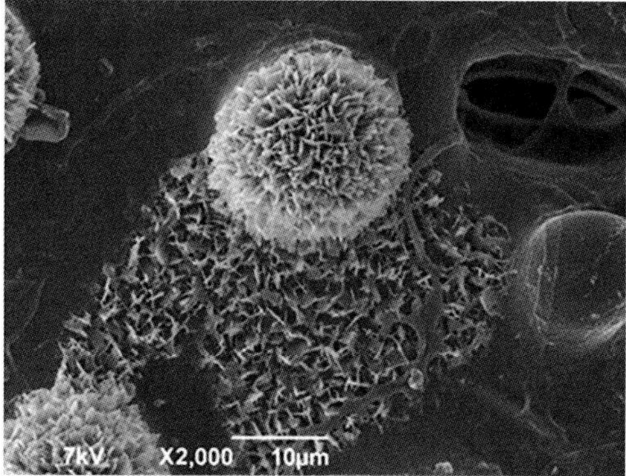

Figure 8.1 Examples of preservation of delicate structures.

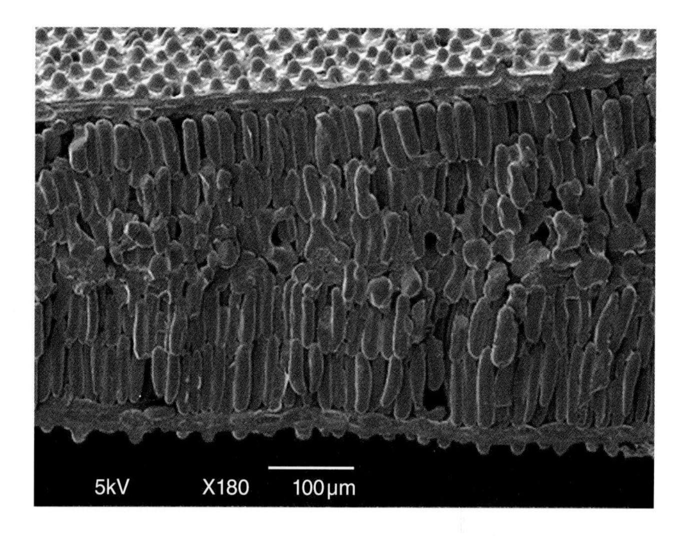

Figure 8.2 Cold fracture through a leaf.

A major advantage of cryo-SEM is the capability to cold-fracture specimens to reveal internal microstructure and gain important information from materials with different dispersion phases, for example oils, polymers, fats and food stuffs (see Chapter 17 in Volume II, Hazekamp and van Ruijven). Figure 8.2 shows a cold fracture through a leaf.

For most protocols, SEM specimens need to be electrically conductive. For this reason, non-conductive and semi-conductive SEM specimens require the deposition of a thin surface layer of metal. This also increases the number of secondary electrons that can be detected from the surface of the specimen in the SEM, and therefore increases the signal to noise ratio. Metal layers need to have a fine grain size and be evenly distributed across the specimen surface.

Cryo preparation techniques for scanning electron microscopy (cryo-SEM) are now considered essential for the successful observation of many wet or 'beam sensitive' specimens. Cryo preparation systems normally include facilities to rapidly freeze and transfer specimens. The cryo preparation chamber includes tools for cryo fracturing, controlled sublimation and specimen coating.

8.2 HISTORY

Initial forays into cryo-SEM were very primitive. Samples were simply plunged into liquid nitrogen and quickly air transferred into the SEM. The SEM was then pumped down and, after waiting for the frost to sublime, a few images were grabbed before the sample warmed up too much. Later, airlocks were added to allow a cleaner transfer and eventually vacuum cryo transfer was added. Today's cryo preparation systems are very different beasts with fully automated cooling systems, integrated turbo pumping systems and automatic sublimation and sputtering control. There are different philosophies as to how and where the systems are cooled.

8.3 TYPES OF COOLING

There are essentially two methods of cooling the cold stage and anti-contaminator (cold trap) in the SEM chamber; braid cooling and cold nitrogen gas cooling. Early systems tended to use braid cooling due to its' inherent simplicity.

8.3.1 Braid Cooling

These systems use copper braids connected to a liquid nitrogen reservoir mounted on the SEM chamber (Figure 8.3). The braids connect directly to the SEM stage and to the anti-contaminator.

The main advantage of this type of system is, potentially, a slightly simpler installation as there is no gas feed through. The large thermal mass means the temperature will be relatively stable but slow to respond. However, the disadvantages are quite numerous:

- Poor base temperature due to the thermal losses across the connections between the braid and other components.
- Poor thermal response due to the mass of the braid. To achieve good cooling a very large diameter braid is needed. This in turn needs space inside the SEM column to route it away from warm surfaces and avoid touches.
- The weight and drag of the braid on the SEM stage can be a problem.
- The large thermal mass makes it hard to control sublimation in the chamber.
- Subliming in the SEM tends to put heat into the braid system, which can affect the temperature of the anti-contaminator and puts heat into the dewar, leading to boiling of the nitrogen.

The fact that there is a large area of cold surface inside the SEM column can help improve the vacuum in the microscope.

8.3.2 Gas Cooling

As cryo-SEM systems developed, gas cooling was introduced (Figure 8.4). Warm, dry nitrogen gas is passed through a heat exchanger to generate cold gas. This cold gas then flows through the cold stage and anti-contaminator.

The advantages of gas cooling are an excellent thermal response as the mass of the cold stage is relatively small. This makes it easy to control sublimation in the SEM chamber and

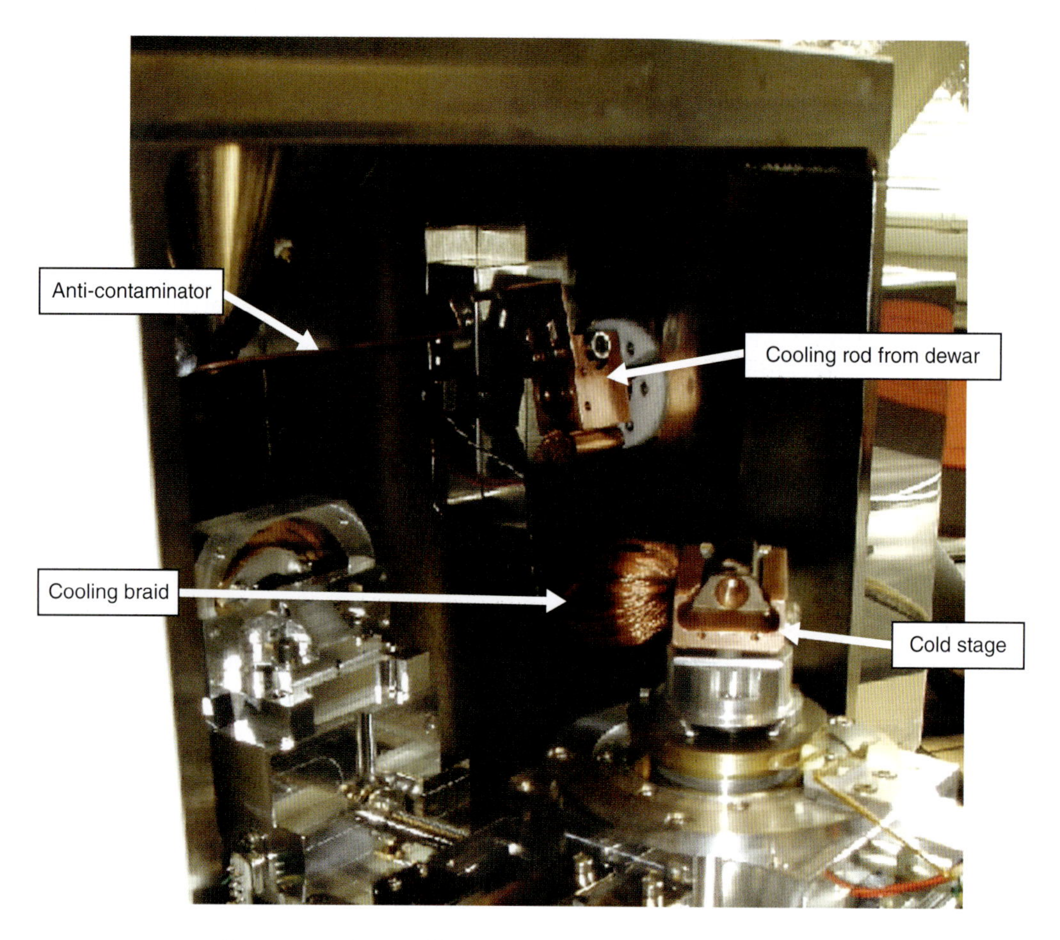

Figure 8.3 A typical braid cooled system (Image Quorum©).

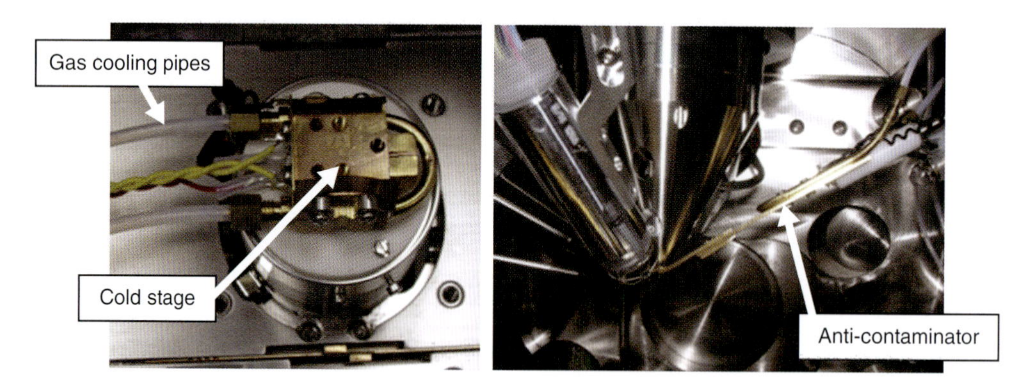

Figure 8.4 A typical gas cooled system (Image Quorum©). The method of controlling the cold gas depends on the system, varying from crude pressure regulators to set a rough temperature and then using a heater to control the fine temperature, to full PID (proportional, integral, derivative) feedback mass flow control (see Theory of PID control, http://www.ni.com/white-paper/3782/en/).

gives a very rapid temperature response. Good gas flow controlled systems can now attain a temperature stability of the order of 0.15 °C in over an hour. Routing of the cooling pipes is very adaptable, making installation simpler. Temperatures are much less influenced by pipes touching warm surfaces. The low mass stage and lightweight tubing leads to less load on the SEM stage and the cold stage can be parked inside the SEM chamber when not being used in cryo mode. By using two separate cooling circuits, the temperature of the SEM cold stage and the anti-contaminator can be driven completely independently.

Gas cooling also lends itself to the use of off-column dewars.

The main disadvantage of gas cooling, especially on early systems, is vibration if the cooling circuit is poorly designed, the gas flow is set too high or the gas flow control circuit is not optimal. This has largely been eliminated on the latest generation of gas cooled systems.

The water content of the supplied nitrogen gas needs to be low to prevent freeze-ups of the cryo tubes. Some systems are now supplied with a gas dryer to counteract this problem.

8.4 LOCATION OF THE PREPARATION CHAMBER

There are basically two choices when it comes to positioning the specimen preparation chamber: on column, where the chamber is bolted directly to the SEM, or off-column, where the preparation chamber is remote from the microscope.

8.4.1 On-Column Preparation Chamber (Figure 8.5)

Why choose on-column?

- Ease of use
  - A single airlock operation brings the specimen into a high-vacuum environment where it stays for the rest of the preparation and observation process.
  - It is easier to check uncoated specimens before subliming or coating as they can quickly and easily be transferred onto the SEM cold stage to be imaged and then returned to the preparation chamber to be processed simply by opening the isolation valve to the SEM column.
  - This method has an obvious speed advantage and there is a reduced risk of contamination as the specimen is not going from high vacuum to low vacuum in an airlock and back again.
  - If a specimen is found to need extra coating or to be fractured at a lower point it is simply withdrawn into the preparation chamber, processed and returned to the SEM.

8.4.2 Off-Column Preparation Chamber (Figure 8.6)

Why choose off-column?

- There is probably less mass hanging on the microscope column and fewer connections need to be made to the microscope.
  - However, there is still a large cooling dewar with boiling nitrogen attached to the column.

Figure 8.5 A typical on-column preparation chamber (Image Quorum©).

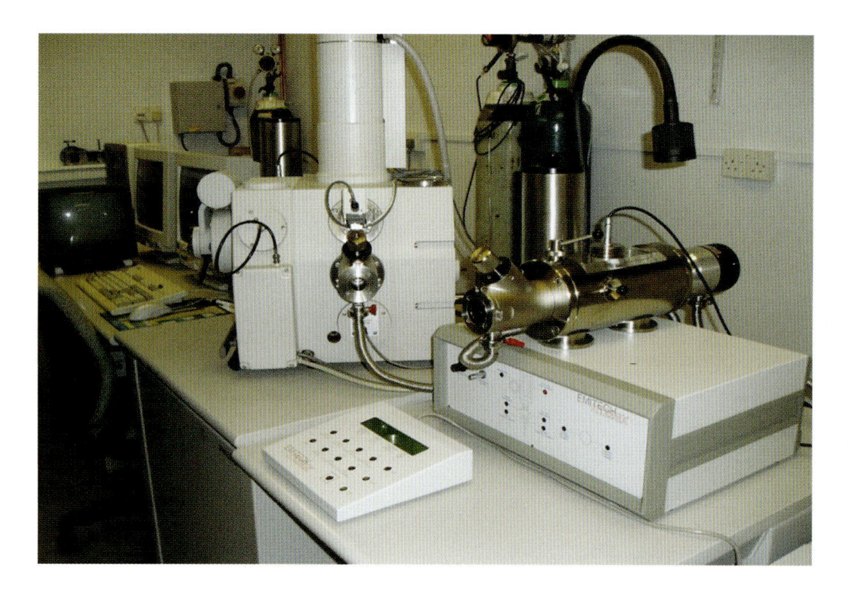

Figure 8.6 A typical off-column preparation chamber (Image Quorum©).

- Most of this type of system are braid cooled so there is the mass of the cooling system to consider.
- Adding extra coating or deeper fracture means removing the specimen from the microscope and transferring across to the preparation chamber through a double airlock operation, with the inherent contamination risks and time constraints.

- If space around the column is very restricted or if extra detectors impede the mounting of an on-column preparation chamber.
- Sometimes the user may want to interface to an existing coating system to save cost.

  - This is largely historical as most cryo-SEM preparation systems have their own coating built in.
  - It is possible to use the same transfer and preparation units on more than one microscope. All that is required is a second airlock, cold stage and anti-contaminator.
- As the preparation chamber is not attached to the SEM it can be larger (to allow extra facilities, such as electron beam evaporation)'

8.5 LOCATION OF THE COOLING DEWAR

Again, the user has the choice of two locations for positioning the cooling dewar.

8.5.1 On-Column Cooling (Figure 8.7)

This consists of a liquid nitrogen dewar mounted directly on a spare port on the SEM column. Whether this then connects to the cold stage via braids or has nitrogen gas passed through and an in-built heat exchanger, it will have a limited hold time unless the dewar is very large, which is often not practical. This means that regular refills are required and are easy to forget when observing an interesting sample, resulting in the sample subliming.

Boiling nitrogen on-column can lead to vibration being seen on the image.

One advantage is that the room footprint required by the system is smaller and space in not needed for an off-column dewar.

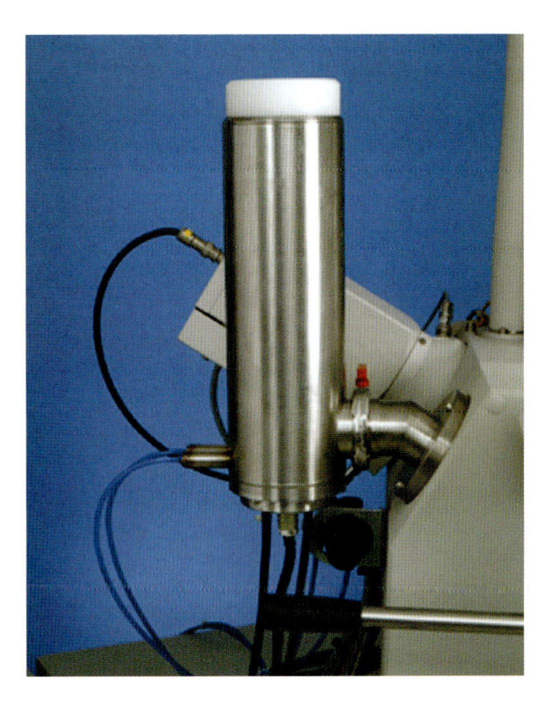

Figure 8.7 A typical on-column cooling dewar (Image Quorum©).

Figure 8.8 A typical off-column cooling system (Image Quorum©).

8.5.2 Off-Column Cooling (Figure 8.8)

By moving the cooling system off-column much larger dewar can be used and hold times of up to 24 hours are achievable, which is very useful for long cryo-FIB experiments. The system is virtually 'fill and forget' as no nitrogen needs to be added during a typical day's run. Some systems even use a single port for cooling and specimen transfer, which means an extra port is available for other accessories.

There is no boiling nitrogen on the SEM column, removing this source of vibration, and less mass on the SEM column. This cooling method is easier to fit to 'busy' microscopes as the port used for the cooling can be almost anywhere on the SEM column.

One disadvantage of off-column cooling is that more floor space is needed for the dewar.

8.6 SAMPLE PREPARATION

Prior to transfer into the preparation chamber the sample will need to be rapidly frozen to preserve it in an 'as close to life-like state' as possible. Whatever freezing method is chosen there will always be limitations in the quality and depth of good preservation and freezing. This depends on the natural water content of the specimen, whether it contains any natural cryo-protectants and the inherent thermal conductivity. Once a thin surface layer has been frozen, heat has to be drawn through this layer, which is a poor thermal conductor, to cool the inner portion of a sample. Thus the rate of cooling drops and larger and larger ice crystals will be formed the deeper into the specimen one goes.

Different cryogens and methods have been used over the years to try and improve the depth of good freezing.

8.7 FREEZING MEHODS

There are several methods available for preparing frozen specimens, each with its; own limitations in depth of good preservation, complexity and cost.

8.7.1 Slushed Nitrogen Freezing

Most cryo preparation systems are supplied as standard with the most basic means of freezing specimens as standard. This method uses slushed nitrogen to freeze the specimen.

Normal liquid nitrogen is at its boiling point at ambient temperature and pressure. This means that when an object is plunged into it, the liquid around it vapourises immediately and forms a layer of nitrogen gas around the specimen. This gas layer is a poor thermal conductor and results in slow freezing. This effect is known as the Leidenfrost phenomenon. To overcome this, the pressure above the liquid nitrogen is reduced using a vacuum pump, causing forced evaporation. The nitrogen cools to its melting point of −210 °C and eventually solidifies. When the system is subsequently brought back to atmospheric pressure the solid nitrogen reverts to a slush. When the sample is plunged into this slushed nitrogen, the nitrogen 'wets' the surface of the sample and it cools much more rapidly than it would in boiling nitrogen.

8.7.2 Propane Jet Freezing (Moor, Kistler and Müller, 1976)

In propane jet freezing a specimen is sprayed by jets of liquid propane, which is cooled by liquid nitrogen (Figure 8.9). The specimen is held between two copper specimen holders. A depth of 10 to 20 μm of good preservation has been claimed. By using cryo-protectants this can be improved but there is the risk of introducing artefacts and rather negates the point of cryo preparation. Because of the explosive nature of the cryogen this method has fallen out of popularity in recent years.

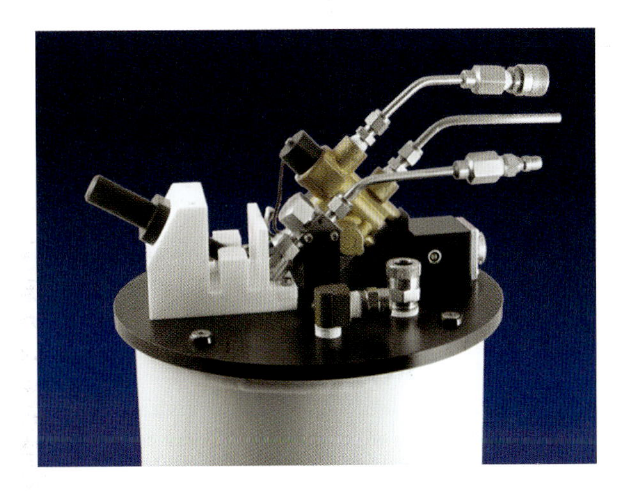

Figure 8.9 A propane jet freezer (image courtesy of RMC-Boeckeler Instruments).

Figure 8.10 A basic plunge freezer (Image Gatan©).

8.7.3 Ethane Plunging (Dubochet *et al.*, 1988)

Plunge freezing a sample into a small pot of liquid ethane surrounded by a bath of liquid nitrogen is a popular method for small samples, usually 3 mm grids or planchettes (Figure 8.10). Initially, this was done by hand but the current range of plunge freezers available on the market are all sophisticated machines with temperature and humidity control. Cooling rates of 10^4 °C/s can be achieved.

8.7.4 Slam Freezing (Dempsey and Bullivant, 1976)

In cold metal block freezing the specimen is rapidly slammed on to a cold metal surface (usually polished silver or copper) that is cooled by liquid nitrogen (Figure 8.11). Specimens can be preserved up to a depth of 10 to 15 µm but the area closest to the metal block may suffer damage.

8.7.5 High-Pressure Freezing (Dahl and Stachelin, 1989)

High-pressure freezing is similar in some ways to propane jet freezing (see Figure 8.12). Just before freezing the sample is pressurised to 2100 bar, at which pressure the melting point of water drops to −21 °C. The water is also many times more viscous than at ambient pressure. It is then frozen with jets of liquid nitrogen. High-pressure freezing allows specimens up to 0.2 mm thick and up to 1 mm^3 to be vitrified at a low freezing rate of 600–1000 °C/s without the need for cryo-protectants.

By freezing samples at regular time intervals and storing under liquid nitrogen until they can be imaged, it is also possible to do time resolved experiments.

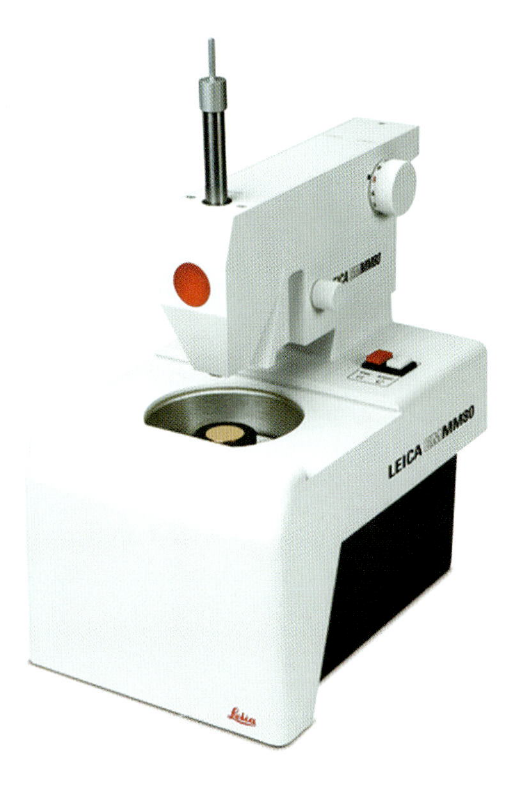

Figure 8.11 A typical slam freezing system (Image Leica©).

Figure 8.12 A typical high-pressure freezer (Image B. Humbel, Lausanne).

8.8 MOUNTING METHODS

There are many ways to mount an SEM specimen on to a stub or shuttle, depending on the geometry and consistency. Most manufacturers supply a large variety of specimen holders. Below are a few simple examples.

8.8.1 Surface Mounting

8.8.1.1 For Flat Samples (Leaves, etc.)

Roughen the stub surface with fine abrasive paper, place a small amount of mounting media (mixture of 50% cellulose acetate/50% colloidal graphite) on to the stub and lay the specimen on top of the mounting media (Figure 8.13).

Cellulose acetate is a natural plastic which is manufactured from natural cellulose. Cellulose acetate membrane filters are hydrophilic, have good solvent resistance, high physical strength and can be sterilised by all methods. When freezing in liquid nitrogen it does not crack. By mixing with graphite, it becomes electrically conductive and also easy to see, which is important when mounting specimens.

8.8.2 Edge Mounting

For Edge Observations and Fracture

Roughen the surfaces of a stub with fine abrasive paper (Figure 8.14). Mount the specimen on edge secured with mounting media or mount the specimen standing on edge in one of the machined slots, secured with mounting media. This is important if the specimen is to be fractured.

8.8.3 Filter Mounting

8.8.3.1 For Liquid Suspensions

Secure a section of filter paper to the stub with four droplets of mounting media (Figure 8.15). Secure the cellulose acetate membrane filter to filter paper base using mounting media. Pipette liquid suspension on to the membrane filter.

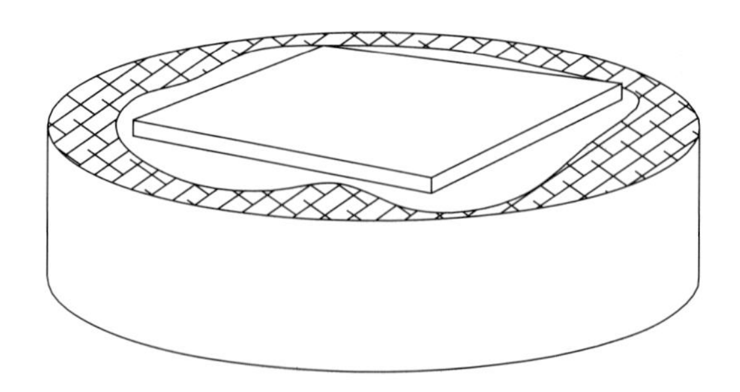

Figure 8.13 Flat specimen preparation.

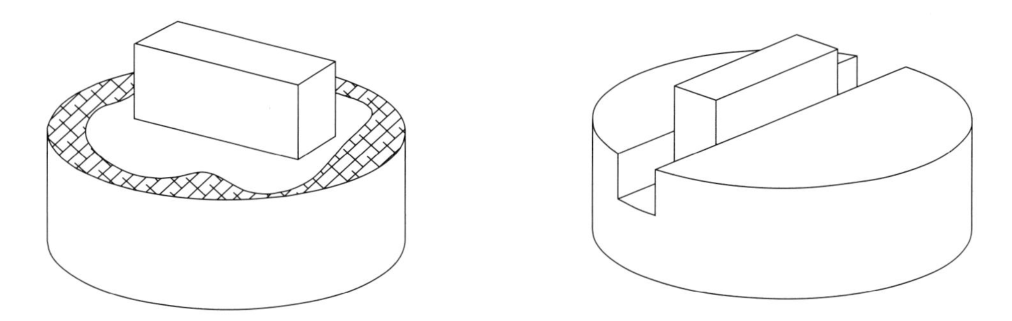

Figure 8.14 Edge mounting preparation.

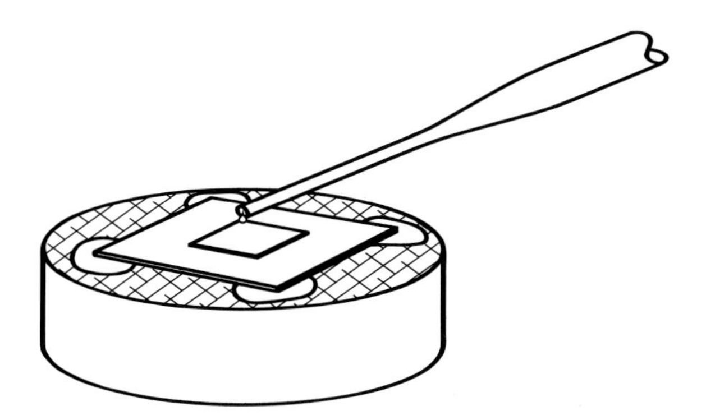

Figure 8.15 Filter mounting preparation.

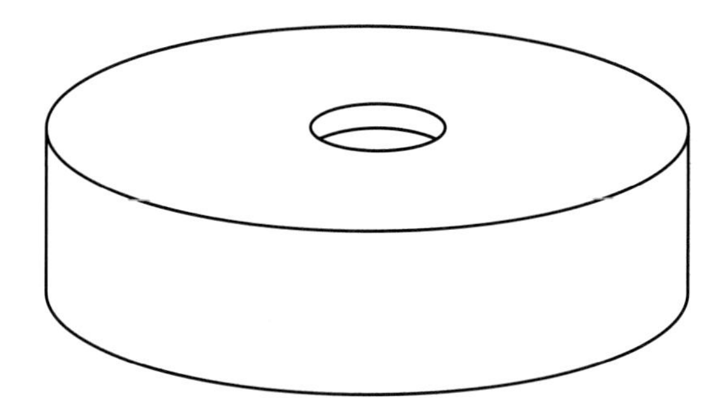

Figure 8.16 Hole mount preparation.

8.8.4 Hole Mounting

8.8.4.1 For Emulsions and Liquids (i.e. Oil, Toothpaste, Water-Based Specimens)

Use a hole or holes drilled in the stub to locate thicker emulsions (Figure 8.16).

8.8.5 Liquid Film Mounting

Liquid specimens can be thinly spread on a plain stub that has been roughened with fine abrasive paper (Figure 8.17).

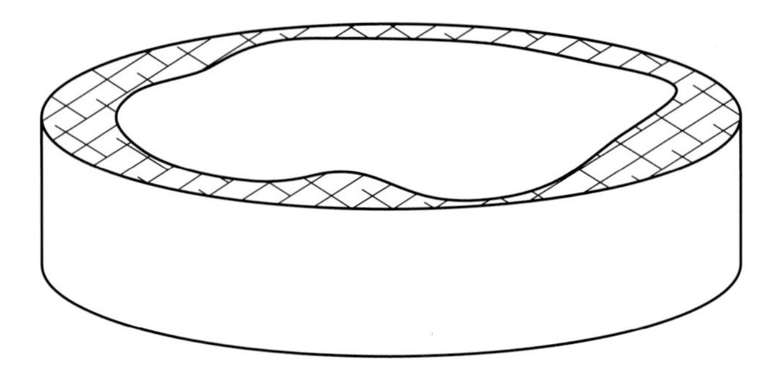

Figure 8.17 Liquid preparation.

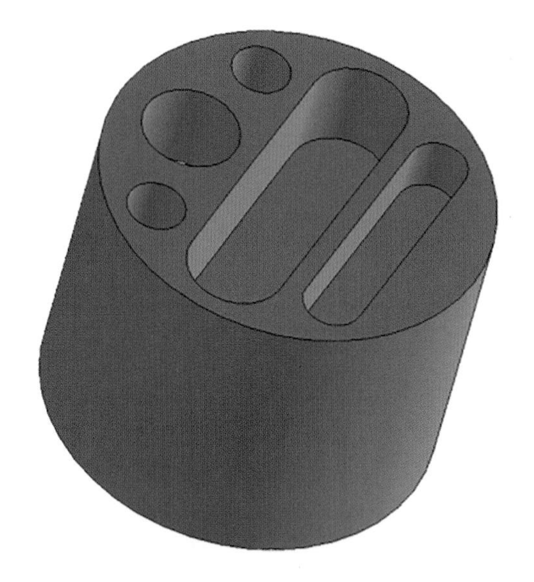

Figure 8.18 Universal specimen stub.

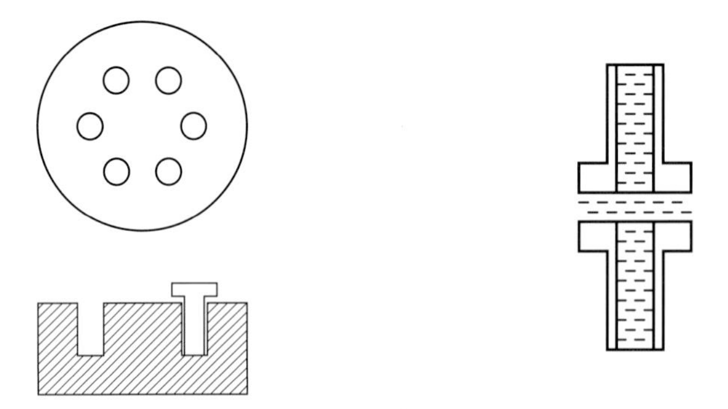

Figure 8.19 Rivet mounting preparation.

Figure 8.20 Selection of shuttles.

8.8.6 Rivet Mounting

Use rivets for liquids and when specimens need to be frozen off the stub to achieve fast freezing rates. The rivet is placed in a hole and filled with liquid prior to freezing, or if the specimen needs to be frozen rapidly off the stub, two rivets are held together, filled with liquid and plunge-frozen prior to mounting in the rivet holder. Rivets will push-fit into holes drilled into standard stubs (Figures 8.18 and 8.19).

Many other mounting stubs and shuttles are available with small vices, spring clamps, planchette holders, cryo-STEM and so on (Figure 8.20).

8.9 FRACTURING

Depending on the type of sample, it is often necessary to fracture through it to reveal an internal structure or to remove material that is not representative of the real sample, having been exposed to atmosphere, damaged during preparation or to reveal internal structures (see Figure 8.21). Most cryo preparation chambers are fitted with one or more means of fracturing the sample prior to imaging, subliming or coating. These are usually cooled knives or blades, either used manually or on a micrometre drive. Although the height of the fracture can be set with a micrometre, the specimen will be brittle and will fracture along the line of least resistance, so a perfectly flat, cryo-planed surface is not usually achievable.

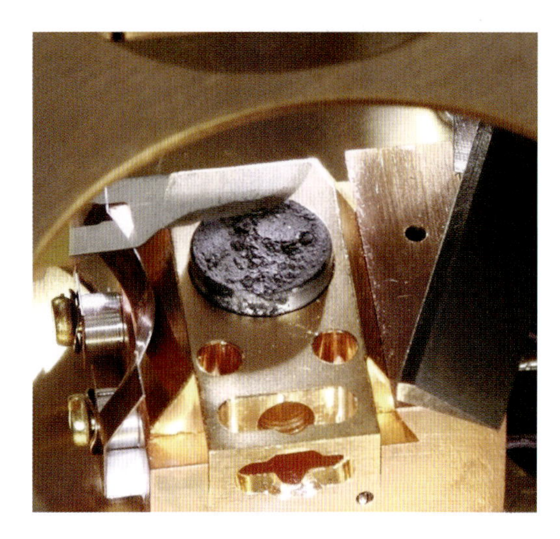

Figure 8.21 Fracturing knife (Image Quorum©).

8.10 SUBLIMATION

Once a fresh surface has been exposed or to 'clean up' non-fractured specimens it is often necessary to sublime the surface to reveal some topography. This involves raising the temperature of the sample in a carefully controlled way. As the temperature increases water sublimes from the surface and is trapped on the cold trap, revealing some of the structure of the sample. Sublimation should be done sparingly to avoid creating artefacts. It can also reveal freezing artefacts by making eutectic patterns visible in areas of high water content. As ice crystals form, solutes tend to be forced to the grain boundaries, which appear as 'walls' or 'curtains' in the image (see Figures 8.22, 8.23 and 8.24).

On most modern cryo preparation systems this process is fully automated and the sublimation profiles can be saved for future use. This gives reproducibility in specimen preparation even if the samples are imaged several months or even years apart.

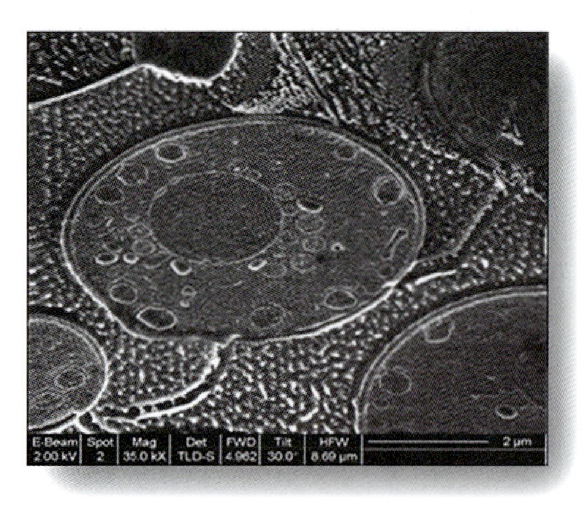

Figure 8.22 Yeast cells after cryo FIB and sublimation.

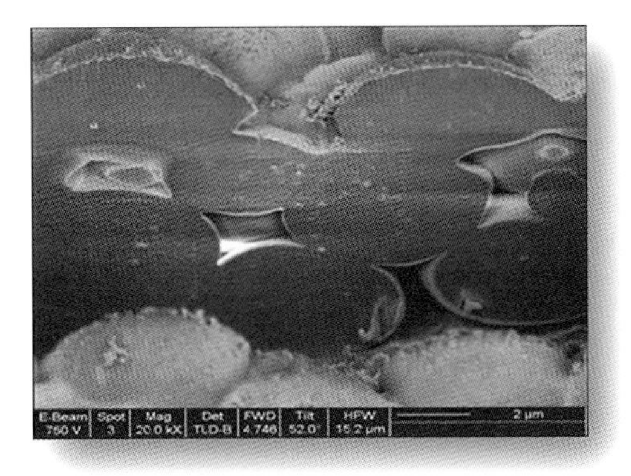

Figure 8.23 Yeast after cryo FIB (Images **FEI/Quorum©**).

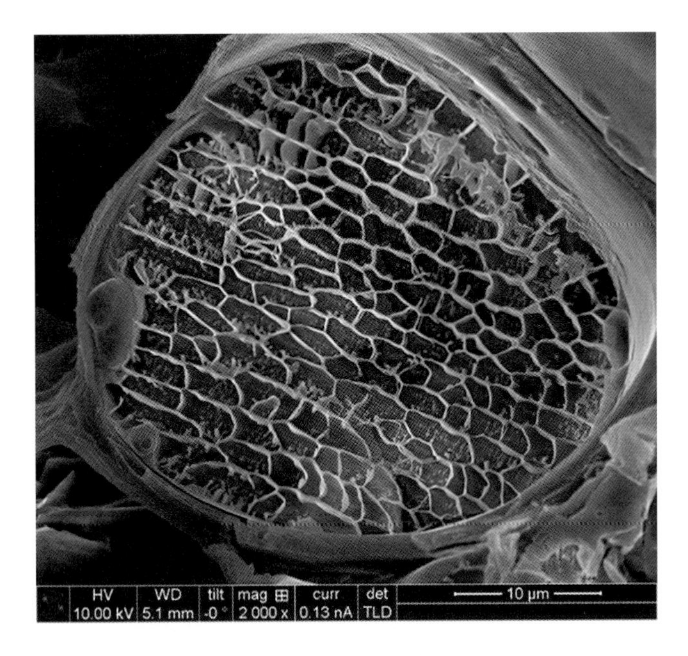

Figure 8.24 Example of over sublimation and eutectics (Image FEI/Quorum©).

Ideally, the sublimation profile should look like a square wave, that is, it rises to the set temperature instantly and cools after the sublimation time similarly. This cannot be achieved physically, so the system needs to get as close as possible to this. During any time taken for the sample to cool from the set point to around −110 °C it is still subliming.

8.11 COATING

Following sublimation, the specimen will usually need to be coated with a thin conductive film of metal or carbon to prevent charging under the electron beam.

8.11.1 Metal Sputtering

Sputter coating in scanning electron microscopy (SEM) is the process of laying down an ultra-thin coating of electrically conducting metal such as gold (Au), platinum (Pt), chromium (Cr) or iridium (Ir) on to a non-conducting or poorly conducting specimen.

Sputter coating prevents charging of the specimen, which would otherwise occur because of the accumulation of static electric fields due to the electron irradiation required during imaging. It also increases the number of secondary electrons that can be detected from the surface of the specimen in the SEM and therefore increases the signal to noise ratio. Sputtered films for SEM typically have a thickness range of 2–20 nm.

Magnetron sputtering using a crossed-field electromagnetic configuration keeps the ejected secondary electrons near the cathode (target) surface and in a closed path on the surface (Figure 8.25). This allows a dense plasma to be established near the sputter target surface. The ions that are accelerated from the plasma do not sustain energy loss by collision before they bombard the sputter target.

For electron microscopy (EM) specimen coating, the magnetron sputtering head design ensures that minimal heat energy (electrons) reach the specimen surface. This is important as it reduces heat damage to the specimen and is a significant factor in ensuring the grain size within the sputtered film is optimally small – essential for high-resolution field emission scanning electron microscopy (FE-SEM).

Sputter coating with metals such as platinum, gold–palladium or chromium is widely used for depositing fine grain, high-resolution films on to (FE-SEM) specimens (Figure 8.26). Chromium oxidises on contact with air, which can present specimen storage problems. For this reason, iridium (Ir) sputter coating is increasingly preferred by many workers.

8.11.2 Carbon Coating

The thermal evaporation of carbon (C) is widely used for preparing specimens for electron microscopy (EM). A carbon source in the form of woven fibre is mounted in a

Figure 8.25 Sputtering (Image Quorum©).

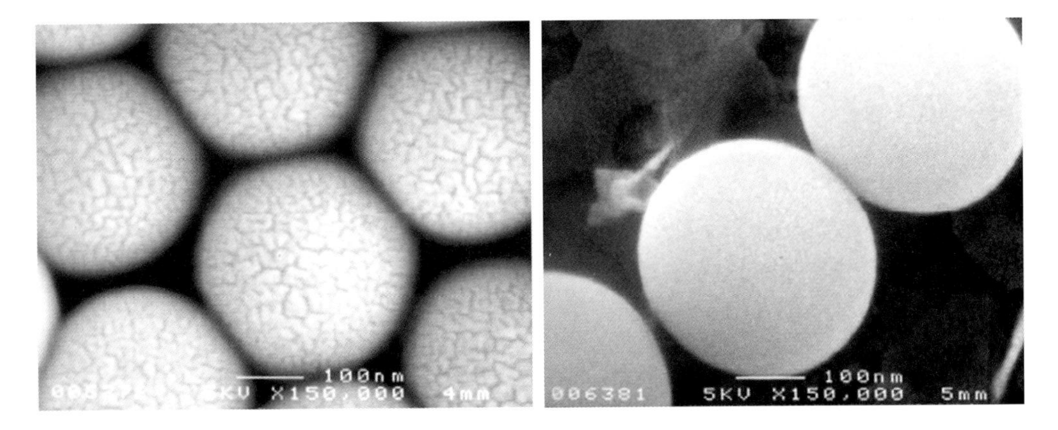

Figure 8.26 Comparison of Au/Pd (left) and Pt (right) coatings (image Quorum).

vacuum system between two high-current electrical terminals. When the carbon source is heated to its evaporation temperature, a fine stream of carbon is deposited on to specimens.

The main applications of carbon coating in EM are making scanning electron microscopy (SEM) specimens conductive for subsequent examination by X-ray microanalysis and being used as specimen support films on transmission electron microscopy (TEM) grids.

Carbon fibre is mainly used in electron microscopy (EM) to produce thin, electrically conductive coatings on specimens. Carbon fibre can be used for transmission electron microscopy (TEM) applications, but carbon rod is normally preferred due to superior control of the evaporation process.

8.12 MORE ADVANCED TECHNIQUES AND EQUIPMENT

As cryo-SEM and cryo-FIBSEM have matured, more advanced hardware has been developed to allow more complex analysis than just regular imaging. Improved specimen handling and workflow solutions are appearing that reduce the number of operator interventions and reduce the risk of contamination or de-vitrification.

8.12.1 CryoFIB Lift-Out and On-Grid Thinning

Many labs are now working with cryo lift-out and on-grid thinning to produce thin lamella that are then transferred to a cryo-TEM (Figures 8.27 and 8.28). A braid connected to the cryo preparation system cold finger cools the needle used for cryo lift-out (Figure 8.29). Once prepared, the lamella is reverse-transferred out of the FIB and into the loading station, where it is transferred to a cryo-TEM carrier and then to the cryo-TEM.

On-grid thinning is ideal for cells that have been cultured on the grids and that are relatively easy to find in the image. Lamella lift-out is more suited to samples where the object of interest has to first be found within the bulk of the sample and then material removed to get to it.

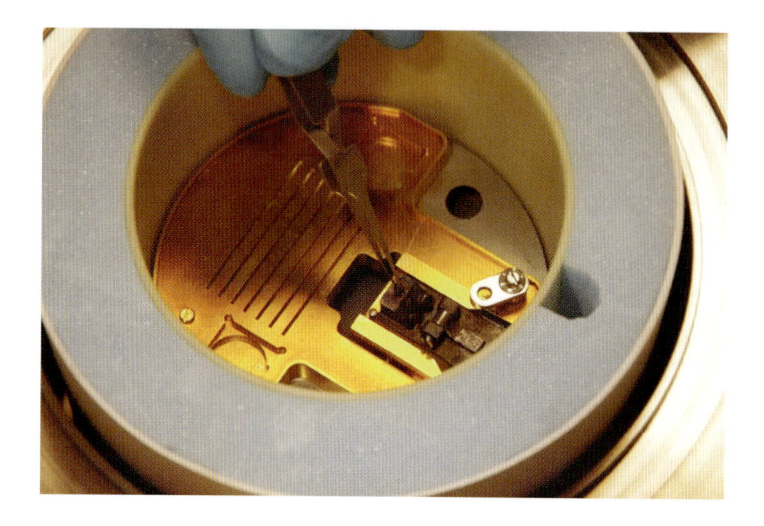

Figure 8.27 Transferring a lamella.

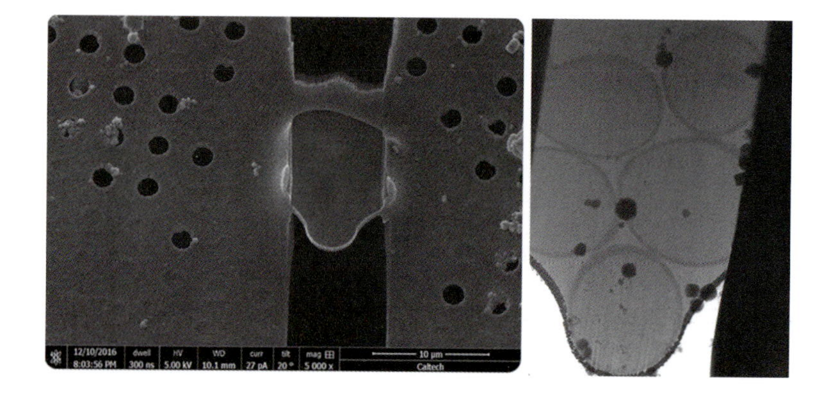

Figure 8.28 Example of on-grid thinning (Jensen Lab, CalTech).

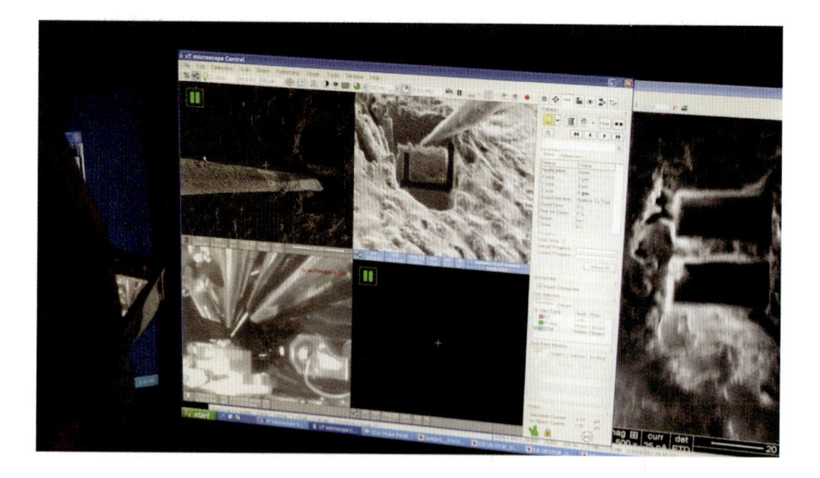

Figure 8.29 Example of cryo lift-out (Kourkoutis Lab, Cornell).

8.12.2 Cryo Rotate Stages

To facilitate production of lamella it is advantageous to have a rotate stage (Figure 8.30) to allow fine control of angle and also to be able to mill both sides of the lamella whilst viewing it from the front. A rotate stage also allows a specific feature to be aligned as desired.

8.12.3 SEM Stage Bias

Another recent development in SEM techniques is the ability to apply a bias voltage to the specimen stage (Figure 8.31). This effectively reduces the landing voltage of the electrons on the sample. Depending on the brand of microscope used this can be up to 5 kV.

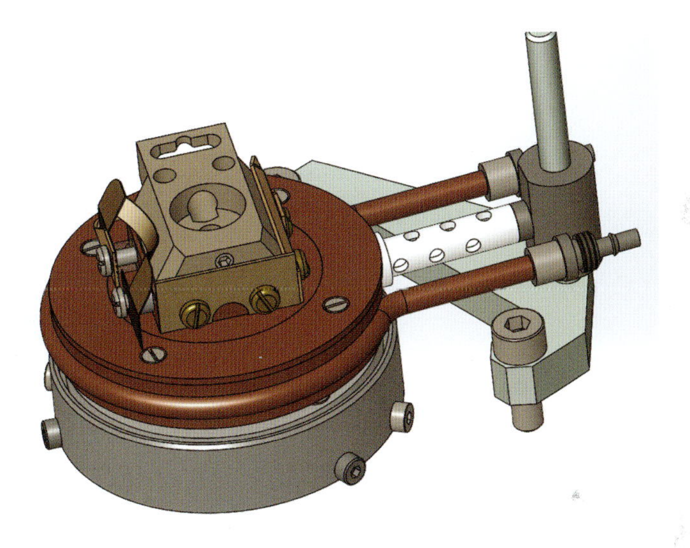

Figure 8.30 Example rotate stage.

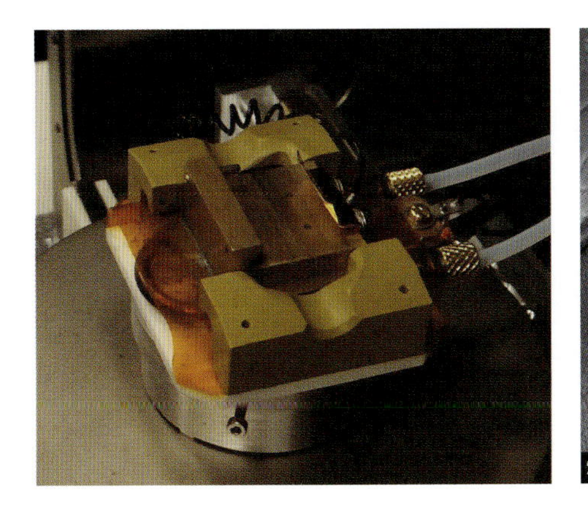

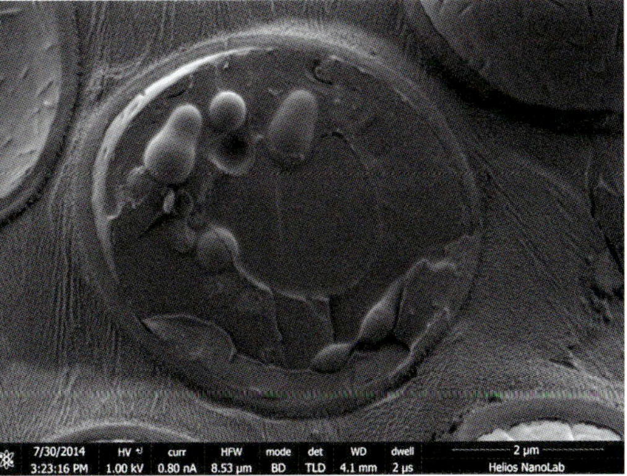

Figure 8.31 Example 5 kV BD compatible stage and yeast sample taken with 4 kV stage bias (B. Humbel, Lausanne).

New cryo-stage designs were needed to allow such high voltages to be used without risking flashovers to the stage heating and temperature measurement circuits.

8.12.4 Cryo-STEM in SEM

Modern FEGSEMs have increased in resolution in recent years to the extent that some cryo-STEM imaging is now viable (Figure 8.32). Stages and shuttles have been developed for most SEMs to allow this to be done.

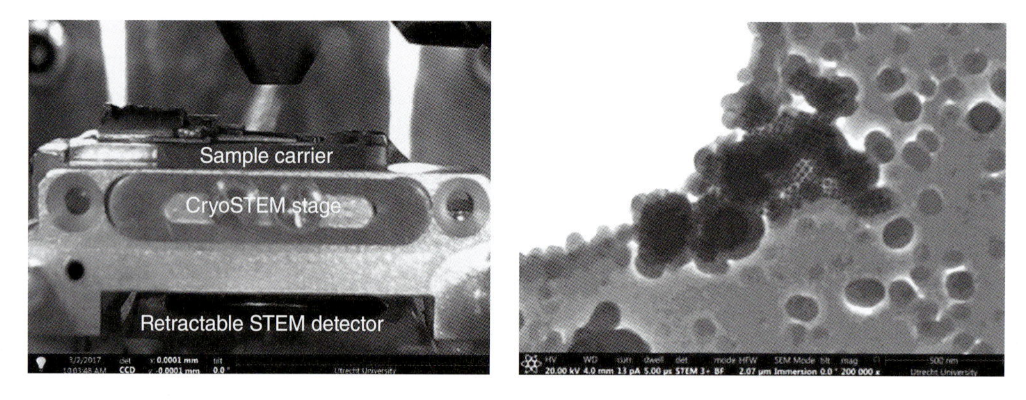

Figure 8.32 A cryo-STEM stage from Quorum with example test image (Matthijs de Winter, Utrecht).

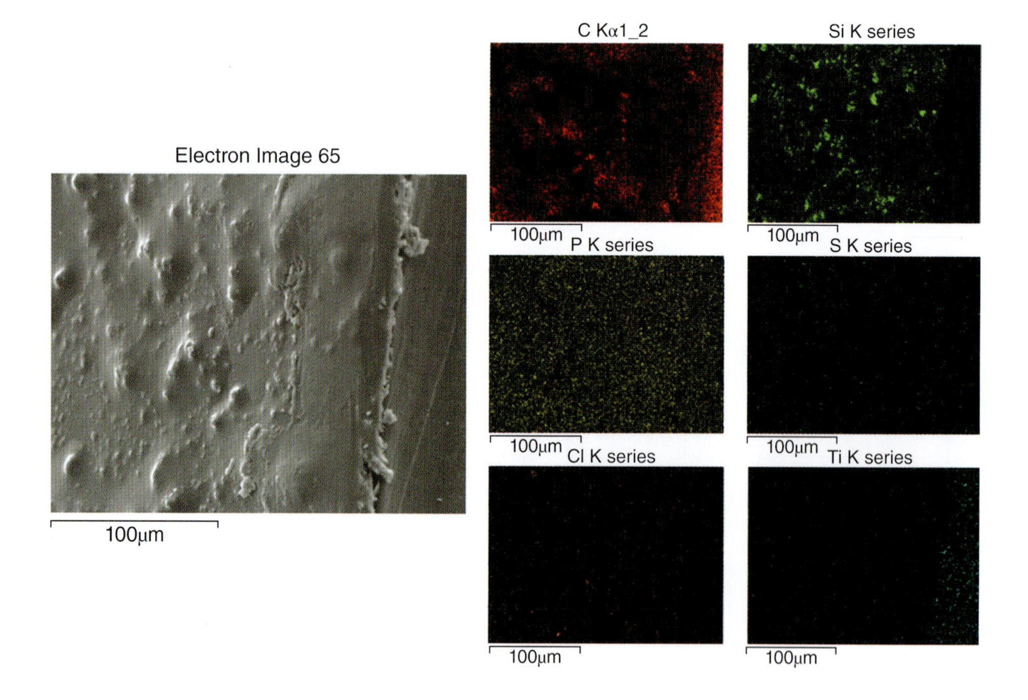

Figure 8.33 Cryo-EDS of a self-adhesive label (Quorum/JEOL UK).

8.12.5 Cryo-EDS

Another area where hardware has improved dramatically and now allows data collection where it was previously impossible is in cryo-EDS. The new generation of large area detectors with their much more efficient acquisition combined with a good, stable cold stage means that it is now possible to obtain X-ray maps of beam-sensitive samples. This applies to both biological samples, such as heavy metal absorbing plants, to polymer and other beam-sensitive materials samples. Figure 8.33 shows an X-ray map from a multi-layer self-adhesive label. The TiO_2 whitener can be seen in the polymer paper and the Si in the varnish layer is also clearly defined. Previous attempts to visualise the polymer paper resulted in it melting.

8.13 CONCLUSION/SUMMARY

Developments in cryo-SEM technologies mean that this is now a relatively simple and quick method for observing wet and beam sensitive samples in as close to 'life-like' state as possible. Where samples have delicate structures, waxy surface coatings or any form of surface secretions cryo-SEM is the only method of observation.

As long as the limitations of freezing depth are taken into account and the user knows how to identify artefacts reliable imaging can be achieved,

REFERENCES

Dahl, R. and Staehelin, A. (1989) High-pressure freezing for the preservation of biological structure: Theory and practice. *Journal of Electron Microscopy Technique*, 13 (3), 165–174.

Dempsey, G.E. and Bullivant, S. (1976) A copper block method for freezing non-cryoprotected tissue to produce ice-crystal-free regions for electron microscopy. *J. Microsc.*, 106, 251–271.

Dubochet, J., Adrian, M., Chang, J.-J., Homo, J.-C., Lepault, J., McDowall, A.W. and Schulz, P. (1988) Cryo-electron microscopy of vitrified specimens. *Quarterly Review of Biophysics*, 21, 129–228.

Moor, H., Kistler, J. and Müller, M. (1976) Freezing in a propane jet. *Experientia*, 32, 805.

9

Cryo-SEM Specimen Preparation Workflows from the Leica Microsystems Design Perspective

Guenter P. Resch
Nexperion e.U. – Solutions for Electron Microscopy, Wien, Austria

9.1 INTRODUCTION

The analysis of frozen hydrated biological specimens in the scanning electron microscope (cryo-SEM) offers a number of advantages over the investigation of samples at room temperature prepared by conventional approaches. First and foremost, observation of specimens in their natural, hydrated state helps to eliminate artefacts typically associated with the chemical fixation and drying/dehydration steps of conventional preparation techniques: the fixation is ideally reduced to a single physical fixation step, instead of a slow and selective chemical procedure, (literally) freezing biological processes and all cellular components in an instant, allowing faithful investigation of structures with a few nanometres resolution. Beyond cellular ultrastructure, this likewise applies to water, whose distribution is of great interest in plant biology, and inorganic elements of tissue electrolytes amenable to microanalytical studies (McCully, Canny and Huang, 2009; McCully *et al.*, 2010).

During the whole preparation and imaging protocol for cryo-SEM, hydrated specimens need to be protected from a number of detrimental influences. (1) The freezing step has to be performed in a way that prevents or minimises the formation of destructive ice crystals, ideally yielding a sample embedded in amorphously frozen (vitrified) water. (2) The specimens have to be protected from warming arising from sources in the environment or due to irradiation that could lead to recrystallisation. Ideally, the sample is kept below the recrystallisation temperature of pure water at approximately −140 °C, but exposure to elevated temperatures for a limited duration, for example for freeze etching at

Biological Field Emission Scanning Electron Microscopy, First Edition.
Edited by Roland A. Fleck and Bruno M. Humbel.

−110 to −90 °C under high vacuum, is acceptable. Natural cryoprotectants in the specimen such as elevated concentrations of salts or sugars, artifical cryoprotectants, the slow recrystallisation rate of water and the mechanism of devitrification, which is less destructive than the primary formation of ice crystals during freezing (Cyrklaff and Kuehlbrandt, 1994; Dubochet, 2007) limit the damage inflicted. (3) Another important requirement is protection against removal of structural water by uncontrolled sublimation at 10^{-7} mbar, which becomes significant above the condensation/sublimation equilibrium temperature of −120 °C (pure water; see Chapter 12 by Tacke *et al.*). (4) Furthermore, contamination of the cold specimen surface by condensation of humidity or material released from the specimen itself in the process of irradiation has to be prevented, particularly after the surface designated for analysis has been exposed.

This chapter describes how different preparation techniques for cryo-SEM can be combined, how these individual steps are implemented on instruments available from Leica Microsystems and how the sample can be protected between these steps. Accordingly, one major subject of this chapter will be the Leica EM VCT500 vacuum cryo-transfer system linking preparation instruments both to each other and also to the SEM. Many of the underlying concepts about freezing, coating, specimen transfer and microscopy and the related instruments presented here have their roots in the laboratories of Hans Moor, Heinz Gross and Martin Müller at the ETH Zurich, Switzerland, and were later commercialised by the Balzers Union or Bal-Tec before the acquisition by Leica Microsystems in 2008.

9.2 SPECIMEN FIXATION

The primary challenge when freezing specimens in an aqueous environment is the formation of ice crystals, causing segregation of cellular content into a water phase and a eutectic phase that contains biological material excluded by the ice crystals, leading to the degradation of cellular structure. Different strategies are used individually or in combination to prevent the formation of crystalline ice and to achieve vitrification: (1) infiltration with cryoprotectants in conjunction with chemical pre-fixation, (2) freezing of very small masses in cryogens that ensure a very high cooling rate or (3) freezing at high pressure.

As the use of chemical fixatives and cryoprotectants (for a protocol, see Severs, 2007) is, however, problematic to the physiology of the cell and hence conflicting with optimal structural preservation, we will focus on methods that allow freezing of native biological samples. Depending on the nature of the sample (type, geometry, region of interest, etc.) different methods requiring different instrumentation are available. Regardless of the freezing technique used, one important consideration depending on required orientation and on what follow-up processing technique will be used is how the fresh or frozen sample can be mounted on a carrier support.

At times it is not possible or not even required by the objective of the study to achieve vitrification. In these cases, the resolution that can be obtained with cryo-SEM will be defined by the size of the ice crystals. Furthermore, practical constraints such as the pathogenicity of a specimen (Wild, Kaech and Lucas, 2012) can dictate the use of chemical pre-fixation, even in cases when the technology to vitrify the native sample is available.

9.2.1 Ambient Pressure Freezing Methods

Freezing biological samples at ambient pressure requires a cooling rate in the order of -10^5 °C/s, depending on the water content, to achieve true vitrification (Frederik and

Hubert, 2005). Achieving these rates throughout the whole specimen requires very small samples (one dimension smaller than a few μm) and cryogens or cooling elements making immediate and direct contact with the specimen and providing high thermal conductivity.

At present, the most popular ambient pressure freezing technique is immersion freezing (plunge freezing), where a thin sample on a thin carrier is plunged into a cryogen such as liquid ethane or liquid propane (Adrian *et al.*, 1984), or a mixture of both (Tivol, Briegel and Jensen, 2008). Both very basic setups of makeshift stands and Styrofoam boxes/metal cups for the cryogens as well as semiautomatic and/or environmentally controlled instruments from various manufacturers (see Frederik and Hubert, 2005, and for a review of current instruments, see Dobro *et al.*, 2010), including the EM GP/EM GP2 from Leica Microsystems (Resch *et al.*, 2011), can be used for immersion freezing. Primarily, this technique is used for preparation of frozen hydrated specimens on EM grids for cryo-transmission EM; the successful combination with cryo-SEM is shown in Ito *et al.* (2015).

Further ultrafast freezing techniques of significance in the past include propane jet freezing (Moor, Kistler and Müller, 1976), spray freezing (Bachmann and Schmitt-Fumian, 1973) and metal mirror/slam freezing by impact against a cold polished metal block (van Harreveld and Crowell, 1964).

9.2.2 High Pressure Freezing

Freezing native samples exceeding a few micrometres in thickness at ambient pressure leads to considerable ice crystal formation and distortion of the biological material. High pressure freezing (HPF) (Moor and Riehle, 1968; Müller and Moor, 1984) was developed to overcome these limitations in specimen thickness and the requirement of cryoprotectants for thicker samples. The high pressure prevents the expansion of water when freezing, lowers the freezing point and retards the ice crystal nucleation point (Moor, 1987) with the effect peaking at about 2100 bar (2.1×10^8 Pa). In this manner, HPF allows freezing of native samples in an aqueous environment of up to approximately 200 μm in thickness with minimal or no ice crystal damage, with the maximum thickness depending on the carrier system used and the composition of the specimen. For HPF, it is essential to carefully synchronise the pressure build-up with the onset of rapid cooling a few milliseconds later; cooling the specimen below the freezing point before fully pressurised would reduce or eliminate the beneficial effect, and an unfrozen specimen exposed to high pressure for excessively long periods will exhibit artefacts. Besides specimen preparation for cryo-SEM, high pressure frozen samples are typically processed by freeze substitution for room temperature TEM analysis (van Harreveld and Crowell, 1964) or cryosectioning for cryo-TEM (Dubochet *et al.*, 1983; Al-Amoudi *et al.*, 2004).

Leica Microsystems has a long-standing history in development and production of high pressure freezers. The EM PACT and the EM PACT2 were two instruments with separate pressurisation and cooling systems based on the design by Studer, Graber and Eggli (2001), while the EM HPF (Studer *et al.*, 1995) and the EM HPM100 developed at the ETH Zurich, Switzerland, both used LN_2 at 2100 bar for pressurisation and cooling in an open system and supported significantly larger specimen carriers. In 2015, Leica introduced their new high pressure freezer 'EM ICE' (Figure 9.1a), largely based on the operating principle of the HPM100. It uses a very similar specimen cartridge system (Figure 9.1b), while the time from finishing to loading the cartridge to freezing could be reduced to one second, allowing the wet specimen to be processed faster and to resolve dynamic events better. Due to a redesign of the high pressure chamber and the valve system compared to early models, alcohol as a pressure/temperature (p/T) synchronisation fluid is no longer required

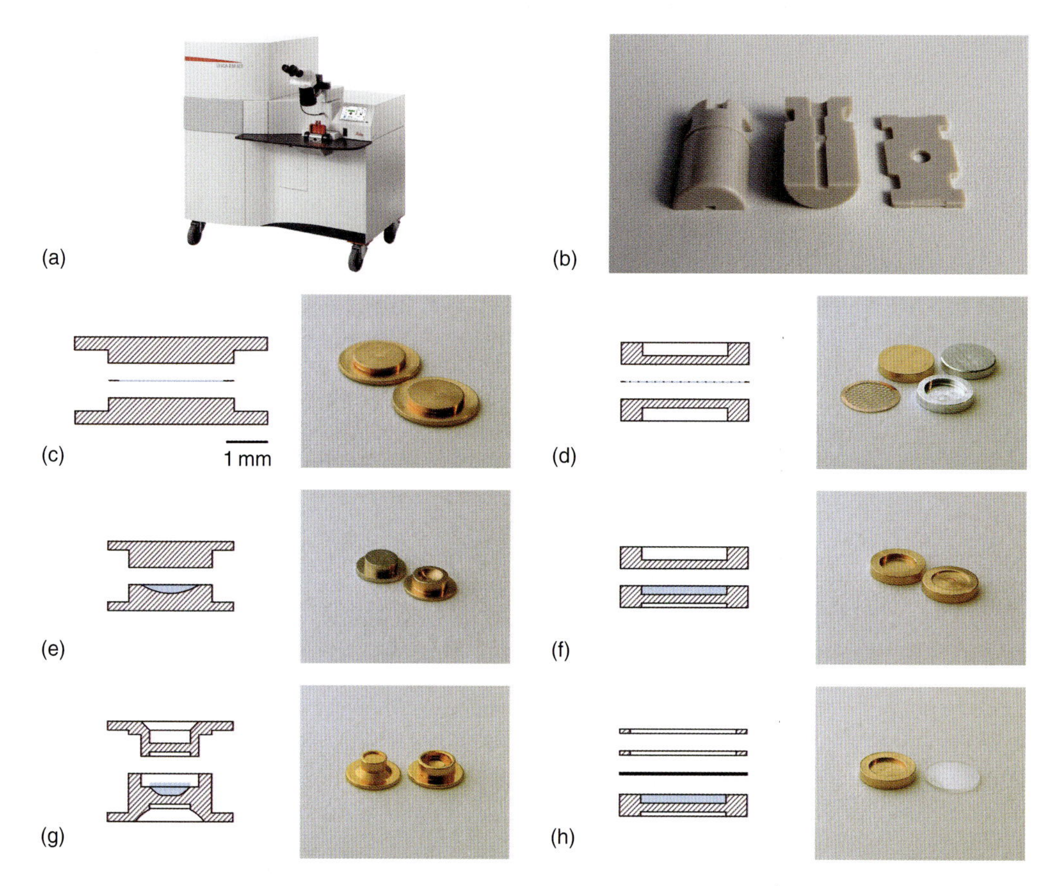

Figure 9.1 The Leica EM ICE high pressure freezer (a) and a selection of consumables typically used for cryo-SEM applications. (b) The upper/lower halves and the middle plate of the 3 mm flat specimen system, produced from insulating polyether ether ketone. The carrier sandwich containing the specimen is inserted into the central hole of the middle plate (right). At the element in the centre, the channel directing the flow of pressurised LN_2 on to the carrier is easily visible. (c–h) Mechanical drawings and photographs of specimen carriers. The technical drawings are to scale, with the volume occupied by the sample highlighted in blue. (c) The 4.6 mm FF system, (d) sandwich of two flat specimen holders and a TEM grid according to Kaech and Ziegler (2014) with aluminum, gold-plated copper and scored carriers depicted, (e) 3.0 mm FF carriers with and without recess, (f) A and B type flat specimen carriers combined, (g) interlocking FF carriers and (h) a flat specimen carrier closed with a transparent sapphire disc for light stimulation. The required spacer rings are not shown on the last photograph.

on the HPM100 and ICE, and the corresponding option has been removed on the latter instrument.

Furthermore, the ICE features a smaller footprint, a lower LN_2 consumption of 30 l/day, a faster turnaround time of 1 minute, automatic sorting of samples into three compartments and an optimised user interaction as compared to its predecessor.

The light stimulation option, which enables the visualisation of highly dynamic processes such as neurotransmission with millisecond precision by HPF in conjunction with optogenetics (Watanabe, Davis and Jorgensen, 2014), was already employed on a prototype Leica EM PACT2 (Watanabe *et al.*, 2013) and available for the HPM100 as an experimental accessory. By tighter integration into the ICE, it now allows a more accurate synchronisation between the light stimulation process and cryofixation. The sample can be stimulated with different wavelengths, both in the visible range (450, 523, 597 and 660 nm) as well as with UV (365 nm). Furthermore, the ICE allows direct electrical stimulation of the specimen – typically neurons or neural tissue – by electrical impulses just before freezing. These pulses are designed to work with millisecond precision and are transferred via a special middle plate that contains a printed circuit board.

While the freezing process itself is largely automated and hence relatively straightforward, a number of steps ahead is crucial to the success of the experiment, including the choice of the optimal specimen carrier for HPF. This decision is influenced by the specimen type, the geometry of the specimen and, last but not least, by the follow-up procedures where good accessibility and a stable mount of the specimen can be essential. As a general rule, the carrier with the thinnest possible cavity should be chosen to increase the probability of obtaining a specimen well frozen in its entirety. Particularly for tissues, this obviously requires a compromise between aiming for a thin specimen for HPF and the mechanical damage inflicted when dissecting a specimen down to a small size. In this context, McDonald *et al.* (2010) recommend the use of carriers in unorthodox configurations, with grids as spacers, etc.

If the freeze fracture (FF) technique is to be employed later, it is essential to keep the specimen sandwich produced in HPF closed until the actual fracture step to prevent contamination. This also precludes the use of any mechanical force that might break the sandwich open too early ('pre-fracture'). On HPF instruments where the use of alcohol as a *p*/*T* synchronisation fluid is optional (HPM100), it is recommended to refrain from using alcohol to prevent the sandwich from freezing on to the holder, requiring undue force to release it (Kaech and Ziegler, 2014).

Different holder and carrier systems are available for and are being used on different HPF instruments. The following overview and Figure 9.1 will introduce the carrier systems most suitable for cryo-SEM applications available for the ICE, which are also backwards-compatible with the HPM100 and HPM010 (Bal-Tec, now ABRA Fluid AG, Widnau, Switzerland).

The gold-plated copper specimen carriers of the 4.6 mm FF system (red middle plate and half cylinders; Figure 9.1c) (Fukazawa *et al.*, 2009) have a flat top and are used with the flat tops facing each other. Hence, this type of carrier is mostly suitable for specimens of a semisolid texture that can be spread on the top, but has been used for freezing tissues as well. To reduce the risk of pre-fracture, it is recommended that a ring of double-sided sticky tape be placed on to the first half before applying the specimen (Fukazawa *et al.*, 2009) and to fix the second carrier in place with that sticky tape. Punches and tape to produce a 0.5 mm wide ring come with the kit system from Leica Microsystems. The same concept of carrier exists for the 3.0 mm FF system (green middle plate and half cylinders) and is being used for similar applications. Using double-sided adhesive tape to prevent pre-fracture is challenging due to the small size of the carriers, but has already been demonstrated (Möbius *et al.*, 2010).

Thicker specimens, such as pieces of tissue, can be frozen in a 3.0 mm FF carrier with a 300 μm indentation on top. These carriers are either used paired with a flat half (Figure 9.1e) or – if absolutely required by the size of the sample – another indented carrier. The issue with pre-fracture due to lateral movement of the carriers against each other was addressed with a new 'interlocking' type of 3 mm FF carriers in 2015 (Kaech, Canny and Huang, 2014): the asymmetrical halves (Figure 9.1g; order numbers 16771874 and 16771875) lock into each other in such a way that separation is only possible by pulling them apart, but not by a lateral movement. Furthermore, the mass of these carriers was reduced with indentations at the back of both halves, aiming for faster cooling rates of the specimen as opposed to the standard 3.0 mm flat specimen carriers. Shadowing and SEM imaging might be influenced adversely, however, due to very steep fracture faces observed in these specimens and the interlocking rim completely surrounding and projecting above the specimen (Andres Kaech, personal communication).

Alternatively, the 3 mm flat specimen carriers (beige middle plate and half cylinders), well established in freeze substitution workflows, can be used for FF samples. They are available in aluminium or gold-plated copper and two versions: 'A' with a 100 and a 200 μm indentation and 'B' with a flat face and a 300 μm recess on the other side. Kaech and Ziegler (2014) describe a method on how to freeze cell suspensions by using a TEM grid serving as a spacer between the flat faces of two B type carriers (Figure 9.1d), based on an earlier design by Semmler *et al.* (1998). This thin layer of specimen allows a very high freezing rate; even samples with a high water content can be frozen with good preservation. Essential steps in this protocol are scoring of both flat surfaces to prevent fracture at this interface and a modification of the hole in the middle plate to allow the slightly larger grids to fit in. In other studies (Walther and Müller, 1999; Walther, 2003), tissue was frozen in the 100 or 200 μm cavity of flat specimen carriers (3 or 6 mm diameter available from Leica Microsystems; Figure 9.1f) and processed for cryo-SEM.

Flat specimen carriers can also be used for experiments in combination with the ICE's light stimulation system: instead of a second metal carrier, the flat specimen carrier is closed with a sapphire disc (Schwarb, 1990) and the stack in the middle plate is filled with spacer rings (see Figure 9.1h). This approach requires the use of at least one transparent polycarbonate half cylinder and has already been used successfully in combination with cryo-SEM on sun screen (Pum, Tomova and Mimietz-Oeckler, 2015).

As important as the choice of the right carrier is the correct filling of these carriers. Gas inclusions, both endogenous as well as from incomplete filling of the carrier with liquid, have to be avoided at any cost: they act as insulators and their collapse during pressurisation can damage the sample. In this context, it is important to use a medium ('filler') to replace the water around the sample and one that is physiologically compatible with the specimen. Typically used fillers for HPF – some of which also function as cryo-protectants – include 10–20% bovine serum albumin, yeast paste, 20% dextran and 1-hexadecene (Studer, Michel and Müller, 1989), each with its particular advantages and drawbacks. An interesting novel approach is 2-methylpentane, as introduced by Harapin *et al.* (2015), which can be removed by sublimation at cryogenic temperatures to expose the sample. For practical advice on specimen and carrier handling for HPF, see McDonald *et al.* (2010) and Kaech and Ziegler (2014). Frozen samples can be stored for an unlimited period of time in LN_2 or transported in a dry shipper (McCully *et al.*, 2009).

9.3 THE VACUUM CRYO-TRANSFER SHUTTLE

As long as the specimen surface to be investigated later in the SEM is not exposed, for example inside a sandwich of closed HPF carriers, the cooled samples are not very sensitive, and no dedicated transfer system is required as long as all transfers are performed quickly and with pre-cooled tools. Once the sample has been fractured or sectioned, however, the surface is very sensitive to contamination by (water vapour) condensation, obscuring fine structural details, and unintentional sublimation due to insufficient cooling. To protect the specimen during transfer from preparation instruments to the microscope, high vacuum cryopreparation chambers directly attached to the SEM are commercially available.

While separated from the microscope chamber with a gate valve, they can be used for different preparative steps like fracturing, etching or coating.

The workflow solution available from Leica Microsystems uses a different approach, a mobile high vacuum cryo-shuttle system interconnecting many of their preparation instruments and linking to analytical instruments such as cryo-SEM, FIB, cryo-TEM and cryo-CLEM. This VCT ('versatile cryo-transfer' or 'vacuum cryo-transfer') shuttle was developed originally in a collaboration of Roger Wepf and Bal-Tec (Ritter *et al.*, 1999; Wepf *et al.*, 2004) and commercialised as VCT100. The whole system included three parts: the shuttle itself, the dock on a preparation instrument and/or the microscope, and the cryo-stage in the SEM, which will be described in a section at the end of this chapter. In 2015, this model was complemented and later replaced with an improved version, the EM VCT500, which is illustrated and explained in detail in Figure 9.2. Amongst other

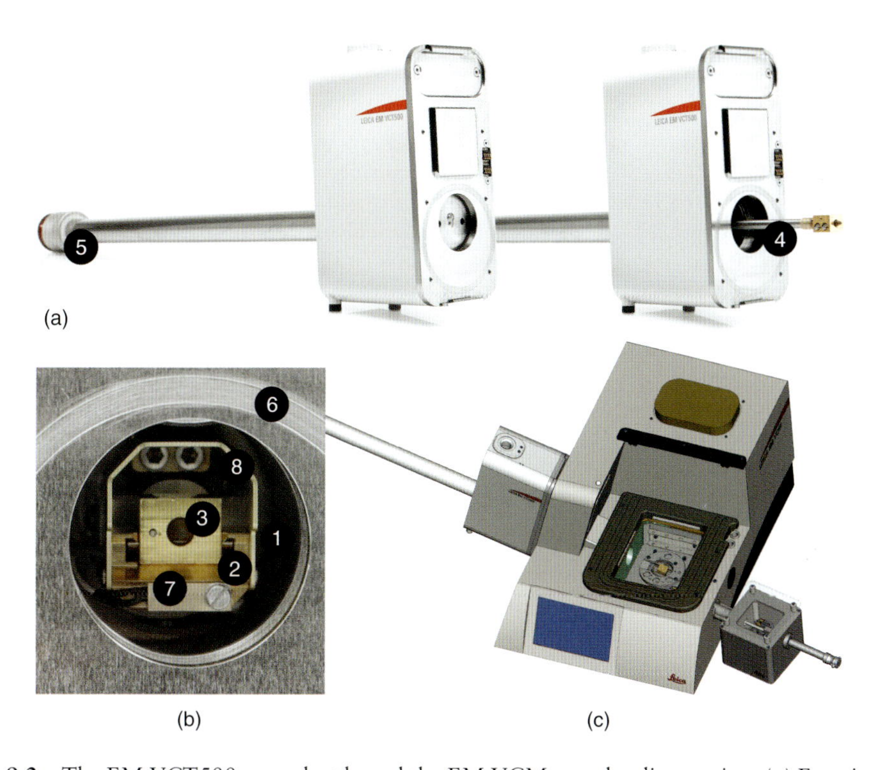

Figure 9.2 The EM VCT500 cryo-shuttle and the EM VCM cryo-loading station. (a) Exterior views of the VCT500 shuttle with the vacuum chamber closed (left), the vacuum chamber open and the manipulator rod extended (right). (b) A view into the vacuum chamber of the shuttle with the high vacuum gate valve open. (c) The EM VCM loading station with a VCT500 shuttle and an EM CRYO-CLEM transfer shuttle docked. Another port at the VCM is available for docking a TEM cryo-transfer holder (not shown); the plastic 'breath guard' protecting the LN_2 bath has been omitted in this graphic for clarity. The VCT500 shuttle uses a high vacuum chamber (1) with a transfer stage (2) to receive the VCT holder carrying the specimen (3) mounted on the manipulator rod (4). This rod can be moved forward and backward by a magnetically coupled handle (5) to insert or retract the VCT holder from instruments, and is rotated with the same handle to clamp or release the holder. When the holder is retracted (a, left), the shuttle chamber is closed with a high vacuum gate valve driven by magnetic coupling to a motor unit. When the holder is connected to a VCT dock at an instrument (not shown), a high vacuum seal (6) is used to seal off the airlock. Inside the shuttle, the holder is clamped to an actively cooled stage at below −178 °C with a built-in temperature sensor (7) and surrounded by an LN_2 cooled anti-contamination shield (8).

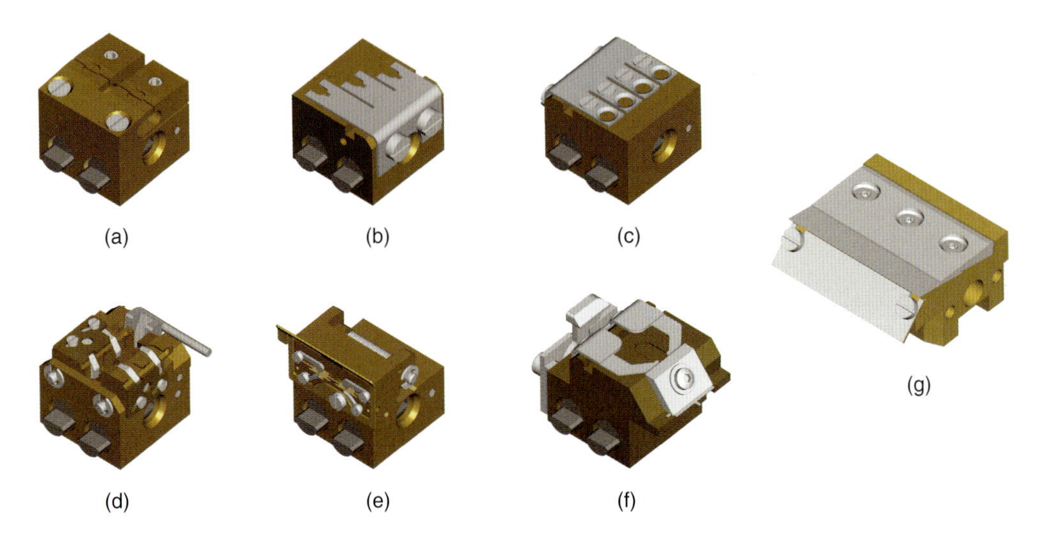

Figure 9.3 The 18 mm × 18 mm square specimen holders compatible with the EM VCT100/500 and other instruments in the Leica workflow. (a) Clamp holder for 3.0 mm diameter flat specimen carriers (order number 16771613); similar holders exist for round 2.8 and 6.0 mm carriers. (b) Specimen holder with a retaining spring for three 3.0 mm × 0.8 mm FF carriers (order number 16771616). This holder is unlocked with a special tool. (c) Specimen holder 16771621 for four 3 mm grids, which is opened and closed with a special tool. (d, e) Holders for two 4.6 mm × 0.6 mm specimen carrier sandwiches to perform a mirror fracturing/double replica technique: (d) the short working distance version 16771618 and (e) the standard version 16771617. (f) The EM TIC3X/VCT cryo-sample holder for HPF 3 mm flat specimen carriers in the horizontal position, order number 16770272. A similar holder was used by Chang and Joester (2015). In addition to (a) to (f), serial models for other applications and customised holders for special applications such as STEM or synchotron measurements are available. (g) A complete knife holder assembly for the EM ACE900, order number 16771780. In the instrument, the knife holder is used upside down (see Figure 9.5 later).

improvements, this new version now uses motor-driven, magnetically coupled valves and elimination of the feedthroughs previously required for valve movement helps to maintain a more stable vacuum in the shuttle.

To accommodate different types of specimen carriers for numerous applications, the shuttle uses square 18 mm × 18 mm specimen holders (Figure 9.3) featuring different clamping mechanisms for specimens. This format is compatible with many current Leica instruments, and also with legacy devices such as the BAF060, the VCT100 or the MED020. A selection of VCT holders used in the Leica cryo-SEM workflow is shown in Figure 9.3. By rotation of the asymmetrical manipulator rod carrying the holder, they can be switched between a released position and firmly clamped to the stage of the different instruments.

Via a dedicated dock (see also Figure 9.6 later), the VCT shuttle can be mechanically and electrically connected to a number of Leica instruments for specimen preparation, electron microscopes and other third party instruments (Figure 9.4). The dock communicates with the shuttle electronically and accommodates pumping and venting lines for the airlock. In contrast to its predecessor VCT100, control of the VCT500 shuttle is integrated into the user interface of compatible systems such as VCM, ACE600 and ACE900. It allows control of the venting, pumping and purging of the airlock with dry N2 gas, pumping and closing the valves; a readout of pressure (via an integrated vacuum gauge) and temperature can be displayed on the host system. An adapter plate allows the VCT100 shuttle to be connected to the VCT500 dock, but not vice versa.

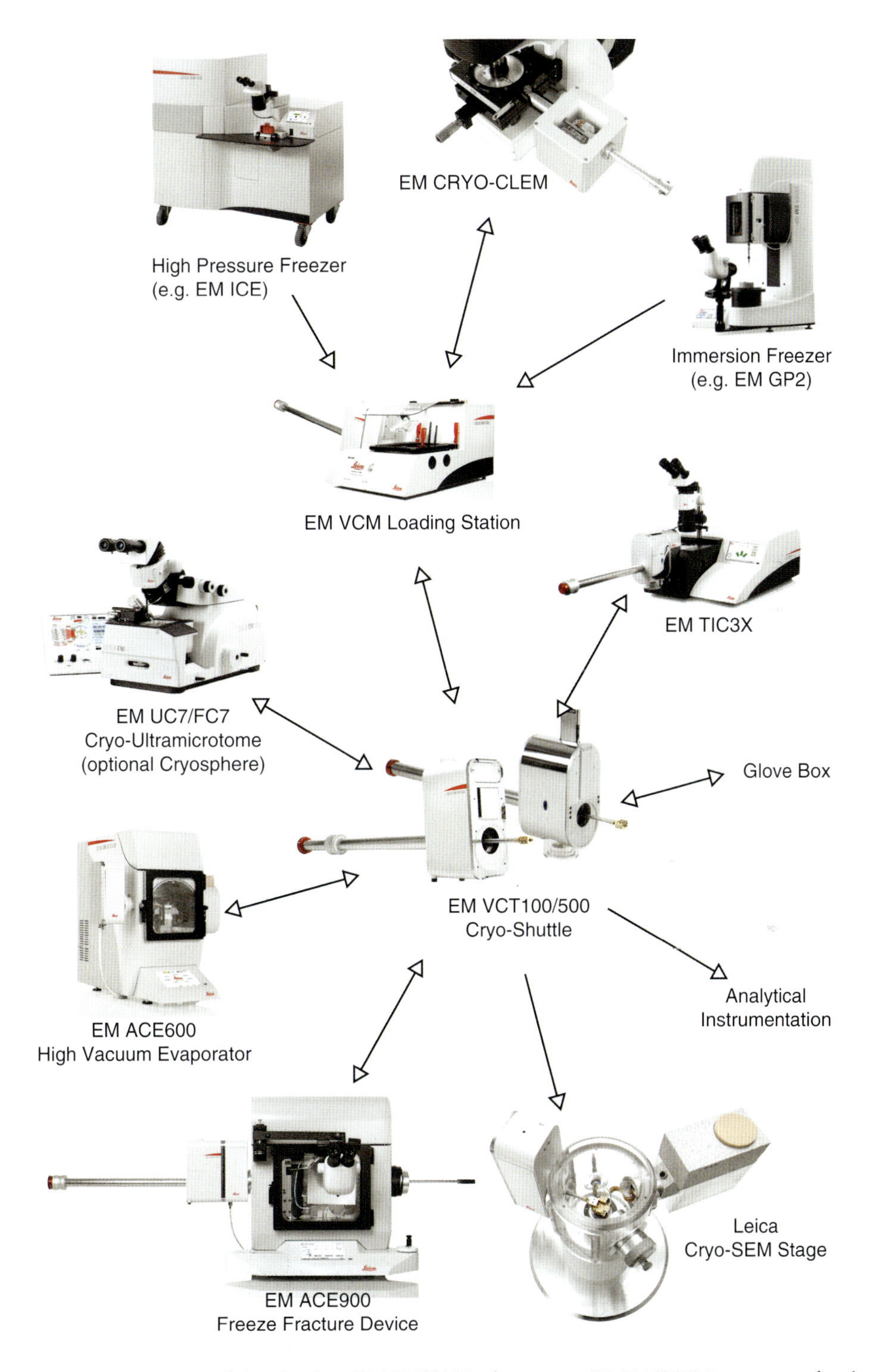

Figure 9.4 Connectivity of the obsolete EM VCT100, the current EM VCT500 cryo-transfer shuttle and the EM VCM loading station to current preparation instruments manufactured by Leica Microsystems. Legacy instruments that the VCT100 shuttle can connect to (SCD500, MED020, BAF060) are not shown.

Anti-contaminators or cryo-shields are essential in instruments such as the cryo-shuttle and in cryo-microscopes to prevent contamination of the specimen surface. Typically, these anti-contaminators are larger in size and colder in temperature (close to the LN_2 temperature) than the sample itself, making them the most probable target for condensation. As opposed to the older version, the 75 ml LN_2 reservoir of the VCT500 shuttle is filled from the top of the device and has been redesigned so that even a small quantity of remaining cryogen can provide full cooling capacity to both the new, cooled stage and the anticontamination system shielding the specimen. One key parameter of a transfer shuttle that does not have its own pumping system and only a limited reservoir of cryogen is the time the specimen can be safely maintained in the shuttle, for example when transferring the shuttle from one instrument to another in a different location. In-factory measurements on the cooled VCT500 shuttle demonstrate that 60 minutes after a last refill with LN_2 and detaching the shuttle from an actively pumping high vacuum system, the vacuum was $< 10^{-6}$ mbar and the temperature at the transfer stage was at −178 °C (Thomas Pfeifer, personal communication).

Owing to the different sizes of SEM chambers, the distance the holder has to travel from the interior of the shuttle to the cryo-stage of the microscope differs greatly between instruments. To account for these differences, the VCT500 shuttle is available in two sizes (L and XL) with different lengths of the manipulator rod.

To load cryo-fixed specimens into the shuttle, Leica is using the EM VCM ('vacuum cryo-manipulation') workstation (see also Figures 9.2 and 9.4). This tool integrates an LN_2 bath for manipulation of frozen samples with automated refilling, positions for two VCT holders and a tool dryer. It includes a dock for the VCT500 and optionally offers connectivity to TEM side entry cryo-transfer holders and the Leica EM CRYO-CLEM system (Figure 9.2c). For a typical cryo-SEM workflow, samples from high pressure freezing are attached under LN_2 to a VCT sample holder (Figure 9.3), which is then attached to and retracted into the shuttle. After (optional) pumping the shuttle to a medium vacuum using the scroll pump, the holder can be transferred to the next instrument without the risk of contamination or warming.

Besides transferring frozen hydrated specimens, the VCT shuttle system is also suitable to transfer samples that are sensitive to environmental conditions, such as oxygen or humidity, safely between preparation instruments and the SEM.

9.4 FREEZE FRACTURE AND FREEZE ETCHING

Unless the surfaces to be studied in cryo-SEM are naturally exposed (e.g. the surface of leaves), bulk frozen hydrated specimens have to be mechanically opened up to make their inner part accessible for analysis: one approach is the freeze fracture (FF) technique, where parts of the specimen are cleaved under high vacuum (Moor *et al.*, 1961). The course of the fracture plane can hardly be predicted: it preferentially follows an inner, hydrophobic area of the lipid bilayers. In a complex system such as a cell, the plane of fracture frequently switches from one bilayer to another, producing a highly corrugated surface with much structural detail. To prevent fracture along the specimen/carrier interface, mechanical or chemical modifications of the specimen carriers are used to improve adhesion of the sample to the carrier.

Typically, fracturing is carried out in the low 10^{-7} mbar vacuum range in an oil-free vacuum system to prevent contamination, with the sample held at −120 °C at an equilibrium of condensation and evaporation of water (see Chapter 12 by Tacke *et al.*) and warmer than the cryo-shield. It should also be noted that the sample temperature influences the fracture plane, with lower temperatures leading to smaller membrane patches, but more detail (Fukazawa *et al.*, 2009). The fracture is either performed with an LN_2 cooled microtome blade passing through the sample ('knife fracture') or by pulling two carriers apart ('tensile fracture' or 'sandwich fracture'). If a spring-loaded, hinged mechanism is used for a tensile fracture ('mirror fracturing'; see Figure 9.3d and e), both halves of the sandwich are retained, enabling the investigation of corresponding positions. These spring-loaded mechanisms are, however, limited in force. In order to separate strongly adhering sandwiches (interlocking carriers, sticky tape) other means such as pulling them apart with the fracture knife have to be used.

Typically, the FF device and the high vacuum coater (see Section 9.6 below) are one combined instrument to protect the specimen in between these steps. If the specimen is metal coated without any freeze etching, the fracture process is often carried out while the metal evaporation has already been started, reducing the risk of uncontrolled sublimation and contamination.

Two instruments from Leica's current product line can be employed for FF: either the EM ACE900 (project name 'EM BAF900'), the newest dedicated FF device, or the EM ACE600, a high vacuum evaporator that can be fitted with FF tools. Both instruments were designed with modularity, a small footprint, a high degree of automation, easy cleaning and an effective vacuum system in mind. To improve the effectiveness of the vacuum system, the size of the chamber was kept small and the number of feedthroughs and dynamic gaskets, for example to motors, was reduced by placing these components entirely into the chamber.

While the ACE600 is smaller and can be flexibly adapted for other applications, the ACE900 offers a number of specific advantages for FF. These include a better end vacuum with a possibility to change the electron guns without breaking the vacuum, and a hard metal knife with a more flexible movement along three axes that can be changed while also keeping the unit evacuated. This dedicated freeze fracture device also features a more effective and faster cooling system with superior anti-contamination via a full chamber cryo-shield. Moreover, the specimen temperature is lower and the cryo-stage supports rotation and low angle coating (see Table 9.1 and Figure 9.5a).

The general workflows on both instruments are very similar. Cooling the instrument (stage, knife, cryo-shield) is started no earlier than in the 10^{-6} mbar range. To assist with reaching the working vacuum quickly and maintaining it during the whole procedure, all transfers on the ACE900 should be performed via air locks or gate valves (see above) and the vacuum should be broken only in exceptional situations, for example for cleaning.

Various older instruments used an N_2 counterflow to keep the cooled and vented chamber free of contamination during specimen transfer. In contrast, the frozen sample on the VCT holder can be inserted into the ACE600/900 vacuum chamber either via a load lock (VCT500 dock required for ACE600) or – preferably – via the VCT100 or 500 shuttle (ACE600 and 900), which connects to the dock mounted on the left side of the instrument: after mechanical attachment, the evacuation and purging of the airlock via the ACE pumping system and opening of the valves are operated via the touch screen user interface. To mount the sample on to the stage, the VCT manipulator rod is extended under constant observation until the holder sits at the pre-centred stage ready to be clamped (Figure 9.5b). After retraction of the manipulator rod, it is recommended that in order to obtain an optimum vacuum the shuttle should be detached and the VCT dock should be sealed with the metal blind flange. After equilibration at the required temperature, the fracture process can be started.

Table 9.1 A comparison of different vacuum coaters and freeze fracture devices available in the past (EM BAF060) and present (EM ACE200, 600 and 900) from Leica Microsystems. Features significant for freeze fracturing, high resolution coating, integration into a uniform workflow and practical considerations are indicated and discussed in the text

	EM BAF060 Freeze Fracture System	EM ACE200 Low Vacuum Evaporator	EM ACE600 High Vacuum Evaporator	EM ACE900 Freeze Fracture System
Pumping system	Two-stage oil-free; diaphragm pump and 260 l/min turbo molecular drag pump	One-stage, rotary vane pump	Two stage oil-free; diaphragm and 67 l/s turbomolecular drag pump	Two stage oil-free; diaphragm and 300 l/s turbomolecular drag pump
Nominal end vacuum	$<1.5 \times 10^{-7}$ mbar (cooled) according to test instructions	$<7.0 \times 10^{-3}$ mbar	$<7.0 \times 10^{-7}$ mbar (cooled)	$<1.5 \times 10^{-7}$ mbar (cooled)
Motion axes and range of cryo-stage	Rotation: 40–99 rpm; angle: fixed. Coating angle set via gun position on arc from 0–90°	n/a	No rotation, tilt +/–27 °	Rotation 40 to 150 rpm, tilt –65 to +45 °
Stage temperature	< –170 to 30 °C	Room temperature	–170 °C to +20 °C	–185 °C to +20 °C
Cryoshielding of specimen	Sample shutter and knive operated at < –180 °C	n/a	Optional: Via cooled built-in knife	Full chamber cryoshield operated at –185 °C; knive operated at –185 °C
Cryogen source	Pressurised 60l LN2 Dewar	n/a	Optional side-mounted 750 ml LN2 dewar for manual refilling	Two-line pumping system from pressure free 25 l LN2 dewar
Exchangable evaporation sources	2	1	2	2

	EM BAF060 Freeze Fracture System	EM ACE200 Low Vacuum Evaporator	EM ACE600 High Vacuum Evaporator	EM ACE900 Freeze Fracture System
Metal evaporation sources	W cathode electron beam source	Sputter head	Sputter head, W cathode electron beam source	W cathode electron beam source
Carbon evaporation sources	W cathode electron beam source	Carbon thread	Carbon thread, carbon rod, W cathode electron beam source	W cathode electron beam source
Knives	Built-in W knife with rotation	n/a	Basic fracturing tool (exchangeable door) and/or built-in W knife with swinging motion	Built-in W knife with rotation-like motion, lateral and height adjustment; exchange via load lock
Interlock to VCT100/VCT500	Optional	No	Optional	Optional
Exchange of sample and accessories (knive, evaporators, …)	Sample: load lock, e-beams: load lock. Knive: fixed	Via vented chamber	Sample: VCT; Accessories: via vented chamber	Via load locks and/or VCT
Footprint (basic unit)	0.86 m^2	0.13 m^2	0.16 m^2	0.50 m^2
Volume (basic unit)	1.20 m^3	0.06 m^3	0.10 m^3	0.32 m^3

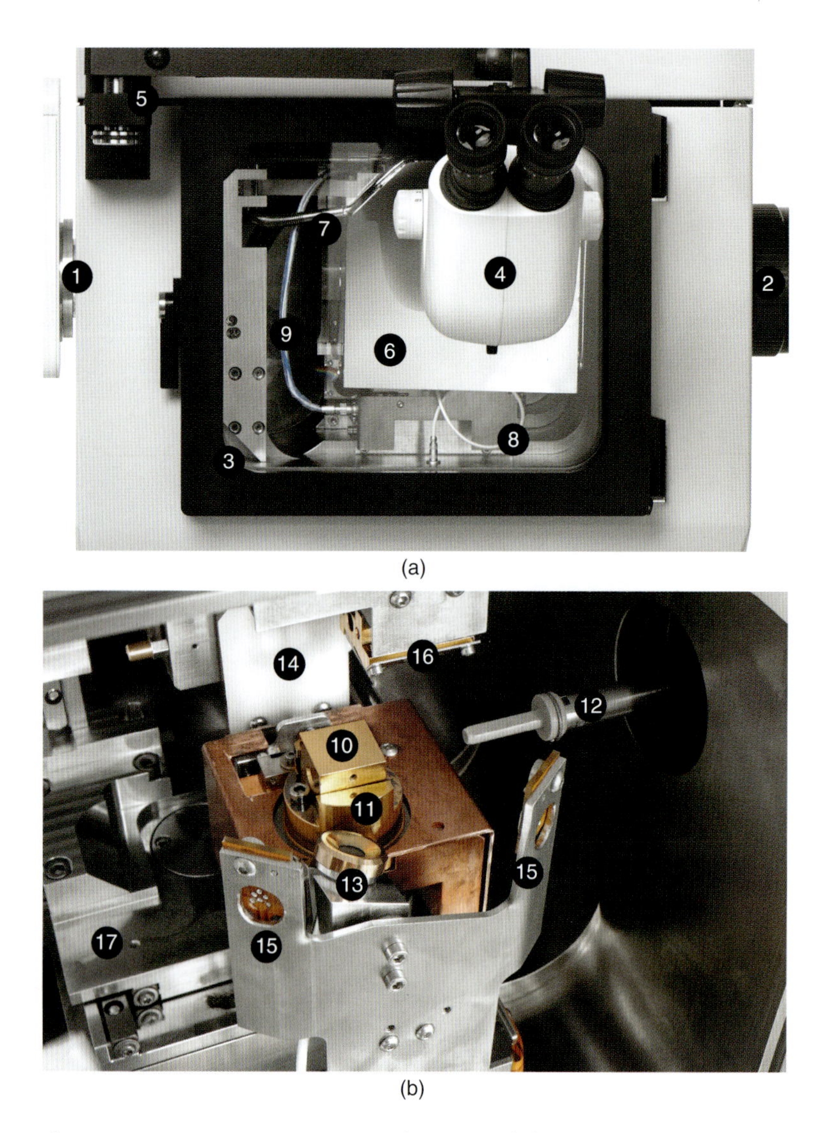

Figure 9.5 The Leica EM ACE900: (a) an outside view and (b) a view into the open vacuum chamber with the cryo-shield tilted away, showing the following components: (1) EM VCT500 dock for specimen transfer under vacuum and with cooling, (2) load lock for specimen insertion with cooling, (3) high vacuum chamber glass door, (4) the stereo microscope including (5) a mechanism for height adjustment, (6) the cryo-shield surrounding the specimen and (7) LN_2 cooling lines, (8) the electrical connection to the quartz film thickness monitor and (9) to the rotary stage. Inside the chamber, an 18 mm × 18 mm VCT specimen holder (10) is mounted on the (11) rotary cryo-stage and can be inserted or removed with (12) the manipulator rod of the load lock. The sensor of the quartz film monitor (13) is located in close proximity to the specimen, which is illuminated via (14) LEDs behind a diffusor from the back and (15) clear LEDs from the front. The fracturing knife (16) is mounted on the knife stage (17) supporting the cutting motion.

On the ACE900, a tungsten knife (Figure 9.5–16) is available for manipulating the sample. It can be operated either manually with buttons controlling x (horizontal position on the knife) and z (height) and a hand wheel controlling the cutting cycle. An automatic mode with a pre-defined feed and speed, and savable positions are also available. On the ACE600, either a 'basic fracturing tool' – a cooled scalpel integrated into the front door – and/or a tungsten knife (1 axis), which can be swung manually over the sample and lifted during the back-stroke, are available.

For knife fracture, either of the tungsten knives or the basic fracturing tool is used under visual observation, using either the magnifying glass at the ACE600 or the stereo microscope of the ACE900 (Figure 9.5a). If flat specimen carriers or dedicated FF carriers are to be used for tensile fracture, they are transferred to the freeze fracture device on a carrier like those shown in Figure 9.3a or b; the top carrier can be hit off by pushing it with the knife or the basic fracturing tool (Kaech and Ziegler, 2014). If FF carriers were used for HPF and loaded into the spring-loaded tensile fracture holders (Figure 9.3e and f), their opening mechanism can be triggered with either the basic fracturing tool or the knife.

9.4.1 Freeze Etching

Additional structural information from freeze fractured specimens can be revealed by controlled sublimation of the ice, a process referred to as "freeze etching" or "partial freeze drying". This uncovering of additional detail can lead to a better insight into the three-dimensional architecture of the specimen. An accurate control of the physical conditions during freeze etching (temperature, pressure) determining the sublimation rate is essential for an optimal result. Typically, freeze etching is performed at around −110 to −90 °C in the range of 10^{-7} mbar pressure, yielding maximum theoretical etching rates of 0.2–14 nm/s for pure water and significantly slower for biological specimens (Umrath, 1983). Depending on the required depth, the sample is held at this temperature for a number of minutes with the cold knife suspended above the specimen acting as a cold trap in the vicinity of the sample. It should be noted that for most glycerol-treated specimens, etching does not reveal any additional details and is redundant as glycerol cannot be removed by vacuum sublimation (Severs, 2007).

To make freeze etching a straightforward and reproducible process, once key parameters have been established, dedicated freeze etching programs can be set up on the ACE600 and ACE900. These programs allow the temperature gradient to/from the etching temperature, the etching temperature itself and the process time for etching while the pumping system is constantly working at full power to be defined. On the ACE900, it is possible to define several consecutive conditions for etching in more than one step.

9.5 CRYO-PLANING

The highly complex course of the fracture through the specimen in FF may impede interpretation and quantitative analysis. This applies particularly when the membrane itself is not the focus of the study but other cellular features underrepresented in fractures. For these applications, the interior of a bulk sample can be exposed by cutting a flat blockface from the cryo-sample, a technique referred to as cryo-planing. As compared to FF, cryo-planing allows a better targetting of the area of interest and – if more than one face is created – also analysis of a structural feature from different perspectives (Rensing, 2012) (see Figure 9.7b later). Furthermore, serial cryo-planing has been demonstrated by Huang *et al.* (1994).

Different tools have been used for cryo-planing including microtomes, ultramicrotomes (Walther and Müller, 1999; Ito *et al.*, 2015), diamond knives built into FF devices (Walther, 2003), focused ion beams (FIB) in dual beam microscopes (Heymann *et al.*, 2006) and broad ion beams (BIB) (Chang and Joester, 2015).

In the Leica specimen preparation workflow, the EM UC7 ultramicrotome with the EM FC7 cryo-setup is used for numerous applications, such as cryo-sections of sucrose infiltrated material for immunolabelling ('Tokuyasu method'; Tokuyasu, 1973) or the production of vitreous sections for cryo-TEM from frozen hydrated samples, and also allows cryo-planing for cryo-SEM. As surface contamination is generally a major issue in cryo-sectioning techniques, a 'Cryosphere' is available for the FC7. This transparent acrylic glass cover, enclosing the whole microtome while still allowing to operate it via ports at the front helps to reduce humidity inside to <10% by a constant overpressure of dry N_2 gas generated from evaporating LN_2.

For cryo-planing, the sample is transferred to the FC7 chamber under LN_2. If the commonly used round 2.8, 3.0 and 6.0 mm flat specimen carriers were used for HPF, they are kept closed in a sandwich until mounted in the microtome holders (chucks). To eliminate the need to trim away a part of the metal HPF carrier, special vertical holders for the microtome are available for this application (Ito *et al.*, 2015; order numbers 16706047, −48 and −49), allowing the flat specimen carriers (except membrane carriers) to be stably mounted with a sectioning plane parallel to the carrier. Alternatively, metal tubes (Shimoni and Müller, 1998; Studer, Graber and Eggli, 2001) can be used for HPF and subsequent cryo-planing (Walther and Müller, 1999).

Typically, the sample at around −150 °C is trimmed with cooled standard tools and planed by sectioning with a feed of 50 nm with a dry cryo-diamond knife (Robert Ranner, personal communication) until the position and shape required have been obtained.

For safe transfer of the planed specimen from the microtome to the next stage of preparation, a special version of the FC7, internally referred to as 'FC7T', is available, including a port and dock for the VCT100/500 shuttle on the right side. The shuttles can also be used in conjunction with the Cryosphere: in this case, the dock is mounted at the Cryosphere and a connection tube protects the exposed sample between the gate valve of the dock and the FC7. When the shuttle is used, the specimen is mounted on to a pre-cooled VCT holder parked in the microtome's cryo-chamber, which is subsequently fixed on the tip of the manipulator rod and retracted into the VCT. Without its own pumping system on the FC7, both the docking of the VCT shuttle (to allow opening of the gate valve) as well as undocking are carried out with the shuttle vented.

If very large areas or samples including hard and brittle materials need to be planed, an alternative to planing with a diamond knife or using a cryo-FIB is BIB milling. Leica's triple Ar^+ ion beam milling system EM TIC3X and its predecessor Bal-Tec TIC020 have been designed with mainly materials applications in mind, but can be adjusted to a diverse selection of material properties. As shown by Chang and Joester (2015) on yeast cells and animal tissues, it can also be employed for milling of frozen hydrated specimens.

This workflow demonstrated on a prototype instrument is now available via options to the TIC3X: the instrument can be fitted with an optional dock for the VCT500 shuttle, allowing the safe transfer of (high pressure) frozen hydrated specimens mounted on dedicated VCT holders including the milling masks (Figure 9.3f) into the instrument and of milled specimens out of it. If required, the frozen specimen can be prepared using a cryo-diamond saw operating the LN_2 loading station of the VCT (VCM) in advance.

The holders are clamped on to the TIC3X cooling stage, keeping both the sample holder and the milling mask at temperatures down to −160 °C. Under these conditions, the sample is ion beam milled until the required depth has been obtained. A dedicated cryo-shield is not required in this application, as potential surface contamination is removed instantaneously by the BIB. According to Chang and Joester (2015), a planed area of 700 000 μm^2 can be achieved within 2 hours without obvious signs of recrystallisation.

Freeze etching as discussed in the previous section can also be employed on planed block-faces produced in either way to reveal additional detail, and has been described as particularly useful for very smooth planed samples (Huang *et al.*, 1994).

9.6 COATING

Both the secondary electron (SE) as well as the backscattered electron (BSE) detectors can be used for signal generation in cryo-SEM. Uncoated cryo-SEM specimens exhibit little contrast, are unstable in the beam and suffer from charging build-up on the non-conductive specimens. Typically, these problems are addressed using a double layer coating. The frozen hydrated sample is coated with a layer of metal, providing a signal and minimising the penetration depth of the beam as well as also improving electrical and thermal conductivity. To further reduce beam damage and charging by increased heat and charge dissipation, and for improved mechanical stability, the metal layer is backed with a carbon layer evaporated on to the sample. The high quality of the metal coat deposited by electron beam evaporation or sputtering (see Chapter 12 by Tacke *et al.*) is essential for the resolution that can be obtained, especially in conjunction with FEGSEM. It has to faithfully reproduce the natural surface profile, be homogenous and featureless, and at the same time deliver optimal contrast restricted to the surface of the specimen.

Metal coating – typically 2.5 to 3.0 nm platinum (Pt) for optimal contrast – can be performed with and without rotation of the specimen stage at different fixed incident angles, typically between 25° and 45°, or with the guns constantly being moved between different angles during shadowing (double axis rotary shadowing DARS; Hermann *et al.*, 1988). Each of these different approaches reveals different aspects of the specimen and have to be chosen according to the question at hand. The subsequent layer of 4 to 10 nm carbon is deposited on to the sample at 45° to 90°.

To measure the amount of material deposited on the sample, an oscillating quartz film thickness monitor (Sauerbrey, 1959) is considered standard. The thickness of a known material deposited on to this sensor can be calculated from the change in oscillations with increasing thickness, and due to the known geometric relationship between sensor, source, and sample, the amount on the sample can be computed (Zeile, 2000).

Heating of frozen hydrated specimens under the intense beams of metal and carbon evaporation is a major issue and has to be taken into account in the construction of the electron guns (Moor, 1970) and the workflow. In practice, this includes parameters such as the aperture sizes used for the guns, the power with which they are operated (i.e. the coating rate), and optimised sample shuttering.

Alternatively, the replicas produced can be used for inspection in a TEM. Replicas from bulk biological specimens must be cleaned of biological material, however, using detergent, household bleach, chromic acid or sulphuric acid (for a protocol, see Severs, 2007). In addition to morphological information, replicas allow investigation of the spatial distribution of membrane proteins using freeze fracture replica immunolabelling (Fujimoto, 1995) but preclude – in contrast to cryo-SEM, where the bulk of the specimen is retained – analytical studies.

The EM ACE600 high vacuum evaporator and the EM ACE900 FF device – introduced in the section on freeze fracture (see Figure 9.5 and Table 9.1) – are also used for coating in the Leica cryo-SEM preparation workflow. The ACE600 operates in the 10^{-7} mbar range with basic cryoprotection while the ACE900 employs a full chamber cryo-shield for optimal protection of the specimen in the low 10^{-7} or high 10^{-8} mbar range (Table 9.1). On both instruments, cryo-SEM samples are coated with W cathode electron beam guns. Alternatively, a sputter head and carbon thread/rod sources are available for other applications on the ACE600.

The electron beam gun modules for the ACE 600/900 and the electronics driving them were completely redesigned from scratch, with easier maintenance and improved stability as compared to previous models in mind. They are no longer located inside the vacuum chamber, but attach as outside modules to the top of the chamber. This was required due to the considerably smaller vacuum chambers on both instruments while maintaining a well-proven working distance. At the same time, this new design allows easy detachment, maintenance and exchange without opening the chamber front door, and on the ACE900 even without venting or warming up the chamber due to a gate valve between the gun and chamber. For both instruments, very similar gun modules are being used, with the ACE900 guns being shorter to account for the aforementioned gate valves. Through small apertures and other design features, the heating of a cooled specimen could be limited to a few degrees (Gisela Höflinger, personal communication). Related design features limit the contamination by deposition of evaporated material in the gun to a much smaller area than with the earlier models, making them easier to clean.

The guns used for evaporation of carbon (C) and metal are almost identical, besides the collet receiving the rod: the C gun receives a 3.0 mm spectrographically pure carbon rod while, in the metal gun, a metal insert is fixed into a recess at the end of a 2.0 mm carbon rod, either by using a slightly deformed insert jammed into the recess or glueing it. At the time of writing this text, the only type of metal insert tested with the new electron guns was Pt, with tests on more materials pending.

In contrast to older models (Bal-Tec BAF300 and 400, Bal-Tec/Leica EM BAF060), the evaporation angle is no longer set by mounting the guns at different positions on an arc inside the vacuum chamber; with the ACE600/900 fixed position guns, the coating angle is now set by tilting the motorised specimen table.

One outstanding feature of the EM ACE series of coaters is that it allows the definition of program 'sequences', designed to allow highly reproducible coating of the specimen (and in the case of protocols, not including the freeze fracture step, even unattended operation on the ACE600). Each sequence can contain a number of procedures such as degasing for removal of residual contamination in the carbon rods and subsequent pumping before the actual evaporation, evaporation and wait times. For the individual evaporation procedures, in turn, parameters such as source and material, static or dynamic coating (stage) angle, rotation of the stage (if available), evaporation power/speed and final thickness of the coat can be pre-defined.

Following coating, the sample is transferred to the cryo-SEM (or other analytical instrumentation) maintaining vacuum and cooling in the VCT shuttle. Samples unloaded via the (vented and only transiently cooled) load lock can only be used for TEM replicas due to the rapid contamination of the exposed sample surface.

9.7 THE LEICA EM VCT500 CRYO-SEM SET

As in the previous steps, the sample has to be protected in the SEM from sublimation and recrystallisation by cooling and from surface contamination by appropriate shielding. The VCT cryo-SEM set available from Leica Microsystems is based on the original design by Wepf *et al.* (2004) and is implemented identically for the VCT100 and VCT500. It consists of an external 700 ml LN_2 Dewar, a cryo-stage that is mounted on top of the manufacturer's room temperature SEM stage, a microscope-specific cryo/anti-contamination shield in the chamber, a dock to attach the VCT cryo-shuttle, an oil-free scroll pump for the dock's airlock and an external controller and operating panel with a touch screen (see Figure 9.6).

Once the microscope's specimen chamber has been evacuated, cooling is initiated by filling the Dewar with up to 4 litres of LN_2, which will suffice for a complete cool-down and fill-up of the system. Via a connection with flexible copper bands (Figure 9.6), the stage is cooled, while thermal isolation against the (warm) microscope is maintained via insulating feet. The minimum temperature specification is dependent on the SEM chamber and the cooling bands used: $< -140\,°C$ is guaranteed for all configurations, while some can maintain the stage at even lower temperatures. Comparable configuration-specific differences apply for

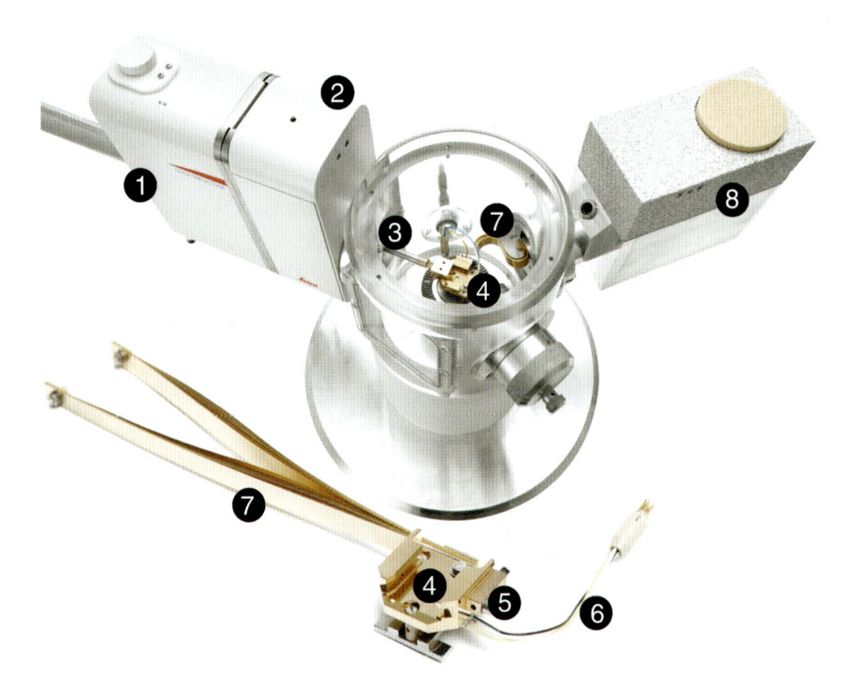

Figure 9.6 The EM VCT500 cryo-SEM set mounted on a mock-up SEM (top) and the actual cryo-stage (bottom). The two components are not shown to scale. The following details can be identified: (1) the docked VCT500 shuttle, (2) the VCT500 dock permanently installed at the microscope, (3) the partially extended manipulator rod, (4) the cryo-stage, (5) the temperature sensor, (6) the wiring for the temperature sensor and the heating element, (7) the flexible copper bands for cooling and (8) the Dewar. The Dewar's vacuum isolation is pumped via the SEM chamber, unless the 'low vacuum version' with its own pumping system has been installed. Not shown: the scroll pump and cryo/anti-contamination shield.

the cool-down time from room temperature to working temperature. A heating element and a temperature sensor integrated into the stage allow regulation of the temperature to the required value and to hold it with 0.1 °C precision. Depending on the stage temperature and aforementioned instrument-specific factors, one LN_2 filling of the Dewar lasts for about 3 hours. To support long-time observations, an automatic refilling system for the Dewar is available.

To insert the frozen hydrated specimen on to the pre-cooled cryo-stage, the SEM stage is moved to a pre-calibrated loading position, the VCT shuttle is attached to the dock, the airlock is pre-pumped via the scroll pump and (optionally) purged; subsequently the VCT holder is inserted under monitoring with a live camera and clamped. After completion of the transfer, the 5 kg VCT shuttle is detached from the microscope to resume pumping or for a bakeout at +110 °C via an internal heating element.

The Leica cryo-stage and the VCT100/500 dock have been integrated into instruments of all major manufacturers. Two available chamber or door ports with a minimum diameter of 40 mm are required for installation; one of these ports has to be oriented directly towards the specimen position of the chamber. The length of the manipulator rod of the shuttle has to be chosen according to the chamber size of the microscope.

The cryo-stage system does not impact the performance of the microscope in a negative way. Leica Microsystems guarantees the manufacturer specified resolution. Equipped with a dedicated Dewar, operating modes such as low vacuum can be used continuously, albeit different limitations for the minimum operating temperature of the stage apply. Due to the flexibility of the copper bands connecting the cryo-stage and Dewar, the microscope's original stage movements in x, y, z and tilt are not restricted; only rotation is limited. Beam deceleration is supported up to a maximum 5 kV.

To bring the stage up to room temperature after use, the VCT Dewar can be heated using an external heating element and a bake-out of the stage at +55 °C can be performed. To switch back to the room temperature stage, the system is designed to allow disconnection of the cryo-stage and dismantling of the anti-contaminator within less than 10 minutes, provided the system is already at room temperature and vented. The Dewar, the cooling shield and the VCT dock will remain in place.

Ideal results produced with the VCT100/500 shuttles/stages are shown along with artefacts typical for cryo-SEM, such as contamination and beam damage, in Figure 9.7. The use of these shuttles is, however, not limited to transferring ready-to-analyse samples into SEMs; they can also be connected to other analytical instrumentation.

9.8 SUMMARY

Most competing instruments for cryo-SEM preparation are based on a high vacuum chamber attached to a side port of the microscope and separated via a gate valve. They are equipped with tools for temperature control, fracturing, etching and coating of the specimens. The shuttle solution described here initially developed by Wepf *et al.* (2004) and advanced by Leica Microsystems offers a number of benefits, while maintaining the frozen-hydrated specimen in a safe environment. The physical separation of the specimen preparation from microscopy allows a more efficient usage of valuable beam time at the microscope, improves throughput in specimen preparation and analysis and offers a higher degree of flexibility and quality due to the wide range of dedicated preparation instruments and techniques available. These instruments – high vacuum coaters, microtomes, ion

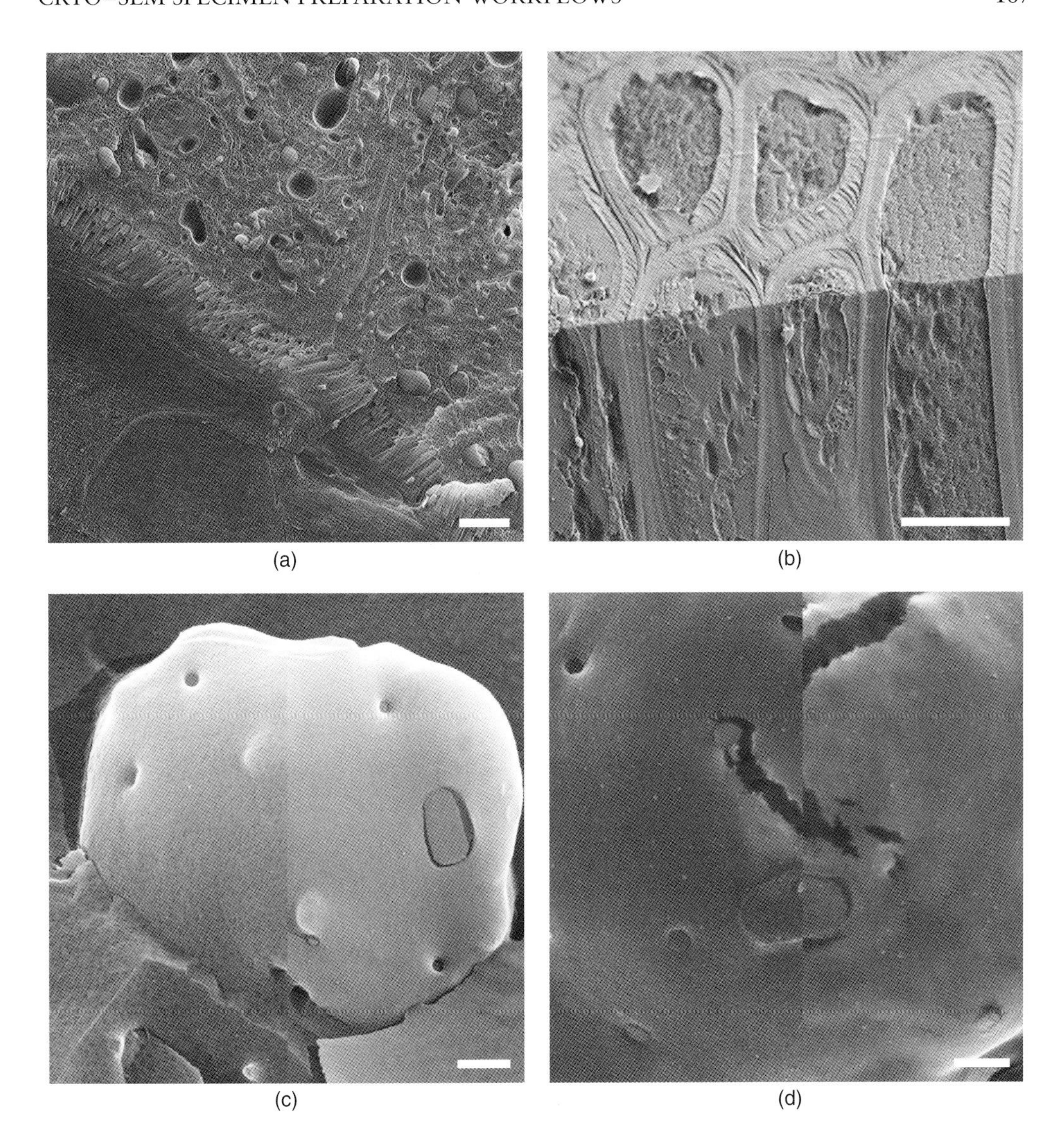

Figure 9.7 Results and artefacts in cryo-SEM. (a) High pressure frozen, freeze fractured, etched and metal/carbon coated specimens from the intestine of a *Drosophila* larva, visualised with in-lens cryo-SEM. The intestinal lumen is oriented towards the bottom left, while endothelial cells with the typical brush border can be seen in the top right. For details on specimen preparation, see Kaech (2016). (b) Poplar xylem prepared using HPF, cryo-trimming at −150 °C to a pyramid shape to reveal more than one face, cryo-planing at the same temperature and metal/carbon coating. This sample was visualised in cryo-SEM (secondary electron detector); for details on specimen preparation and imaging see Rensing (2012). (c) A sample of freeze fractured human platelets, exhibiting ice condensation on the surface, is visible as dark spots during the first scan (left part of image). After several scans of the same region (right part), the contamination was visibly reduced. Courtesy of Andres Kaech, Center for Microscopy and Image Analysis, University of Zurich, Switzerland. (d) Progression of beam damage in a sample of human platelets over several scans of the same region (left part versus right part). Courtesy of Andres Kaech. Scale bars: (a) 1 μm, (b) 10 μm, (c, d) 200 nm.

beam millers, the VCT shuttle, etc. – are optimised for the critical parameters to yield high resolution cryo-SEM data and address the issues of contamination and sublimation encountered in cryopreparation.

ACKNOWLEDGEMENTS

The author would like to thank Leander Gaechter, Andreas Hallady, Gisela Höflinger, Ian Lamswood, Dr Saskia Mimietz-Oeckler, Renate Neuberg, Dr Thomas Pfeifer, Robert Ranner, and Dr Cveta Tomova, all from Leica Microsystems at the time of writing, for their time and the insight provided. The feedback from and discussion with Dr Andres Kaech, University of Zurich, Dr Paul Walther, University of Ulm, Alexander Rosenthal, Microscopy Improvements and the editors is highly appreciated. Leica Microsystems Vienna is acknowledged for permission to use product photos and graphics.

REFERENCES

Adrian, M. *et al.* (1984) Cryo-electron microscopy of viruses. *Nature*, 308, 32–36.

Al-Amoudi, A., Chang, J.J., Leforestier, A., McDowall, A., Salamin, L.M., Norlén, L.P., Richter, K., Blanc, N.S., Studer, D. and Dubochet, J. (2004) Cryo-electron microscopy of vitreous sections. *EMBO J.*, 23 (18), 3583–3588.

Bachmann, L. and Schmitt-Fumian, W.W. (1973) Spray-freezing and freeze-etching, in *Freeze-Etching, Techniques and Applications*, Société Française de Microscopie Électronique, Paris, pp. 73.

Chang, I.Y.T. and Joester, D. (2015) Cryo-planing of frozen-hydrated samples using cryo triple ion gun milling (CryoTIGMTM). *Journal of Structural Biology*, 192 (3), 569–579.

Cyrklaff, M. and Kuehlbrandt, W. (1994) High-resolution electron microscopy of biological specimens in cubic ice. *Ultramicroscopy*, 55 (2), 141–153.

Dobro, M.J. *et al.* (2010) Plunge freezing for electron cryomicroscopy, in *Methods in Enzymology*, Elsevier Inc., pp. 63–82.

Dubochet, J. (2007) The physics of rapid cooling and its implications for cryoimmobilization of cells. *Methods in Cell Biology*, 2007 (79), 7–21.

Dubochet, J., McDowall, A.W., Menge, B., Schmid, E.N. and Lickfeld, K.G. (1983) Electron microscopy of frozen-hydrated bacteria. *J. Bacteriol.*, 155 (1), 381–390.

Frederik, P.M. and Hubert, D.H.W. (2005) Cryoelectron microscopy of liposomes. *Methods in Enzymology*, 391, 431–448.

Fujimoto, K. (1995) Freeze-fracture replica electron microscopy combined with SDS digestion for cytochemical labeling of integral membrane proteins. Application to the immunogold labeling of intercellular junctional complexes. *J. Cell Science*, 108, 3443–3449.

Fukazakwa, Y. *et al.* (2009) SDS-digested freeze-fracture replica labelling (SDS-FRL), in *Handbook of Cryo-Preparation Methods for Electron Microscopy* (eds. A. Cavalier, D. Spehner and B.M. Humbel), CRC Press, Boca Raton, London, New York, pp. 567–586.

Harapin, J. *et al.* (2015) Structural analysis of multicellular organisms with cryo-electron tomography. *Nature Methods* July, 12 (7), 634–636.

Hermann, R., Pawley, J., Nagatani, T. and Müller, M. (1988) Double-axis rotary shadowing for high-resolution scanning electron microscopy. *Scanning Microscopy*, 2, 1215–1230.

Heymann, J.A.W. *et al.* (2006) Site-specific 3D imaging of cells and tissues with a dual beam microscope. *Journal of Structural Biology*, 155 (1), 63–73.

Huang, C.X., Canny, M.J., Oates, K., McCully, M.E. (1994) Planing frozen hydrated plant specimens for SEM observation and EDX microanalysis. *Microsc. Res. Tech.*, 28 (1), 67–74.

Ito, Y. *et al.* (2015) Development of a cryo-SEM system enabling direct observation of the cross sections of an emulsion adhesive in a moist state during the drying process. *Microscopy*, 64 (6), 459–463.

Kaech, A. (2016) *Application Note: Cryo-SEM/Drosophila Larva*, Leica Microsystems, Vienna.

Kaech, A. and Ziegler, U. (2014) High-pressure freezing: Current state and future prospects, in *Electron Microscopy: Methods and Protocols, Methods in Molecular Biology*, vol. 1117, *Methods in Molecular Biology* (ed. J. Kuo), Humana Press, Totowa, NJ, pp. 151–171.

Kaech, A., Zumthor, J.P. and Hehl, A. (2014) *Application Note - Giardia lamblia*. Leica Microsystems, Vienna.

McCully, M.E., Canny, M.J. and Huang, C.X. (2009) Invited review: Cryo-scanning electron microscopy (CSEM) in the advancement of functional plant biology. Morphological and anatomical applications. *Functional Plant Biology*, 36 (2), 97–124.

McCully, M.E. *et al.* (2010) Cryo-scanning electron microscopy (CSEM) in the advancement of functional plant biology: Energy dispersive X-ray microanalysis (CEDX) applications. *Functional Plant Biology*, 37 (11), 1011–1040.

McDonald, K. *et al.* (2010) 'Tips and tricks' for high-pressure freezing of model systems, in *Methods in Cell Biology*, Elsevier Inc., pp. 671–693.

Möbius, W. *et al.* (2010) Electron microscopy of the mouse central nervous system, in *Methods in Cell Biology*, Elsevier Inc., pp. 475–512.

Moor, H. *et al.* (1961) A new freezing-ultramicrotome. *The Journal of Biophysical and Biochemical Cytology*, 10, 1–13.

Moor, H. (1970) High resolution shadow casting by the use of an electron gun. *Proceedings of the 7th International Congress on Electron Microscopy*, 1, 413–414.

Moor, H. (1987) Theory and practice of high-pressure freezing, in *Cryotechniques in Biological Electron Microscopy* (eds R.A. Steinbrecht and K. Zierold), Springer-Verlag, Berlin, pp. 175–191.

Moor, H. and Riehle, U. (1968) Snap-freezing under high pressure: a new fixation technique for freeze-etching, in *Proceedings of the Fourth European Reg. Conference on Electron Microscopy*.

Moor, H., Kistler, J. and Müller, M. (1976) Freezing in a propane jet. *Experientia*, 32 (No. 6).

Müller, M. and Moor, H. (1984) Cryofixation of thick specimens by high pressure freezing, in *The Science of Biological Specimen Preparation*, SEM.

Pum, D., Tomova, C. and Mimietz-Oeckler, S. (2015) High Pressure Freezing with Light Stimulation – Cryo-SEM analysis of UV light stimulation sun screen lotion. Accessed December 18, 2015, http://www.leica-microsystems.com/science-lab/high-pressure-freezing-with-light-stimulation/.

Rensing, K. (2012) LSN Application Note: Superior Cryopreparation of Biological and Industrial Samples – High Pressure Frozen and Cryo Planed Biological Samples, Leica Microsystems, Vienna.

Resch, G.P. *et al.* (2011) Immersion freezing of biological specimens: rationale, principles, and instrumentation. *Cold Spring Harbor Protocols*, 2011 (7), 778–782.

Ritter, M., Henry, D., Wiesner, S., Pfeiffer, S. and Wepf R. (1999) A versatile high vacuum cryo-transfer for cryo-FESEM, cryo-SPM and other imaging techniques. *Microsc. Microanal.*, 5 (Suppl. 2: *Proceedings*).

Sauerbrey, G. (1959) Verwendung von Schwingquarzen zur Wägung dünner Schichten und zur Mikrowägung. *Z. Physik*, 155, 206.

Schwarb, P. (1990) Morphologische Grundlagen zur Zell-Zell Interaktion bei adulten Herzmuskelzellen in Kultur. PhD Thesis, Swiss Federal Institute of Technology, Zurich.

Semmler, K., Wunderlich, J., Richter, N. and Meyer, H.W. (1998) High-pressure freezing causes structural alterations in phospholipid model membranes. *Journal of Microscopy*, 190, 317–327.

Severs, N.J. (2007) Freeze-fracture electron microscopy. *Nature Protocols*, 2 (3), 547–576.

Shimoni, E. and Müller, M. (1998) On optimizing high-pressure freezing. From heat transfer theory to a new microbiopsy device. *Journal of Microscopy*, 192, 236–247.

Studer, D., Graber, W. and Eggli, P. (2001) A new approach for cryofixation by high-pressure freezing. *Journal of Microscopy*, 203(January), 285–294.

Studer, D., Michel, M. and Müller, M. (1989) High-pressure freezing comes of age. *Scanning Microscopy Supplement*, 3, 253–269.

Studer, D., Michel, M., Wohlwend, M., Hunziker, E.B. and Buschmann, M.D. (1995) Vitrification of articular cartilage by high-pressure freezing. J. Microsc., 179 (Pt 3), 321–332.

Tivol, W.F., Briegel, A. and Jensen, G.J. (2008) An improved cryogen for plunge freezing. *Microscopy and Microanalysis: the official journal of the Microscopy Society of America, Microbeam Analysis Society, Microscopical Society of Canada*, 14 (5), 375–379.

Tokuyasu, K.T. (1973) A technique for ultracryotomy of cell suspensions and tissues. *J. Cell Biol.*, 57 (2), 551–565.

Umrath, W. (1983) Berechnung von Gefriertrocknungszeiten fuer die elektronenmikroskopische Praeparation. *Mikroskopie (Wien)*, 40, 9–37.

van Harreveld, A. and Crowell, J. (1964) Electron microscopy after rapid freezing on a metal surface and substitution fixation. *Anat. Rec.*, 149, 381–385.

Walther, P. (2003) Recent progress in freeze-fracturing of high-pressure frozen samples. *Journal of Microscopy*, 212 (1), 34–43.

Walther, P. and Müller M. (1999) Biological ultrastructure as revealed by high resolution cryo-SEM of block faces after cryo-sectioning. *Journal of Microscopy*, 196, 279–287.

Watanabe, S., Davis, M.W. and Jorgensen, E.M. (2014) Flash-and-freeze electron microscopy: Coupling optogenetics with high-pressure freezing, in, eds. *Nanoscale Imaging of Synapses: New Concepts and Opportunities, Neuromethods*, vol. 84, Neuromethods (eds. U.V. Nägerl and A. Triller), Springer, New York, pp. 43–57.

Watanabe, S. *et al.* (2013) Ultrafast endocytosis at *Caenorhabditis elegans* neuromuscular junctions. *ELife,* 2, e00723.

Wepf, R. *et al.* (2004) Improvements for HR- and Cryo-SEM by the VCT 100 high-vacuum cryotransfer system and SEM cooling stage. *Microscopy and Microanalysis*, 10 (Suppl. 2), 970–971.

Wild, P., Kaech, A. and Lucas, M.S. (2012) High-resolution scanning electron microscopy of the nuclear surface in Herpes Simplex Virus 1 infected cells, in *Scanning Electron Microscopy for the Life Sciences* (ed. H. Schatten), Cambridge University Press, Cambridge, pp. 115–136.

Zeile, U. (2000) *Coatings and Shadow Casting Techniques for Electron Microscopy and Improvements in Coating Quality*, Balzers.

10

Chemical Fixation

Bruno M. Humbel[1,5], Heinz Schwarz[2], Erin M. Tranfield[3] and Roland A. Fleck[4]

[1]*Electron Microscopy Facility, University of Lausanne, Switzerland*
[2]*Max Planck Institute for Developmental Biology, Tübingen, Germany*
[3]*Instituto Gulbenkian de Ciência, Oeiras, Portugal*
[4]*Centre for Ultrastructural Imaging, King's College London, UK*
[5]*Imaging Section, Okinawa Institute of Science and Technology, Onna-son, Okinawa, Japan*

10.1 INTRODUCTION

There is a wealth of information about the chemical fixation of biological samples. Most of these studies were performed on a single sample, such as bacteria, or cultured cells like fibroblasts, or on model substances such as protein solutions. Therefore, there are many examples where one study contradicts another study. In the light of this we believe it is of great importance to revisit the individual so-called 'standard protocols' in every electron microscopy lab. Often these protocols were introduced with a certain purpose and over time the protocol has become the holy grail, not to be altered, not to be questioned, without the present user knowing why. Very frequently over the course of time, small, unnoticed changes have occurred in the protocol so that the presently used 'standard protocol' is not the one originally introduced to the lab. As scientists we need to understand the chemistry we are using and the artefacts we are inducing in our sample. This knowledge allows the assessment of to what level of resolution the images may be subsequently analysed and interpreted. Artefacts affecting the structures of interest can be minimised through a combination of understanding fixation chemistry and careful experimental design. However, it is important to emphasise from the beginning that chemical fixation is a technique full of artefacts. At best artefact formation can be carefully controlled so that the existing artefacts do not hinder the ability to answer the scientific question.

Herein, we try to give information we think is important for sample preparation but readers are strongly encouraged to read the excellent treatises by M.A. Hayat (Hayat, 2000,

Biological Field Emission Scanning Electron Microscopy, First Edition.
Edited by Roland A. Fleck and Bruno M. Humbel.

1986, 1981) and Gareth W. Griffiths (Griffiths, 1993) as well as the original publications cited therein and in this chapter.

10.2 ALDEHYDES

10.2.1 Formaldehyde (FA)

Formaldehyde (Walker, 1944) was discovered and synthesised by August Wilhelm Hofmann in 1869 (Hofmann, 1869). He oxidised 'Holzgeistdämpfe' vapours of wood alcohol, that is methanol, over a glowing spiral of platinum wire and found the oxidised product formaldehyde. This compound was first used as a fixative for tissue in 1893 (Blum, 1893). Today formaldehyde is prepared by depolymerising para-formaldehyde. For a buffered 2% formaldehyde solution 2 g of para-formaldehyde is suspended in 45 ml of bidistilled water, warmed to about 50 °C and stirred vigorously in a closed Erlenmeyer flask for about 30 minutes. Then 20 μl of 1 M sodium hydroxide solution is added. If the para-formaldehyde is not completely dissolved after 10 minutes, the hydroxide step is repeated. After depolymerisation is completed, bidistilled water is added to make a final solution volume of 50 ml. Then 50 ml of buffer, either 400 mM HEPES or 200 mM phosphate buffer at pH 7.2 is added (see Robertson, Bodenheimer and Stage, 1963). The final pH should be checked to be pH 7.2. It is critically important to understand that the active fixative is formaldehyde and not its polymer para-formaldehyde (see Figure 10.1). Therefore when using formaldehyde as a fixative it should be correctly called formaldehyde.

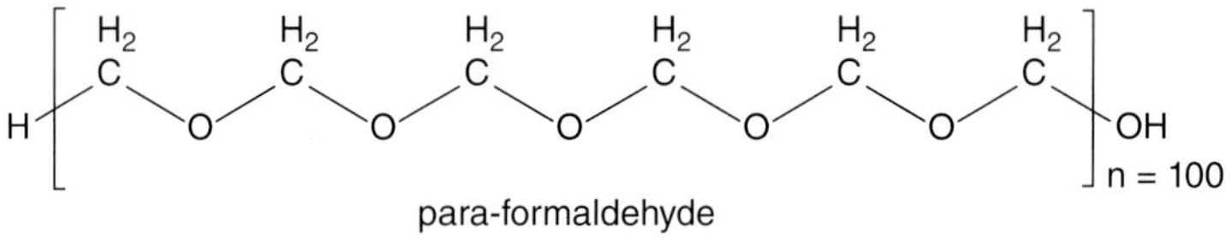

H2 C O H2 C O H2 C O H2 C O H2 C O H2 C OH n = 100 H
para-formaldehyde

Figure 10.1 Para-formaldehyde. Redrawn from Kiernan (2000).

Formaldehyde is a gas that is dissolved in water (see Figure 10.2). Thus storage containers need to be sealed otherwise the fixative will slowly evaporate, decreasing the concentration.

The reaction of formaldehyde is very complex and is described in great detail by Jones (1973). It reacts with unsaturated fatty acids and several amino acids, for example arginine, lysine, terminal amino groups, histidine, proline, tryptophan, serine, threonine, cystein, phenylalanine, tyrosine and peptide bonds (Jones, 1973). Studies by Roozemond (1969) showed that phosphatidyl ethanol amine is less soluble after formaldehyde fixation, whereas the solubility of other lipids may increase (Jones, 1973; Roozemond, 1969). The active component of formaldehyde is the carbonium ion ($^{+}CH_2OH$), which is produced at a lower pH; therefore buffered formaldehyde fixatives are in fact reducing the fixing power (Jones, 1973). Formaldehyde solutions can oxidise to formic acid, leading to a decrease in pH. It is therefore critically important that formaldehyde is used in a buffered solution. More

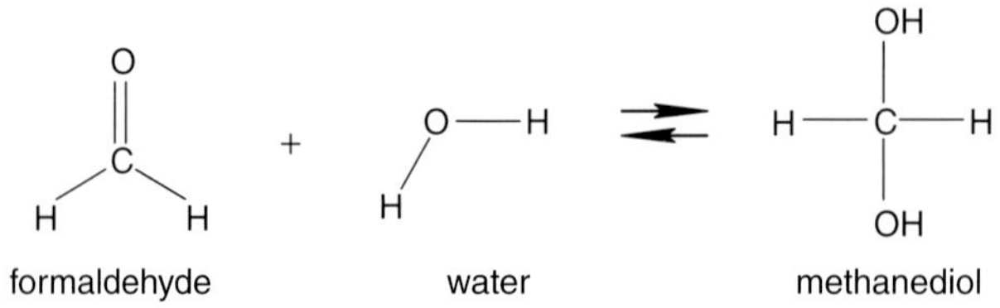

O C H H + O H H ⇌ OH H C H OH
formaldehyde water methanediol

Figure 10.2 Formaldehyde dissolved in water produces methanediol. Redrawn from Kiernan (2000).

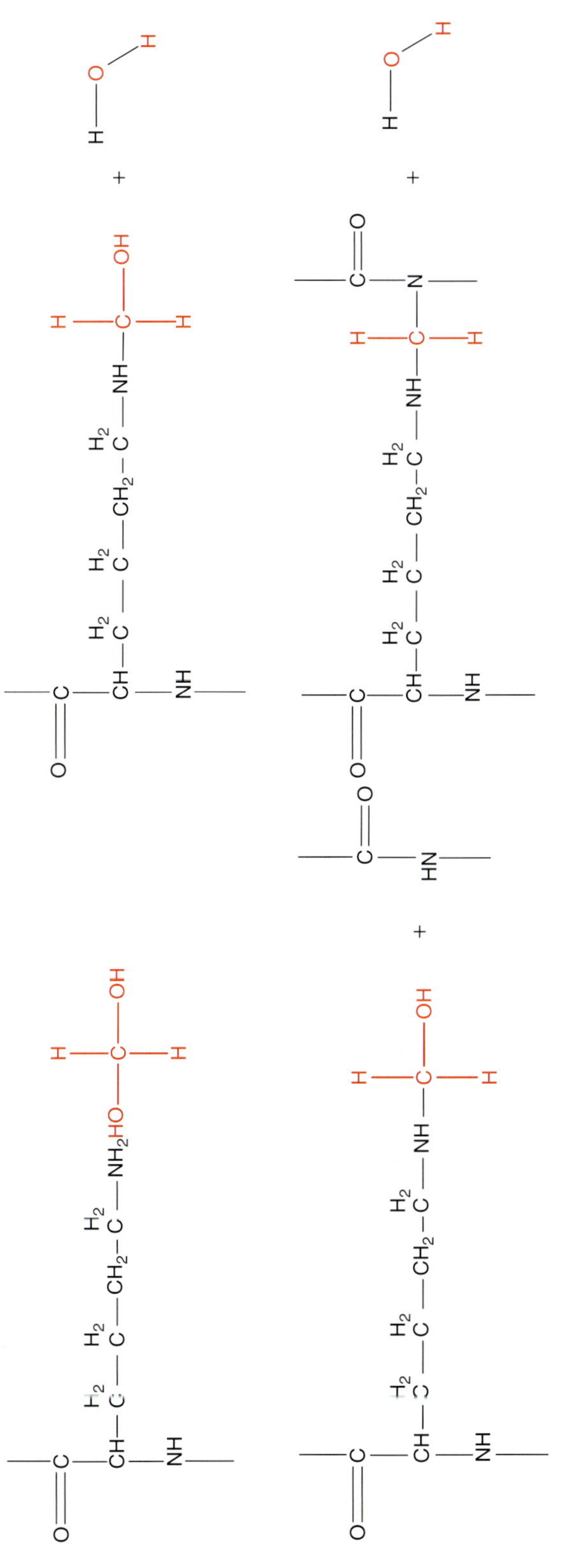

Figure 10.3 The terminal amino acid of lysine (left) is reacting with methandiol (red), which then reacts with the –NH– of the peptide bond (right), releasing two molecules of water. Redrawn from Kiernan (2000).

generally it is assumed that buffered formaldehyde cross-links the amino group of the side chain of lysine to the peptide nitrogen (Kiernan, 2000) (see Figure 10.3). However, the reaction is very slow although the penetration of the small molecule into tissue is fast. Formaldehyde can react with unsaturated fatty acids (Jones, 1973) but this has no influence on the mobility of membrane proteins (Jost, McMillan and Griffith, 1983). Jones suggested that this reaction can lead to the loss of lipids during fixation.

In his treaty Jones suggests replacing formaldehyde with acrolein as an everyday fixative. He also assumes that acrolein is a more powerful lipid fixative than glutaraldehyde (Jones, 1973).

In most electron microscopy labs formaldehyde is still highly valued as a fixative for immunolabelling studies. The limited power to perfectly preserve the cellular fine structure has the advantage that the proteins of interest, the antigens, are more accessible for antibodies, making localisation studies more successful. Even though it is a weak fixative it is an effective fixative. As an example, the ultrastructure of liver can be nicely preserved using 4% formaldehyde with 5–7.5% sucrose in phosphate buffer (Holt and Marian Hicks, 1961). In this study it was pointed out that other buffers were inferior to phosphate buffer.

It is important not to use formalin (e.g. Merck or Sigma-Aldrich) for electron microscopy. Formalin is a solution of 37% formaldehyde that is stabilised with 10–15% methanol, which will cause cytoplasmic extraction and the degradation of the ultrastructure in biological samples. Beyond the methanol, stabilised formalin often contains a lot of precipitated para-formaldehyde and during long-term storage formaldehyde reacts with itself producing methanol and formic acid (Kiernan, 2000). All combined, these variables make formalin a poor fixative for ultrastructural preservation.

10.2.2 Glutaraldehyde (GA)

Sabatini, Bensch and Barrnett (1962, 1963) studied the quality of fixation of different aldehydes (glutaraldehyde, glyoxal, hydroxyadipaldehyde, crotonaldehyde, pyruvic aldehyde, acetaldehyde, metacrolein, acrolein) and concluded that glutaraldehyde and acrolein showed excellent morphological preservation (Sabatini *et al.*, 1963). Though not further reasoned, maybe because of the better preservation of enzyme activity, they claim that glutaraldehyde gave the best general preservation (Sabatini *et al.*, 1963). This publication led to glutaraldehyde being broadly adopted as a main fixative for electron microscopy. For optimal preservation of the ultrastructure Hündgen suggested to use glutaraldehyde at a concentration of 2% and to add 6.9% sucrose to the buffer (Hündgen 1968). In his contribution to the method CD of the European 3D network of Excellence, Griffiths called glutaraldehyde euphorically the Queen of the fixatives (Griffiths, 2009).

Aldehydes in general bind preferentially to terminal amino groups, essentially to lysine. Glutaraldehyde can also react with tyrosine, tryptophan, phenylalanine, histidine and sulfhydryl groups, and hence cystein and methionine. It seems, however, that lysine is the most important component (Hayat, 1981, and references therein; Peters and Richards, 1977). Aldehydes form a Schiff base with a primary amino group (Jencks, 1964; Qin *et al.*, 2013; Tidwell, 2008). They, however, do not react with nucleic acids, unless denatured, nor carbohydrates or lipids, with the possible exception of phosphatidyl serine and phosphatidyl ethanol amine (Roozemond, 1969; Wood, 1973). This means that if a sample is fixed only with an aldehyde the proteins have been fixed, but the lipids, the carbohydrates and the nucleic acids remain unfixed.

It has been shown that commercial 25% glutaraldehyde, pH 3.1, contains 79% water, 3% glutaraldehyde and 18% derivatives of higher molecular weight that could be broken down

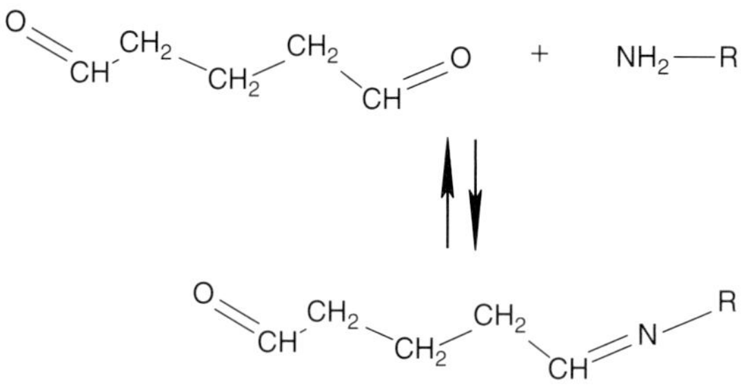

Figure 10.4 At low pH glutaraldehyde exists predominately in its monomeric form and the resulting Schiff's base with a terminal amino group is reversible.

to glutaraldehyde (Peters and Richards, 1977), and hence 18% polymers of glutaraldehyde. At higher pH the generation of higher unsaturated polymers increases until they even precipitate (Peters and Richards, 1977). The quality of glutaraldehyde can also be assessed by photospectroscopy. The monomer absorbs at 280 nm, whereas the polymers at 235 nm (Gillett and Gull, 1972).

At pH 3, the Schiff base is reversible (Peters and Richards, 1977). In other words, at low pH glutaraldehyde does not fix irreversibly and in a satisfactory way (see Figure 10.4).

At higher pH, pH 7.2–7.4, which is usually used for chemical fixation, glutaraldehyde is predominately an α, β unsaturated polymer, which results in a stable Schiff base formation (Hopwood, 1970, 1972; Monsan, Puzo and Mazarguil, 1975; Peters and Richards, 1977; Richards and Knowles, 1968) and in irreversible, permanent chemical fixation (see Figure 10.5). This is why chemical fixation with glutaraldehyde needs to be done in a buffered solution.

10.3 ACROLEIN

The high reactivity of acrolein preserves the submicroscopic organisation of cells adequately but completely destroys enzyme activity (Sabatini *et al.*, 1963).

Acrolein can react with proteins and fatty acids (see Figure 10.6), followed rapidly by a polymerisation process (Jones, 1973). It has been suggested as the better alternative to formaldehyde (Jones, 1973), especially for the fixation of lipids (Feustel and Geyer, 1966). Concentrated acrolein has also successfully been used as a sole substitution fluid during freeze-substitution (B.M. Humbel, unpublished). It worked quickly and resulted in good ultrastructure. However, because of the strong lachrymatory effect and potential health hazards its use has been abandoned after a few tests.

10.4 OSMIUM TETROXIDE

Osmium tetroxide preceded glutaraldehyde as a fixative and was widely used by the pioneers of electron microscopy (Palade, 1952; Porter and Kallman, 1953). Interestingly, in chick embryo fibroblast the extended mitochondrial network could already be appreciated in this early work (Porter, Claude and Fullam, 1945). The fibroblasts were cultivated on grids, osmium tetroxide vapour fixed and air dried.

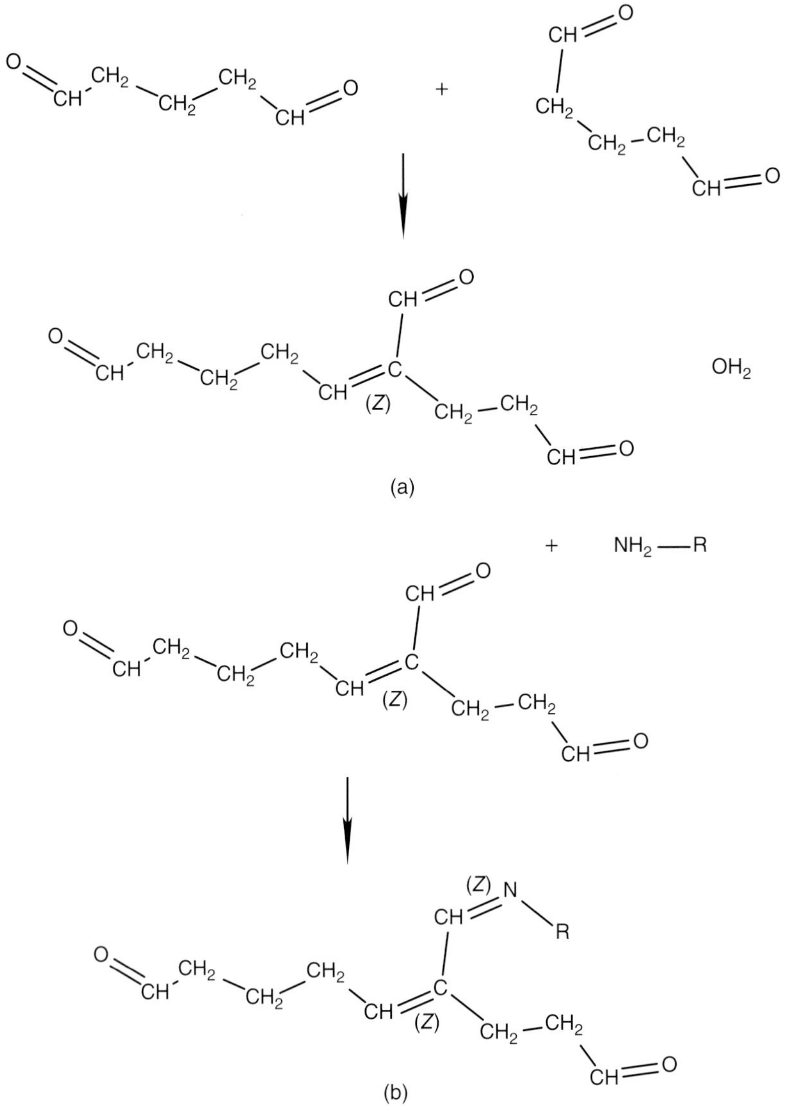

Figure 10.5 (a) At higher pH, for example 7.4, glutaraldehyde polymerises into dimers or higher polymers, where the carboxyl group gets stabilised by the mesomeric effect of the O=C–C=C bond. (b) The resulting Schiff's base is stabilised through the mesomeric effect of the N=C–C=C bond and the reaction becomes stable and is no longer reversible.

According to Behrman (1984) osmium tetroxide reacts with unsaturated fatty acids resulting in the depicted cross-link in Figure 10.7. Korn alternatively suggests a reaction with only one osmium tetroxide molecule (1967). Note that the oxidation state of osmium tetroxide changes from +8 to +6 after the reaction. White *et al.* (1976) studied the reaction of osmium tetroxide by X-ray photoelectron spectroscopy. They deposited cholesterol and osmium tetroxide at −100 °C, then raised the temperature and at −70 °C the reaction started and Os(VIII) became Os(VI), which is in accordance with the chemical reaction depicted in Figure 10.7. In this experiment, at 22 °C all Os(VIII) had become Os(VI). Using

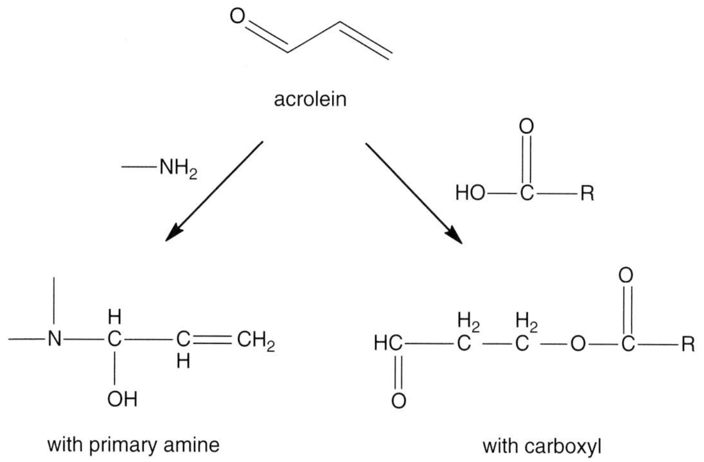

acrolein

with primary amine

with carboxyl

Figure 10.6 Acrolein reactions. Redrawn from Jones (1973).

phosphate buffered 2% osmium tetroxide to fix erythrocyte ghosts for 1 hour at 4 °C resulted in 42% Os(VI), 37% Os(IV) and 21% Os(III), meaning that only 42% of the osmium tetroxide has the fixing configuration depicted. After dehydration with ethanol virtually all osmium has turned into the Os(III) species. Interestingly, the reaction with glutaraldehyde, which precedes the osmium tetroxide treatment, already results in 21% Os(VI), 29% Os(IV) and 50% Os (III) species (White *et al.*, 1976). In other words, only a small percentage of osmium tetroxide is in the VI state that is assumed to be responsible for the fixation, and if osmium tetroxide is mixed with glutaraldehyde in solution, the percentage of Os(VI) is further reduced. The other species are probably responsible for the blackening of the osmium tetroxide.

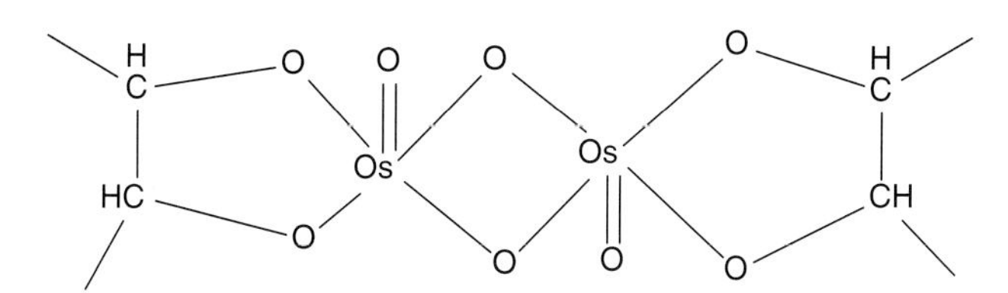

Figure 10.7 Osmium tetroxide reacted with double bonds of two unsaturated fatty acids. Redrawn from Behrman (1984) and White *et al.* (1976).

Osmium tetroxide has a relatively slow penetration rate into tissue. For dense tissues, osmium tetroxide typically does not penetrate faster than 0.5 mm in 1 hour and after 1 hour not a lot of further penetration will be observed. After leaving kidney samples in osmium tetroxide in phosphate buffer at 4 °C for 24 hours, the penetration depth was only 2.4 mm (Hopwood, 1970).

Osmium tetroxide fixation at temperatures above 0 °C can result in proteolysis of proteins (Baschong *et al.*, 1984; Behrman, 1984; Locke, 1994; Maupin-Szamier and Pollard, 1978). This effect can be used to image internal cellular membranes by scanning electron microscopy (e.g. Riva *et al.*, 1999; Tanaka and Mitsushima, 1984). From these results it is legitimate to assume that the lower oxidation state osmium tetroxide species might be responsible for the proteolytic effect. Using osmium tetroxide during freeze-substitution, we realised that the sample blocks only become black above about −30 °C (Humbel and Müller, 1986). This has led us to deduce that at temperatures under −30 °C the majority of the osmium tetroxide is in the fixation configuration. Preliminary experiments done in conjunction with the development of a cryo-fixation, freeze-substitution rehydration protocol for Tokuyasu cryo-sectioning (Van Donselaar *et al.*, 2007) showed that osmium tetroxide used during freeze-substitution fixation has little influence on the efficiency of immunolabelling (Van Donselaar and Humbel, unpublished observation). This observation supports the notion that during the low temperatures of freeze-substitution osmium tetroxide is a fixative rather than a protease.

Hayat and Guiquinta suggested that osmium tetroxide vapour fixation should be applied before starting with the standard chemical fixation protocol to better preserve overall cell components (Hayat, 1981; Hayat and Guiquita, 1970; Seligman, Wasserkrug and Hanker, 1966). Osmium tetroxide vapour fixation was later used to stabilise cryo-fixed and freeze-dried samples, especially in conjunction with element analysis by X-ray microanalysis (Coulter and Terracio, 1977; Edelmann, 1978, 1986; Hayat, 1981). Edelmann and others used osmium tetroxide vapour treatment to stain cryo-fixed, freeze-dried and Lowicryl HM20 embedded dendritic cells *en bloc* (Edelmann, 2002).

10.5 URANYL ACETATE

In 1956 uranyl acetate was introduced as a whole mount electron microscopic stain after osmium tetroxide fixation to increase the contrast, particularly in the cytoplasm and the nucleus (Strugger, 1956). In 1958 it was investigated in detail as an on-section stain by Watson (1958). Since then uranyl acetate is generally used as a stain in electron microscopy either on section (e.g. Locke and Krishnan, 1971) or as whole mount staining before dehydration and embedding (Terzakis, 1968; Locke, 1994; Locke, Krishnan and McMahon, 1971). In the course of improving the preservation of the cellular ultrastructure of bacteria, with special attention on the DNA, and establishing a general, more reliable protocol, uranyl acetate was not only added as a stain but also as a fixative (Ryter *et al.*, 1958; Stoeckenius, 1961). The specificity of the interaction of uranyl acetate and DNA was further investigated and compared with the Feulgen stain (Huxley and Zubay, 1961), a specific light microscopy stain for DNA (Feulgen and Rossenbeck, 1924; Feulgen and Voit, 1924a, 1924b, 1924c), showing that the Feulgen pattern was in accordance with the uranyl pattern. Later Feulgen chemistry has been adapted to specifically stain DNA for electron microscopy (Gautier, 1976).

Uranyl acetate fixes nucleic acids (Ryter *et al.*, 1958; Stoeckenius, 1961) and it is a potent fixative for phospholipids (Silva *et al.*, 1968, 1971; Terzakis, 1968; Ting-Beall. 1980). In bacteria, the addition of a uranyl acetate fixation step before dehydration reduced the loss of phospholipids from 20% to 1.5% (Silva *et al.*, 1971) and in rat brain from 18% to 0.16% (Silva *et al.*, 1968).

10.6 LESS COMMON FIXATIVES

10.6.1 Malachite Green

Malachite green (CAS # 569-64-2), otherwise known as Aniline green, Basic green 4, Benzal green, China green, Diamond green B or Victoria green B is an organic compound produced in a two-step process. The chemical formula is $C_{23}H_{25}N_2Cl$ (see Figure 10.8).

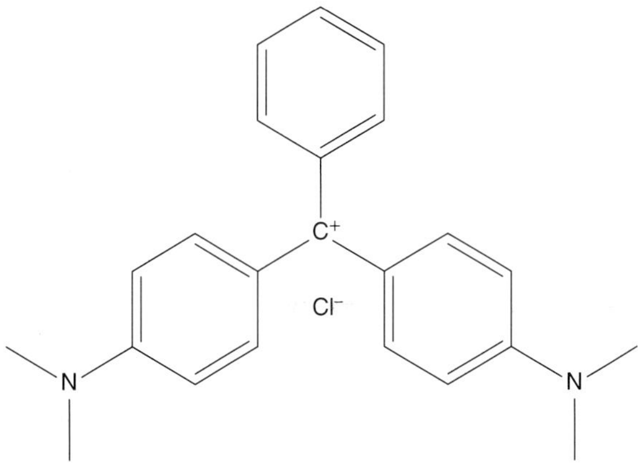

Figure 10.8 Chemical structure of malachite green or N,N,N′,N′-tetramethyl-4,4′-diaminotriphenylcarbenium chloride.

Malachite green is a cationic stain that has been used in light microscopy and electron microscopy sample preparation since the early 1970s.

In vitro chromatographic studies from 1974 revealed that malachite green binds to fatty acids (oleic, myristic, stearic and palmitoleic), fatty aldehydes, phospholipids (phosphatidylcholine, lysolectin, phosphatidylserine and phosphatidylethanolamine), glycolipids and cholesterol but it does not bind to sugars (D-glucose and sucrose) or amino acids (including glycine, alanine, asparagine, L-proline, L-serine, L-histidine, tryptophane) (Teichman, Takei and Cummins, 1974). These data, in combination with the observation by the same group that the addition of malachite green to the glutaraldehyde fixative step retained glutaraldehyde-soluble lipid elements (Teichman, Fujimoto and Yanagimachi, 1972), led to the conclusion that malachite green stains long-chain carbon groups found in many lipids.

The predominant method of use in the literature is the addition of malachite green to the glutaraldehyde fixative step (Teichman *et al.*, 1972). It has been used for immersion fixation as well as perfusion fixation (Pourcho, Bernstein and Gould, 1978).

A detailed study of the application of malachite green in electron microscopy suggested that malachite green adds electron density to ribosomes, to myofilaments and to lipid droplets in a uniform manner. When further enhanced by post-staining with uranyl acetate and lead citrate, tissues fixed with a 3% glutaraldehyde, 0.1% malachine green solution had increased electron density of both membrane and lipids components of the cells when compared to just 3% glutaraldehyde fixed tissues. The older literature uses cacodylate buffer with malachite green, but newer literature has also used 0.1 M PHEM (pH 7.4, see below) buffer (Vale-Costa *et al.*, 2016). Malachite green has also been used in light microscopy and super-resolution studies (Fitzpatrick *et al.*, 2009).

10.6.2 Ruthenium Red

Ruthenium red (CAS# 11103-72-3), otherwise known as ammoniated ruthenium oxychloride, is an inorganic, synthetically produced, cationic dye. It has been used to stain plant material since the 1890s and it was introduced to biological sample preparation for electron microscopy in 1964 by Luft (1964). Ruthenium red solutions are more stable around neutral pH. At low pH the solution will oxidise to ruthenium brown and at high pH the ammonium molecules will decompose (Luft 1971a). The chemical formula is $Cl_6H_{42}N_{14}O_2Ru_3$ (see Figure 10.9).

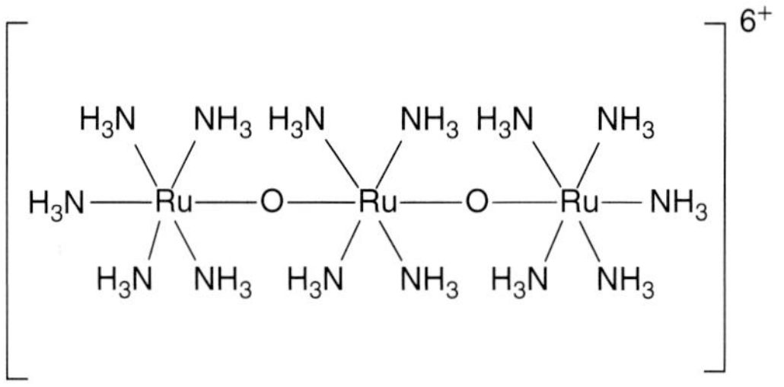

$$\left[\begin{array}{ccccccc} & H_3N\ \ NH_3 & & H_3N\ \ NH_3 & & H_3N\ \ NH_3 & \\ H_3N- & Ru & -O- & Ru & -O- & Ru & -NH_3 \\ & H_3N\ \ NH_3 & & H_3N\ \ NH_3 & & H_3N\ \ NH_3 & \end{array}\right]^{6+}$$

Figure 10.9 Chemical structure of ruthenium red or ammoniated ruthenium oxychloride.

Ruthenium red is frequently mixed into the aldehyde fixation (Luft, 1971a) or with osmium tetroxide in the secondary fixation step (Dykstra and Aldrich, 1978). Ruthenium red is also used in multiple steps within the same preservation protocol (Fassel *et al.*, 1992). Experimental evidence suggests that ruthenium red binds to anionic sites and leaves a fine grained electron density on fatty acids, phospholipids, acidic protein polysaccharides and acidic mucopolysaccharides (Luft, 1971a, 1971b). Negatively charged glycocalyx materials have also been demonstrated by finely grained electron-dense material following the addition of ruthenium red and lysine to glutaraldehyde or a combination of glutaraldehyde and formaldehyde (Fassel *et al.*, 1998). It is important to note that ruthenium red reacts with phosphate-based buffers, so it is necessary to use an uncharged buffering system such as cacodylate buffer (Dykstra and Aldrich, 1978) when adding ruthenium red to a protocol.

The stain should be excluded from cells by intact cell membranes (Luft, 1971b). It is reported that ruthenium red in combination with glutaraldehyde and then osmium tetroxide effectively cross-links and limits the penetration depth into the tissue. Efforts have been made to improve penetration by working with smaller tissues, or even by cutting frozen hydrated sections and staining the 80–100 nm sections with ruthenium red (El-Saggan and Uhrík, 2002).

For maximum staining, the ruthenium red/osmium tetroxide mixture should be freshly prepared and used as soon a possible. When a 3-hour old solution was used a noticeable decrease in surface staining was observed, but there was enhanced penetration of the stain into intercellular regions, suggesting that the older solution may be a useful intercellular tracer (Handley and Chen, 1981).

Ruthenium red mimics calcium and has an effect on the excitation and contraction of muscle fibers (Snowdowne and Howell, 1984; Zacharová *et al.*, 1990). It has been used in many functional assays. However, as it is combined with fixatives in electron microscopy sample preparation, the functional aspects of ruthenium red should be of limited concern. Caution should be taken when using ruthenium red in studies on highly active cells, such as platelets, because morphological changes have been observed in these cases (Dierichs, 1979), presumably due to the interference with calcium and sodium channels in the initial moments of aldehyde fixation.

10.7 MIXTURES OF FIXTURES

10.7.1 Formaldehyde and Glutaraldehyde

Morris J. Karnovsky suggested using a mixture of formaldehyde and glutaraldehyde (Karnovsky, 1965) for better preservation of the fine structure (Hayat, 1981).

Formaldehyde, as a small molecule, penetrates faster and immobilises the structure before gross alterations occur due to a slow fixation process. After the initial reversible formaldehyde fixation the ultrastructure is preserved by the irreversible but slower reaction with glutaraldehyde. The original mixture of Karnovsky is 4% formaldehyde and 5% glutaraldehyde in cacodylate or phosphate buffer at a final pH 7.2 (Karnovsky, 1965). Because of the high osmolarity of 2010 mosmol, today half-strength (2% formaldehyde and 2.5% glutaraldehyde) Karnovsky fixative is most frequently used (Hayat, 1981). We have successfully used even lower concentrations, such 2% formaldehyde and 0.02–0.1% glutaraldehyde (Jan Leunissen, personal communication) for Tokuyasu cryo-sectioning (Tokuyasu, 1973). The chemical reaction of high concentrations of glutaraldehyde exhibit strong autofluorescence that interferes with fluorescence microscopy, but at concentrations of 0.5% or lower the induced autofluorescence is negligible (e.g., Loussert Fonta and Humbel, 2015). As an added note: glutaraldehyde induced and endogenous autofluorescence can efficiently be destroyed using sodium borohydride (Beisker, Dolbeare and Gray, 1987; Clancy and Cauller, 1998; Corrodi, Hillarp and Jonsson, 1964). Some years ago we made the interesting observation that in 2% formaldehyde fixed cells unbound glucocorticoid receptors were washed off during the immunolabelling procedure but were retained when 0.1% glutaraldehyde was added to the fixation protocol (Brink *et al.*, 1992).

10.7.2 Buffered Formaldehyde and Picric Acid

The buffered formaldehyde and picric acid fixative was introduced by Zamboni and colleagues for the fixation of spermatozoa, a sample known for its sensitivity to chemical fixation and for being difficult to preserve (Ito and Karnovsky, 1968; Stefanini, DeMartino and Zamboni, 1967; Zamboni and De Martino, 1967). This fixative is based on Bouin's fixative made with picric acid (Bouin, 1897) cited in (Romeis, 1968). Bouin's solution poorly preserves the cellular ultrastructure but has the advantage that picric acid penetrates very quickly through cell membranes. Zamboni's fixative is buffered and contains 15 ml of a saturated aqueous solution of picric acid and 2% formaldehyde to make 100 ml in 128 mM phosphate buffer at pH 7.3. We used this fixative successfully to prepare another difficult to fix organism, *Dictyostelium discoideum*, for the Tokuyasu cryo-sectioning technique (Humbel and Biegelmann, 1992).

10.7.3 Glutaraldehyde and Osmium Tetroxide

Hasty and Hay noted that glutaraldeyde fixation alone resulted in four classes of membrane artefacts in areas without intramembrane proteins. These are membrane blebs or blisters, detached vesicles, myelin figures and multivesicular protrusions (Hasty and Hay, 1978). Also the mesosomes in bacteria were identified as glutaraldehyde-induced artefacts (Ebersold, Cordier and Lüthy, 1981). These artefacts were attributed to the fact that glutaraldehyde only fixes proteins but not lipids (Jost *et al.*, 1983). They could be avoided by a simultaneous fixation with a mixture of 1–2% glutaraldehyde and 1% osmium tetroxide (Hayat, 1981), a fixation protocol suggested by Trump and Bulger (Franke, Krien and Brown, 1969; Trump and Bulger, 1966). The mixture must be prepared immediately before use and seems to be stable for about 30 minutes at 25 °C (Hopwood, 1970); however, mixing the component and fixation is preferentially done at 0 °C (Franke *et al.*, 1969). The gutaraldehyde–osmium tetroxide mixture also seems to reduce precipitation of the annoying osmium tetroxide crystals (Hayat, 1981). With a mixture of 0.83% glutaraldehyde and

0.67% osmium tetroxide the notoriously difficult to handle slime mould *Dictyostelium discoideum* was successfully preserved with excellent morphology of the highly dynamic Golgi apparatus (Schwarz, 1973).

The downside of the mixture of glutaraldehyde and osmium tetroxide is the depth of penetration into tissue. In 24 h at 4 °C the mixture of glutaraldehyde and osmium tetroxide in phosphate buffer penetrated 0.4 mm whereas osmium tetroxide alone penetrated 2.4 mm into rat kidneys (Hopwood, 1970). Furthermore, in the combined solution the majority of osmium tetroxide has turned into osmium black (White *et al.*, 1976), the non-fixing form of osmium tetroxide.

10.7.4 Osmium Tetroxide–Potassium Ferrocyanide Staining

Potassium ferricyanide, Fe(III), (IUPAC: potassium hexacyanoferrate(III), also known as red Prussiate of Potash) was introduced by Elbers, Ververgaert and Demel (1965) as a potential fixative to preserve phospholipids in brain tissue and by de Bruijn to stain glycogen (1973). Later de Bruijn used potassium ferrocyanide, Fe(II), (IUPAC: potassium hexacyanoferrate(II), also known as yellow Prussiate of Potash) and osmium tetroxide to fix glycogen in liver tissue (De Bruijn, 1995; De Bruijn and den Breejen, 1975). Next to the staining of the glycogen it turned out that this mixture has a stabilising effect on microfilaments and is an excellent stain for membranes (McDonald, 1984), especially for sarcoplasmic reticulum and glycocalyx (White, Mazurkiewicz and Barrnett, 1979). The staining of membranes, however, is variable and depends on the organism (Schnepf, Hausmann and Herth, 1982). Again using X-ray photoelectron spectroscopy, White *et al.* (1979) proposed a staining mechanism. By adding ferrocyanide to osmium tetroxide the components equilibrate into the following products: OsO_4, $Fe(CN)_6^{-4}$, $Fe(CN)_6^{-3}$, $OsO_2(OH)_4^{-4}$ and the labile cyano-bridged Fe–Os species in their VIII, VII and VI oxidation stages (White *et al.*, 1979). They speculate that the greater reactivity of Os (VII and VI) intermediates lead to more Os deposition than OsO_4 alone and that the chelation of Os is responsible for the immobilisation of osmium tetroxide and the more intense staining (White *et al.*, 1979). Osmium tetroxide in combination with potassium ferrocyanide is often called reduced osmium tetroxide in the literature.

10.7.5 Osmium Tetroxide–Thiocarbohydrazide–Osmium Tetroxide (OTO)

Because of the poor contrast of osmium tetroxide fixed cells embedded in epoxy resins, Seligman and co-workers set out to find histological methods to increase the contrast of lipids and discovered the osmophilic reagent thiocarbohydrazide (Seligman *et al.*, 1966). After osmium tetroxide fixation thiocarbohydrazide attaches to the bound osmium and serves as a new binding site for fresh osmium tetroxide. In this way the contrast of osmophylic compounds, especially lipids, could greatly be enhanced. There is evidence that thiocarbohydrazide also reduces the osmium tetroxide artefacts and the application of the OTO technique results in better preservation of actin filaments (Aoki and Tavassoli, 1981) and the organ of Corti (Davies and Forge, 1987). Today, there is a revival of the OTO technique sparked by the needs of the block-face scanning microscopy techniques for volume microscopy.

10.7.6 Osmium Tetroxide and Tannic Acid

Already in the nineteenth century botanists had observed that osmium tetroxide reacted with polyphenols in an intense blue-black colour and Bolles Lee (1887) suggested to use this reaction for histochemistry, which led to a method to stain endothelial cells using osmium tetroxide with pyrogallic acid (Kolossow, 1892). The combination of osmium tetroxide and gallic acid to stain fat droplets and mitochondria for light microscopy was reinvestigated by Wigglesworth (1957). In 1976, Simionescu and Simionescu extensively tested the application of galloylglucoses to improve cell compartment contrast, mainly membrane contrast (Simionescu and Simionescu, 1976a, 1976b). The best and most reproducible effect they got was with the well-defined tannic acid from nutgalls that consisted predominately of penta- and hexagalloyl glucoses (Allepo) or hepta- and decagalloyl glucoses (oriental) (Simionescu and Simionescu, 1976a). They suggest that tannic acid is best applied after aldehyde and osmium tetroxide fixation before dehydration. The tannic acid treatment after osmium tetroxide fixation predominately enhanced the contrast of membranes (Blanchette-Mackie and Scow, 1981; Simionescu and Simionescu, 1976a) but also better preserved the ultrastructure of actin filaments (Maupin and Pollard, 1983) and collagen fibrils (Meek and Chapman, 1986). The addition of pyrogallol found a new application to stain brain tissue for volume microscopy (Mikula and Denk, 2015).

For SEM preparations tannic acid reduced shrinkage (Wollweber, Strache and Goethe, 1980) and made the samples electrically conductive for non-coated scanning electron microscopy (Jongebloed *et al.*, 1999; Murakami *et al.*, 1987).

Tannic acid dissolves easily in acetone and is therefore also used in freeze-substitution applications to enhance membrane contrast and preserve cellular fine details including the cytoskeleton (Akisaka and Shigenaga, 1983; Giddings, 2003; Jiménez *et al.*, 2009; Tranfield *et al.*, 2014; Wilson, Farmer and Karwoski, 1998).

10.8 SUMMARY OF ACTION OF FIXATIVES

Jost *et al.* (1983) examined the efficiency of fixation on the motion of proteins and lipids in membranes. Their results suggest that formaldehyde has no influence, that glutaraldehyde does immobilise proteins but not lipids and that osmium tetroxide greatly reduces the mobility of lipids. From these results it is concluded that formaldehyde is a very weak fixative and glutaraldehyde preserves proteins and osmium tetroxide lipids. As mentioned above, buffered formaldehyde is not an efficient fixative and it retains phosphatidyl ethanol amine as the only lipid (Roozemond, 1969); therefore it has been suggested to replace formaldehyde by acrolein (Jones, 1973).

In summary, proteins are preserved by aldehydes, lipids and fatty acids are stabilised by osmium tetroxide and acrolein, uranyl actetate and malachite green can fix phospholipids and formaldehyde and uranyl acetate are the main fixatives for nucleic acids. To reduce sample artefacts, it is suggested to proceed through the fixation protocol without storage in buffer. When needed the samples can be stored after the aldehyde fixation in 1% formaldehyde, for example for shipping of samples designated for immunolabelling, or overnight storage in 0.5–2% aqueous uranyl acetate.

10.9 BUFFERS

The purpose of a buffer is to counteract pH changes that will occur during the reaction of fixatives with the tissues and counteract any acids formed during the ageing of the fixatives. If the pH changes are not counteracted, then artefacts are likely to occur. Interestingly

Table 10.1 Comparison of different buffers.

Buffer	pKa	pH buffering range
PIPES	6.8	6.1 – 7.5
HEPES	7.55	6.8 – 8.2
Phosphate	7.21	5.8 – 8.0
Cacodylate	6.27	5.0 – 7.2
MES	6.15	5.5 – 6.7

Table extracted from the 2008 handout of AppliChem GmbH.

enough, the expensive and poisonous arsenic containing (Weakley, 1977) sodium cacodylate buffer is still the most used buffer in electron microscopy labs. It first appeared in 1948 to replace phosphate buffer in the pH range of 6.0 to 7.0 to study the kinetics of phosphatases (Plumel, 1948). Soon thereafter it must have found its way into electron microscopy, probably as it does not cause precipitations of osmium crystals. Unfortunately, cacodylate buffer is used at the very edge of its buffering range, which is pH 5.0–7.2. This makes cacodylate buffer a very poor choice (Griffiths, 1993) and it would be much better to replace it with the so-called Good buffers (Good *et al.*, 1966) that have been standard in biochemical labs for a long time. In the 2008 handout of AppliChem it says cacodylate is very toxic and is nowadays usually replace by MES, 2-(*N*-morpholino)-ethanesulfonic acid (AppliChem GmbH, Darmstadt, Germany).

An effective buffer is used close to its pKa within about 0.2–0.3 pH units (Griffiths, 1993; Johnson, 1985), which makes HEPES, *N*-(2-hydroxyethyl)-piperazine-*N′*-ethanesulfonic acid and phosphate buffer a good choice, and PIPES, piperazine-*N,N′*-bis(2-ethanesulfonic acid) (Salema and Brandão, 1973) a reasonable choice (see Table 10.1). There is a combination buffer called PHEM, which is composed of 60 mM PIPES, 25 mM HEPES, 10 mM EDTA and 2 mM $MgCl_2$ (Schliwa, van Blerkom and Porter, 1981). This buffer was initially used for cytoskeletal studies but it works very well for general electron microscopy sample preparation (Vale-Costa *et al.*, 2016; Van Donselaar *et al.*, 2007). The widely used TRIS, tris(hydroxymethyl)-aminomethane, buffer cannot be used for electron microscopy as its primary amino group will react with aldehydes!

10.10 WATER SOURCE

The water source for buffers and fixatives is important to consider. Most people will use purified water for experiments but it is important to understand there are different approaches to remove the impurities from water, which results in different kinds of impurities remaining. Not all kinds of purified water are interchangeable.

To make glass-distilled water, source water is put into a glass container and heated from below. The steam has been collected, cooled and condensed into glass-distilled water. In this case, minerals (such as salts), microorganisms and small particles are removed from the water. However, any compounds volatile at under 100 °C remain in the water. Double distilled water has gone through this process twice.

To make deionised water, source water is processed through an electrically charged resin. Typically both positive and negative charged resins are used, which results in any cations and anions in the water being exchanged with H^+ and OH^- respectively in the resins. The final result is the production of pure water. However, deionisation of water does not remove the uncharged particles including small microorganisms or small compounds that are uncharged and dissolve in water. Furthermore, deionised water will continue to change as it contacts air. Freshly prepared deionised water has a pH of 7, but as soon as it contacts carbon dioxide from the air, the carbon dioxide will dissolve in the water and react with the water to produce H^+ and HCO^{3-}. The end result is that briefly aged deionised water tends to be slightly acidic. However, the pH will be adjusted back to 7.2–7.4 during the preparation of the buffer solution.

To make reverse osmosis water, source water is pushed against a membrane with very small pores that are capable of blocking virtually all particles, bacteria and organics that are larger than 200 daltons. To counteract the osmotic pressure of the water filtration system, hydraulic pressure is used to push the water through the pore at a reasonable rate. The end result is very pure water.

To achieve the best water for a buffer or fixative solution the combination of several of these techniques is suggested. It is also suggested that regular maintenance should be performed on the water purification systems to maintain high quality purification.

10.11 SEM PREPARATION

10.11.1 Critical-Point Drying and Room Temperature Preparation

Water molecules (the major component of biological tissues) are incompatible with vacuum environments and therefore with electron microscopy imaging. Water has a high surface tension and as it transitions from liquid to gaseous phase during drying the forces of surface tension can adversely affect the specimen nano- and microultrastructure. Many strategies to preserve structure and minimise or avoid these forces have been developed to support high resolution SEM imaging. Hydrated samples can be imaged with environmental, variable pressure, scanning electron microscopes (Jornsanoh *et al.*, 2011; Kirk, Skepper and Donald, 2009; Muscariello *et al.*, 2005; Stokes, 2003), discussed in Chapter 20 by Reimer, Eggert and Hohenberg in Volume II, or by cryo-scanning electron microscopy (Walther, 2003; Walther and Müller, 1999; Walther *et al.*, 1995), discussed in Volume II in Chapter 17 by Hazekamp and van Ruijven and Chapter 18 by Walther.

However, the main approach is critical-point drying (CPD), which is a highly effective method to preserve ultrastructure and provide a room temperature stable sample for imaging in the SEM. The sample itself is also normally conductive (due to chemical processing prior to drying), which further aids high-resolution imaging. The 'critical point' is where the physical characteristics of liquid and gaseous states are not distinguishable (Burstyn and Bartlett, 1975). Compounds that at this critical point can be converted into the liquid or gaseous phase without crossing the interfaces between liquid and gaseous avoiding the damaging effects of the forces of surface tension. The critical point of water lies at a point (T_c 374 °C, P_c 229 bar) where biological samples would be damaged beyond recognition prior to drying. Thus an alternative gas:liquid transition system is required. Liquid carbon dioxide (CO_2) is more suitable for biological applications whose critical point lies at 31 °C and 74 bar (Burstyn and Bartlett, 1975). As CO_2 is less miscible with water, for the CO_2 to act as an effective transitional fluid for critical-point drying, the water in biological tissue is best replaced with an organic solvent (e.g. ethanol (T_c 241 °C, P_c 60 bar) or acetone (T_c 235 °C, P_c 46 bar)) miscible in both water and liquid CO_2 (Burstyn and Bartlett, 1975). Other specialist exchange fluids have been tried, for example amyl acetate for pollen (Lynch and Webster, 1975). Once the water in the sample is substituted with the exchange fluid, the exchange fluid is in turn replaced with the transition fluid (the liquid CO_2). Once complete, the CPD chamber is brought to the critical point for CO_2 and the liquid is converted to the gaseous phase by decreasing the pressure at the constant critical point temperature. It is important to consider the risk of some air drying of a specimen during the chemical fixation steps and the CPD loading steps. To mitigate these Baigent (Baigent *et al.*, 1978) developed an apparatus to avoid exposing a specimen to any air transfer steps during processing, thus avoiding any air drying and preserving ultrastructure. The construction and operation of the apparatus are described by Hockley and Jackson (2000) and Hockley, Jackson and Fleck (2001).

An alternative to critical-point drying is air drying of hexamethyldisilazane (HMDS) treated samples (Braet, De Zanger and Wisse, 1997; Heegaard, Jensen and Prause, 1986). Comparing HMDS drying with critical-point drying suggested that the overall morphology for biofilms (Araujo *et al.*, 2003), cell culture cells (Braet et al., 1997; Katsen-Globa *et al.*, 2016) and organs (Bray, Bagu and Koegler, 1993) are of equal quality; only plant material dried through HMDS suffered considerable shrinkage (Bray et al., 1993).

After drying, the samples are metal coated for further stabilisation and reducing charging artefacts. High resolution metal coating is discussed in Chapter 12 by Tacke *et al.* in Volume I.

Dehydration of biological samples can lead to gross alterations of the cellular structure. The most important point is shrinkage (Boyde, 1980; Boyde and Boyde, 1980; Boyde and Maconnachie, 1979, 1980, 1984). The group of Boyde could demonstrate that their test sample, glutaraldehyde fixed mouse limb, shrinks by about 30% in volume during dehydration (Boyde, 1980; Boyde and Maconnachie, 1979) and up to 60% in volume after critical-point drying (Boyde and Maconnachie, 1979). Dehydration in the usual steps of 30%, 50%, 70%, 80%, 90% and 100% of ethanol or in finer steps did not alter the final result. The effect of dehydration could also be shown by MacKenzie (1972) and the group of Gross (Gross, 1987; Wildhaber, Gross and Moor, 1982) by the fact that removing the last 50–30% of water the cellular structures, even on the molecular level, collapsed. Freeze drying is a much better approach, where shrinkage is only in the range of 10% (Boyde, 1980).

10.11.2 Ionic Liquids

The first air- and water-stable ionic liquid, 1-ethyl-3-methylimidazolium tetrafluoroborate, was reported by Wilkes and Zaworotko in 1992 (Arimoto *et al.*, 2008; Wilkes and Zaworotko, 1992) (see Figure 10.10). They have interesting unique properties (Torimoto *et al.*, 2010):

(a) They are molten salts, composed of cations and anions.
(b) They are fluid at room temperature.
(c) They have a high electrical conductivity.
(d) They have a negligible vapour pressure and do not evaporate under vacuum conditions.
(e) They are biocompatible (Tsuda *et al.*, 2011).

These favourable characteristics, electrical conductivity and very low vapour pressure offer an alternative for biological sample preparation for scanning electron microscopy. A biological sample impregnated with hydrophilic ionic liquid should become stable under vacuum conditions, be electrically conductive and possibly show less shrinkage. First experiments with 1-butyl-3-methylimidazolium tetrafluoroborate impregnated seaweed

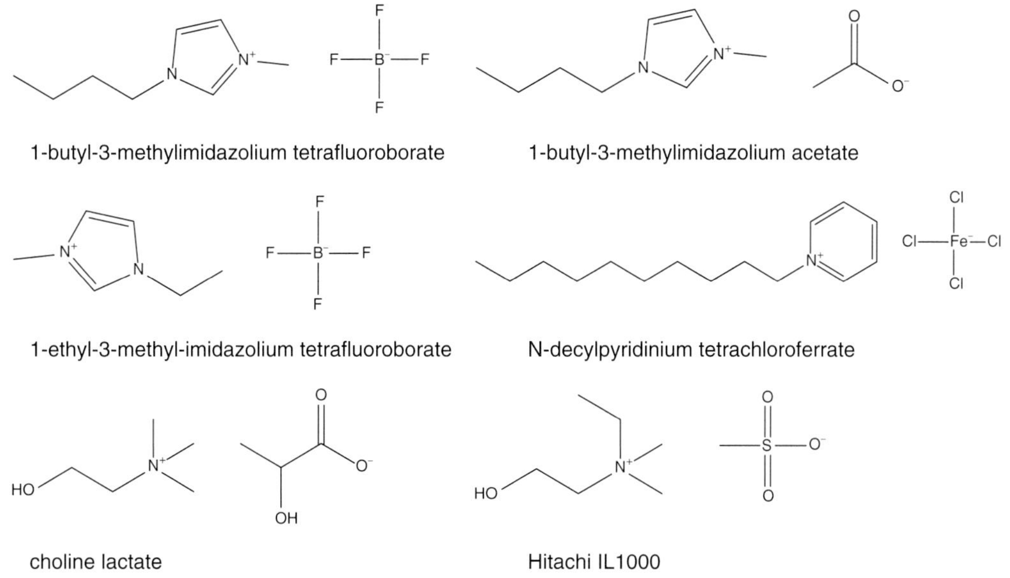

1-butyl-3-methylimidazolium tetrafluoroborate

1-butyl-3-methylimidazolium acetate

1-ethyl-3-methyl-imidazolium tetrafluoroborate

N-decylpyridinium tetrachloroferrate

choline lactate

Hitachi IL1000

Figure 10.10 Hydrophilic ionic liquids: 1-butyl-3-methylimidazolium tetrafluoroborate (Kuwabata *et al.*, 2006; Torimoto *et al.*, 2010); 1-butyl-3-methylimidazolium acetate (Asahi *et al.*, 2015); 1-ethyl-3 methyl-imidazolium tetrafluoroborate (Ishigaki *et al.*, 2011a, 2011b; Wilkes and Zaworotko, 1992); *N*-decylpyridinium tetrachloroferrate (Zhuravlev, Ivanova and Grechishkin, 2015); choline lactate (Asahi *et al.*, 2015; Komai *et al.*, 2014); Hitachi IL1000 (Bittermann, Rodighiero and Wepf, 2016; Joubert *et al.*, 2017; Nimura *et al.*, 2014; Sakaue *et al.*, 2014; Shiono *et al.*, 2014).

were successfully performed by Kuwabata *et al.* (Arimoto *et al.* 2008; Kuwabata *et al.*, 2006), described in more detail by Takahashi, Shirai and Fugi (2013).

In recent years the following ionic liquids have emerged for electron microscopy: 1-butyl-3-methylimidazolium acetate (Asahi *et al.*, 2015), 1-ethyl-3-methyl-imidazolium tetrafluoroborate (Wilkes and Zaworotko, 1992), *N*-decylpyridinium tetrachloroferrate (Zhuravlev *et al.*, 2015), choline lactate (Komai *et al.*, 2014) and the commercially available Hitachi IL1000 (Nimura *et al.*, 2014). Ionic liquids were successfully applied to several biological samples like biofilms of *Streptococcus mutans* (Asahi *et al.*, 2015) and *Aspergillus fumigatus* (Joubert et al. 2017); crustacean (Shiono *et al.*, 2014); *Helicobacter bilis* (Sakaue *et al.*, 2014); pollen (Komai *et al.*, 2014; Tsuda *et al.*, 2011); cell culture cells (Asahi *et al.*, 2015; Bittermann *et al.*, 2016; Ishigaki *et al.*, 2011a, 2011b; Tsuda et al, 2011) and intestinal villus (Tsuda *et al.*, 2011). The hydrophilic liquids seem to be the better choice (Arimoto *et al.*, 2008; Ishigaki *et al.*, 2011a). Compared to critical-point drying, ionic liquid prepared samples show better intracellular morphology (Bittermann *et al.*, 2016).

When using ionic liquids care has to be taken that the liquid is completely removed from the surfaces in order not to obscure fine surface details. It is suggested to use low concentrations of the ionic liquid (Ishigaki *et al.*, 2011a) or to remove it mechanically by centrifugation (Takahashi *et al.*, 2013). Further, at higher magnifications charging can occur, which might be the reason that in current publications no high resolution data are shown. The ionic liquid is certainly an exciting new tool for sample preparation in electron microscopy but more investigations on the type of liquid, the concentration used and the imaging conditions are needed.

10.11.3 Resin

Typically scanning electron microscopy is applied to study surfaces. With the invention of array tomography (Micheva and Smith, 2007), serial block-face scanning electron microscopy (SBF-SEM) (Denk and Horstmann, 2004; Leighton, 1981) and the application of focused ion beam scanning electron tomography in biology (FIB-SEM) (Knott *et al.*, 2008), however, SEMs are now frequently used to image sections or the remaining block-face of resin embedded biological samples. Therefore, it seems necessary to include resins into this chapter.

There are two types of resin used in electron microscopy: epoxy resins (Lee and Neville, 1967) and methacrylates (Newman and Hobot, 2001). Epoxy resins react through their epoxy group and can covalently bind to many components of a biological sample (Causton, 1986), leading to a co-polymer (see Figure 10.11). The potential reaction partners are carboxyl groups of glutamic and aspartic acids and the C terminus of proteins; phenol groups, phenols in plants and tyrosine; amino groups, lysine and N-terminus of proteins; imidazole, histidine; guanidyl groups, arginine; indole groups, tryptophan; amide groups, glutamine and asparagine; hydroxyl groups, serine, theronine and possibly alcohols; and sulfhydryl groups in cystine (Causton, 1986). Therefore epoxy resins can also be used as a fixative (Matsko and Müller, 2005; Sung *et al.*, 1996). Small peptides, such as, for example, the neurotransmitter mesotocin, Cys–Tyr–Ile–Gln–Asn–Cys–Pro–Ile–Gly–NH_2, could not be fixed by glutaraldehyde. Only by embedding into Epon is it possible to localise mesotocin in the pituitary gland of Xenopus (Wang, Humbel and Roubos, 2005). Epon can react with the N-terminus, Cys, Tyr, Gln, Asn, and hence with six of the nine amino acids (Causton, 1986).

Methacrylates polymerise with a radical reaction initiated by a heat, benzoyl peroxide (Franklin, 1984), or UV, benzoin methylether (Carlemalm, Garavito and Villiger, 1982), inducible starter chemical (see Figure 10.12). Therefore the methacrylate components cross-link only with themselves with very little interaction with biological macromolecules (Causton, 1986). The embedding is more like trapping the biological components in the meshwork of the polymer, like fishes in a net.

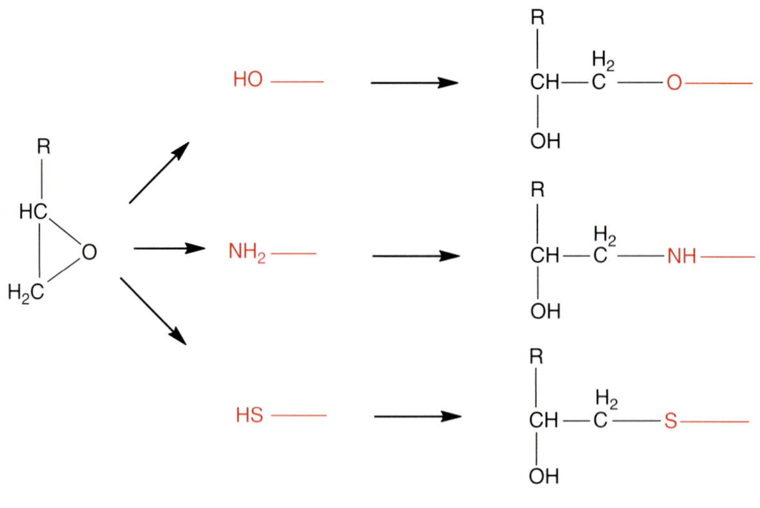

Figure 10.11 Epoxy reaction with hydroxyl, primary amines and sulfhydryl groups. Redrawn from Causton (1986).

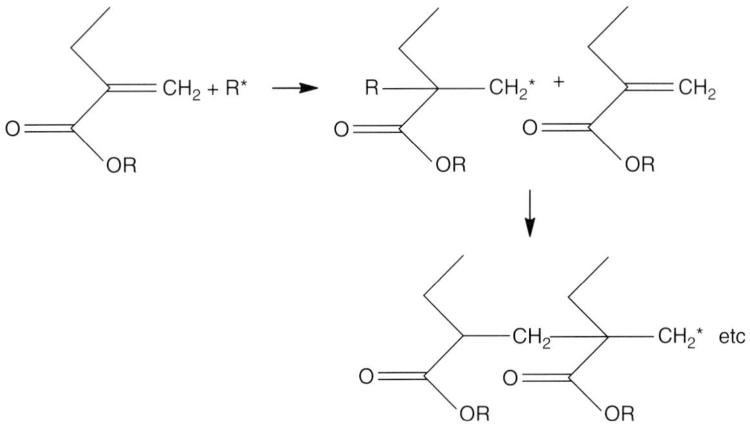

Figure 10.12 R* is the initial radical that is produced by the decay of the initiator of the reaction, benzoyl peroxide or benzoin methylether. Redrawn from Causton (1981).

Widely used epoxy resins are Araldite (Glauert and Glauert, 1958), Epon (Luft, 1961), Spurr's low viscosity resin (Spurr, 1969) and Durcupan (Kushida, 1964, 1965; Kushida, Nagato and Kushida, 1981; Stäubli, 1963).

The methacrylates are Lowicryl (Carlemalm *et al.*, 1982, 1986), LR white (Ellinger and Pavelka, 1985; Newman and Hobot, 1987; Newman and Jasani, 1984) and LR gold (Bendayan, Nanci and Kan, 1987). For a comprehensive study of the different methacrylates, their chemistry and application consult the excellent monograph of Newman and Hobot (2001). Also note that monomers of the resins can be toxic, carcinogenic, irritants of skin and mucosa and environmental pollutants (Causton, 1981; Tobler and Freiburghaus, 1990).

10.12 SOME THOUGHTS ON THE PREPARATION PROTOCOLS FOR VOLUME MICROSCOPY

There are two different approaches for volume microscopy:

1. Imaging the sections, that is serial sectioning and array tomography.
2. Imaging the remaining block face, discarding the sections, that is SBF-SEM and FIB-SEM.

There are no special needs when the sections are imaged as they can be treated and stained like ordinary sections with uranyl acetate and lead citrate (Reynolds, 1963; Venable and Coggeshall, 1965) before imaging in a TEM or SEM (Schwarz and Humbel, 2014).

For block face imaging, however, there are a few constraints: the sample has to be stained throughout the block; it needs a lot of stain, say high concentrations of heavy metals, to allow for fast image acquisition; and it should be electrically conductive to avoid charging. Thick sections used for TEM tomography are somewhere in between as on-section staining is not always successful in staining through a thick resin layer (Peters, Hinds and Vaughn, 1971; Shalla, Carroll and DeZoeten, 1964). We stained a thick Epon section of cells, freeze-substituted without heavy metals, on-section with uranyl acetate and lead citrate. In the tomogram it became obvious that the stain did not penetrate deeply into the section (Tomova and Humbel, unpublished), which is in agreement with previously published observations (Locke, Krishnan and McMahon, 1971).

The first protocols for volume microscopy used standard preparation methods: aldehyde fixation followed by osmium tetroxide and/or osmium tetroxide modified with potassium ferri or ferrocyanide (De Bruijn, 1973, 1995; De Bruijn and den Breejen, 1975) and embedding in resins, for example Durcupan (Knott *et al.*, 2008). Obviously, FIB-SEM tomography is less affected by charging than SBF-SEM; a thin layer of about 5 nm of platinum sputtering is sufficient to prevent major charging effects. It may be speculated that gallium ions are deposited on to the sample during cutting and that this thin layer provides sufficient conductivity.

For SBF-SEM, however, several labs have established their own protocols (Briggman, Helmstaedter and Denk, 2011; Hua, Laserstein and Helmstaedter, 2015; Tapia *et al.*, 2012; Thai *et al.*, 2016; Wilke *et al.*, 2013); see Chapter 23 by Genoud in Volume II. Interestingly, these protocols are very similar, using the same steps and chemicals and differ mainly in incubation time and temperature.

The main element of these protocols is the so-called OTO technique, osmium tetroxide–thiocarbohydrazide–osmium tetroxide introduced by Seligman (Seligman *et al.*, 1966) to enhance the contrast of lipids. Concentrations of osmium tetroxide is 1% or 2% and before thiocarbohydrazide (TCH) treatment, osmium tetroxide is usually reacted with potassium ferri (III) or ferro (II) cyanide (Briggman *et al.*, 2011; Hua *et al.*, 2015; Thai *et al.*, 2016; Wilke *et al.*, 2013). TCH is used at concentrations of 0.1% to 1%. Note that at 24.7 °C the solubility of TCH in water is only 0.55% (Kurzer and Wilkinson 1970) (see Table 10.2). This is probably the reason why Seligman made the 1% TCH solution at 50 °C (Seligman *et al.*, 1966).

In addition, the above-mentioned groups adapted the *en bloc* staining with lead aspartate, as introduced by Judie Walton (1979) to follow the *en bloc* uranyl acetate staining. The Pb^{2+} ions are buffered with aspartate as a weak chelator such that low concentrations of free Pb^{2+} ions are available over a long period of time, resulting in an even staining throughout the tissue block, without the danger of large precipitates that are known for lead citrate staining (Walton, 1979). The original recipe stains for 1 hour at 60 °C. It might be interesting to combine the previously mentioned whole mount uranyl staining (Locke, 1994; Terzakis, 1968) with the whole mount lead staining.

These preparation protocols have become essential to study the connectome (sum of neuronal connections) in the brain. The membranes and the vesicles are clearly and heavily stained, facilitating (semi)automatic segmenting of neurons and synapses (Cardona *et al.*, 2012; Hayworth *et al.*, 2015; Helmstaedter, Briggman and Denk, 2011; Kasthuri *et al.*, 2015; Kreshuk *et al.*, 2015; Mikula, Binding and Denk, 2012; Mikula and Denk, 2015).

Using these elegant protocols, however, the cells look rather empty. Going back to the action of osmium tetroxide, the empty cells might be explained by the proteolytic activity of osmium tetroxide (Behrman, 1984; Maupin-Szamier and Pollard, 1978; Riva et al., 1999; Tanaka and Mitsushima, 1984). Hence, when intracelluar structures need to be studied more reliable protocols, preferably based on cryo-fixation, should be applied (see Chapter 11 by Fleck, Shimoni, Humbel in Volume I). An attempt in this direction has been done by combining cryo-fixation by high-pressure freezing with freeze-substitution with 0.1% glutaraldehyde and 2% osmium tetroxide in acetone ending at 0 °C. Then the substitution fluid was replaced by 1% tannic acid (Simionescu and Simionescu, 1976a) in acetone, again followed by osmium tetroxide in acetone, an OTO technique where T stands for tannic acid (Jiménez et al., 2009). There are many more variations thinkable, for example to combine cryo-fixation and freeze-substitution chemical fixation with a rehydration protocol (Ripper, Schwarz and Stierhof, 2008; Stierhof *et al.*, 2008; Van Donselaar *et al.*, 2007), such as,

Table 10.2 Solubilities of thiocarbohydrazide.

Solvent	Temperature in °C	Solubility, *x* g/100 g
Water	0	0.18
Water	24.7	0.55
Ethanol	24.7	0.26
Chloroform	24.7	0.05
Carbon tetrachloride	24.7	0.03
Hydrazine hydrate	24.7	13.60

From Kurzer and Wilkinson (1970).

for example, potassium ferroyanide and thiocarbohydrazide, which have a low solubility in organic solvents (Kurzer and Wilkinson, 1970) (see Table 10.2).

10.13 CONCLUSION

The moment we want to observe a natural phenomenon we are introducing artefacts by the way we are looking at it and the method and instruments we are using (Vogt, 1972).

It is generally acknowledged that cryo-preparation and cyro-observation methods give the most reliable image of a biological structure. These methods, however, are time consuming and depend on highly sophisticated and expensive equipment. Therefore the so-called 'standard' chemical preparation methods will remain the technique of choice for most electron microscopy labs. This chapter is a wake-up call for microscopists not to blindly follow protocols but to question every single step to prevent mistakes. Much more importantly, knowing the effects of the reagents helps interpreting the results. The chemical reactions discussed are far more complex; here only the basic mechanisms are shown and therefore the rich primary literature should be consulted for full comprehension.

REFERENCES

Akisaka, T. and Shigenaga, Y. (1983) Ultrastructure of growing epiphyseal cartilage processed by rapid freezing and freeze-substitution. *J. Electron Microsc.*, 32, 305–320

Aoki, M. and Tavassoli, M. (1981) OTO method for preservation of actin filaments in electron microscopy. *J. Histochem. Cytochem.*, 29, 682–683.

Araujo, J.C., Téran, F.C., Oliveira, R.A., Nour, E.A.A., Montenegro, M.A.P., Campos, J.R. and Vazoller, R.F. (2003) Comparison of hexamethyldisilazane and critical point drying treatments for SEM analysis of anaerobic biofilms and granular sludge. *J. Electron Microsc.*, 52, 429–433.

Arimoto, S., Sugimura, M., Kageyama, H., Torimoto, T. and Kuwabata, S. (2008) Development of new techniques for scanning electron microscope observation using ionic liquid. *Electrochim. Acta*, 53, 6228–6234.

Asahi, Y., Miura, J., Tsuda, T., Kuwabata, S., Tsunashima, K., Noiri, Y., Sakata, T., Ebisu, S. and Hayashi, M. (2015) Simple observation of *Streptococcus mutans* biofilm by scanning electron microscopy using ionic liquids. *AMB Express*, 5, 6.

Baigent, C.L., Mannweiler, K., Andresen, I., Rutter, G.A. and Neumayer, U. (1978) Verbessertes und vereinfachtes 'Austauschverfahren' zur Präparation von infektiösen Deckglaszellkulturen für vergleichende REM- und TEM-Studien. *Beitr.* Elektronenmikroskop Direktabb. Oberfl., 11, 251–263.

Baschong, W., Baschong-Prescianotto, C., Wurtz, M., Carlemalm, E., Kellenberger, C. and Kellenberger, E. (1984) Preservation of protein structures for electron microscopy by fixation with aldehydes and/or OsO_4. *Eur. J. Cell Biol.*, 35, 2126.

Behrman, E.J. (1984) The chemistry of osmium tetroxide fixation, in *The Science of Biological Specimen Preparation 1983* (eds J.P. Revel, T. Barnard and G.H. Haggis), AMF O'Hare, SEM Inc., IL, pp. 1–5.

Beisker, W., Dolbeare, F. and Gray, J.W. (1987) An improved immunocytochemical procedure for high-sensitivity detection of incorporated bromodeoxyuridine. *Cytometry*, 8, 235–239.

Bendayan, M., Nanci, A. and Kan, F.W.K. (1987) Effect of tissue processing on colloidal gold cytochemistry. *J. Histochem. Cytochem.*, 35, 983–996.

Bittermann, A.G., Rodighiero, S., and Wepf, R.A. (2016) Ionic liquids for biological SEM and FIB/SEM, in *16th European Microscopy Congress*, 2016, Lyon, France.

Blanchette-Mackie, E.J. and Scow, R.O. (1981) Membrane continuities within cells and intercellular contacts in white adipose tissue of young rats. *J, Ultrastruct. Res.*, 77, 295–318.

Blum, F. (1893) Der Formaldehyde als Härtungsmittel. *Zeitschrift für wissenschaftliche Mikroskopie und Mikroskopische Technik*, 10, 314–315.

Bolles Lee, A. (1887) *La Cellule*, 4, 107–133.

Bouin, P. (1897) Etudes sur l'évolution normale et l'involution du tube séminifère. *A. Anat. Micr.*, 1, 225–339.

Boyde, A. (1980) Review of basic preparation techniques for biological scanning electron microscopy, in *Proceedings of the 7th Congress on Electron Microscopy* (eds P. Brederoo and W. Priester), 1980, The Hague, pp. 768–777.

Boyde, A. and Boyde, S. (1980) Further studies of specimen volume changes during processing for SEM: including some plant tissue. *Scanning Electron Microsc.*, 2, 117–124.

Boyde, A. and Maconnachie, E (1979) Volume changes during preparation of mouse embryonic tissue for scanning electron microscopy. *Scanning*, 2, 149–163.

Boyde, A. and Maconnachie, E. (1980) Treatment with lithium salts reduces ethanol dehydration shrinkage of glutaraldehyde fixed tissue. *Histochemistry*, 66, 181–187.

Boyde, A. and Maconnachie, E. (1984) Not quite critical point drying, in *The Science of Biological Specimen Preparation 1983* (eds J.P. Revel, T. Barnard and G.H. Haggis), AMF O'Hare, SEM Inc., IL, pp. 71–75.

Braet, F., De Zanger, R. and Wisse, E. (1997) Drying cells for SEM, AFM and TEM by hexamethyldisilazane: a study on hepatic endothelial cells. *J. Microsc.*, 186, 84–87.

Bray, D.F., Bagu, J. and Koegler, P. (1993) Comparison of hexamethyldisilazane (HMDS), Peldri II, and critical-point drying methods for scanning electron microscopy of biological specimens. Microsc. Res. Tech., 26, 489–495.

Briggman, K.L., Helmstaedter, M. and Denk, W. (2011) Wiring specificity in the direction-selectivity circuit of the retina. *Nature*, 471, 183–188.

Brink, M., Humbel, B.M., de Kloet, E.R. and van Driel, R. (1992) Evidence against the model of nuclear translocation for the glucocorticoid receptor. *Endocrinology*, 130, 3575–3581.

Burstyn, H.P. and Bartlett, A.A. (1975) Critical point drying: application of the physics of the PVT surface to electron microscopy. *Am. J. Phys.*, 43, 414–419.

Cardona, A., Saalfeld, S., Schindelin, J., Arganda-Carreras, I., Preibisch, S., Longair, M., Tomancak, P., Hartenstein, V. and Douglas, R.J. (2012) TrakEM2 software for neural circuit reconstruction. *PLoS ONE*, 7, e38011.

Carlemalm, E., Garavito, R.M. and Villiger, W. (1982) Resin development for electron microscopy and an analysis of embedding at low temperature. *J. Microsc.*, 126, 123–143.

Carlemalm, E., Villiger, W. and Kellenberger, E. (1986) New resins for low temperature embedding that are designed for immunocytochemistry and for observing unstained biological material. *J. Electron Microsc.*, 35, 2167–2168.

Causton, B.E. (1981) Resins: toxicity, hazards and safe handling. *Proc. RMS*, 16, 265–267.

Causton, B.E. (1986) Does the embedding chemistry interact with tissue? in *The Science of Biological Specimen Preparation 1985* (eds M. Müller, R.P. Becker, A. Boyde and J.J. Wolosewick, AMF O'Hare, SEM Inc., IL, pp. 209–214.

Clancy, B. and Cauller, L.J. (1998) Reduction of background autofluorescence in brain sections following immersion in sodium borohydride. *Journal of Neuroscience Methods*, 83, 97–102.

Corrodi, H., Hillarp, N.-Å. and Jonsson, G. (1964) Fluorescence methods for the histochemical demonstration of monoamines. 3. Sodium borohydride reduction of the fluorescent compounds as a specificity test. *J. Histochem. Cytochem.*, 12, 582–586.

Coulter, H.D. and Terracio, L. (1977) Preparation of biological tissues for electron microscopy by freeze-drying. *Anat. Rec.*, 187, 477–493.

Davies, S. and Forge, A. (1987) Preparation of the mammalian organ of Corti for scanning electron microscopy. *J. Microsc.*, 147, 89–101.

De Bruijn, W.C. (1973) Glycogen, its chemistry and morphologic appearance in the electron microscope: I. A modified OsO_4 fixative which selectively contrasts glycogen. *J. Ultrastruct. Res.*, 42, 29–50.

De Bruijn, W.C. (1995) Contrast staining with reduced osmium complexes. *J. Histochem. Cytochem.*, 43, 965–966

De Bruijn, W.C. and den Breejen, P. (1975) Glycogen, its chemistry and morphological appearance in the electron microscope II. The complex formed in the selective contrast staining of glycogen. *Histochem. J.*, 7, 205–229.

Denk, W. and Horstmann, H. (2004) Serial block-face scanning electron microscopy to reconstruct three-dimensional tissue nanostructure. *PLoS Biol.*, 2, 1900–1909.

Dierichs, R. (1979) Ruthenium red as a stain for electron microscopy. Some new aspects of its application and mode of action. *Histochemistry*, 64, 171–187.

Dykstra, M.J. and Aldrich, H.C. (1978) Successful demonstration of an elusive cell coat in amebae. *J. Protozool.*, 25, 38–41.

Ebersold, H.R., Cordier, J.-L. and Lüthy. P. (1981) Bacterial mesosomes: method dependent artifacts. *Arch. Microbiol.*, 130, 19–22.

Edelmann, L. (1978) A simple freeze-drying technique for preparing biological tissue without chemical fixation for electron microscopy. *J. Microsc.*, 112, 243–248.

Edelmann, L. (1986) Freeze-dried embedded specimens for biological microanalysis. *Scanning Electron Microsc.*, IV,1337–1356.

Edelmann, L. (2002) Freeze-dried and resin-embedded biological material is well suited for ultrastructure research. *J. Microsc.*, 207, 5–26.

El-Saggan, A.H. and Uhrík, B. (2002) Improved staining of negative binding sites with ruthenium red on cryosections of frozen cells. *Gen. Physiol. Biophys.*, 21, 457–461.

Elbers, P.F., Ververgaert, P.H.J.T. and Demel, R. (1965) Tricomplex fixation of phospholipids. *J. Cell Biol.*, 24, 23–30.

Ellinger, A., and Pavelka, M. (1985) Post embedding localization of glycoconjugates by means of lectins on thin sections of tissues embedded in LR white. *Histochem. J.*, 17, 1321–1336.

Fassel, T.A., Van Over, J.E., Hauser, C.C., Buchholz, L.E., Edmiston, C.E. and Sanger, J.R. (1992) Evaluation of bacterial glycocalyx preservation and staining by ruthenium red, ruthenium red-lysine and alcian blue for several methanotroph and staphylococcal species. *Cells and Materials*, 2, 37–48.

Fassel, T.A., Mozdziak, P.E., Sanger, J.R. and Edmiston, C.E. (1998) Paraformaldehyde effect on ruthenium red and lysine preservation and staining of the staphylococcal glycocalyx. *Microsc. Res. Tech.*, 36, 422–427.

Feulgen, R. and Rossenbeck, H. (1924) Mikroskopisch-chemischer Nachweis einer Nucleinsäure vom Typus der Thymonnucleinsäure und die darauf beruhende elektive Färbung von Zellkernen in mikroskopischen Päparaten. *Hoppe-Seyler's Z. Physiol. Chem.*, 135, 203.

Feulgen, R. and Voit, K. (1924a) Über den Mechanismus der Nuclealfärbung. I. Über den Nachweis der reduzierenden Gruppen in den Kernen partiell hydrolysierter mikroskopischer Präparate. *Hoppe-Seyler's Z. Physiol. Chem.*, 135, 249–252.

Feulgen, R. and Voit, K. (1924b) Über den Mechanismus der Nuclealfärbung. II. Über das Verhalten der Kerne partiell hydrolisierter mikroskopischer Präparate zur fuchsinschwefligen Säure nach voraufgegangener Behandlung mit Phenylhydrazin. *Hoppe-Seyler's Z. Physiol. Chem.*, 136, 57–61.

Feulgen, R. and Voit, K. (1924c) Über die für die Nuclealfärbung und Nuclealreaktion verantwortlich zu machenden Gruppen. *Hoppe-Seyler's Z. Physiol. Chem.*, 137, 272–286.

Feustel, E.-M. and Geyer, G. (1966) Zur Eignung der Acroleinfixierung für histochemische Untersuchungen. *Acta Histochem.*, 25, 219–223.

Fitzpatrick, J.A.J., Yan, Q., Sieber, J.J., Dyba, M., Schwarz, U., Szent-Gyorgyi, C., Woolford, C.A., Berget, P.B., Waggoner, A.S. and Bruchez, M.P. (2009) STED nanoscopy in living cells using fluorogen activating proteins. *Bioconjugate Chem.*, 20, 1843–1847.

Franke, W.W., Krien, S. and Brown, R.M. (1969) Simultaneous glutaraldehyde-osmium tetroxide fixation with postosmication. *Histochemie*, 19, 162–164.

Franklin, R.M, (1984) Immunohistochemistry on semi-thin sections of hydroxypropyl methacrylate embedded tissues. *J. Immunol. Methods.*, 68, 61–72.

Gautier, A. (1976) Ultrastructural localization of DNA in ultrathin tissue sections. *Int. Rev. Cytol.*, 44, 113–191.

Giddings, T.H. (2003) Freeze-substitution protocols for improved visualization of membranes in high-pressure frozen samples. *J. Microsc.*, 212:53-61.

Gillett, R. and Gull, K. (1972) Glutaraldehyde – its purity and stability. *Histochemie*, 30, 162–167.

Glauert, A.M. and Glauert, R.H. (1958) Araldite as an embedding medium for electron microscopy. *J. Biophys. Biochem. Cytol.*, 4, 191–194.

Good, N.E., Winget, G.D., Winter, W., Connolly, T.N., Izawa, S. and Sing, R.M.M. (1966) Hydrogen ion buffers for biological research. *Biochemistry.* 5, 467–477.

Griffiths, G. (2009) Chemical fixation for electron microscopy, in *EM in Life Sciences* (eds E.V. Orlova and A.J. Verkleij), 3DEM Network of Excellence.

Griffiths, G. (1993) Fine Structure Immunocytochemistry, Springer-Verlag, Berlin, Heidelberg.

Gross, H. (1987) High resolution metal replication of freeze-dried specimens, in *Cryotechniques in Biological Electron Micoscopy* (eds R.A. Steinbecht and K. Zierold), Springer-Verlag, Berlin, Heidelberg, pp. 203–215.

Handley, D.A., Chen, S. (1981) Oxidation of ruthenium red for use as an intercellular tracer. *Histochemistry*, 71, 249–258.

Hasty, D.L. and Hay, E.D. (1978) Freeze-fracture studies of the developing cell surface. II. Particle-free membrane blisters on glutaraldehyde-fixed corneal fibroblasts are artefacts. *J. Cell Biol.*, 78, 756–768.

Hayat, M.A. (1981) *Fixation for Electron Microscopy*, Academic Press, New York, London, Toronto, Sydney, San Francisco.

Hayat, M.A. (1986) *Basic Techniques for Transmission Electron Microscopy*, Academic Press, Orlando, San Diego, New York, Austin, Boston, London, Sydney, Tokyo, Toronto, p. 411.

Hayat, M.A. (2000) *Principles and Techniques of Electron Microscopy. Biological Applications*, 4th edn, Cambridge University Press, Cambridge.

Hayat, M.A. and Guiquita, R. (1970) Vapor fixation prior to fixation by immerson for electron microscopy. *Proceedings of the 7th International Congress on Electron Microscopy*, p. 391.

Hayworth, K.J., Xu, C.S., Lu, Z., Knott, G.W., Fetter, R.D., Tapia, J.C., Lichtman, J.W. and Hess, H.F. (2015) Ultrastructurally smooth thick partitioning and volume stitching for large-scale connectomics. *Nat. Methods*, 12, 319–322.

Heegaard, S., Jensen, O.A. and Prause, J.U. (1986) Hexamethyldisilazane in preparation of retinal tissue for scanning electron microscopy. *Ophthalmic Research*, 18, 203–208.

Helmstaedter, M., Briggman, K.L. ande Denk, W. (2011) High-accuracy neurite reconstruction for high-throughput neuroanatomy. *Nature Neuroscience*, 14, 1081–1088.

Hockley, D. and Jackson, M. (2000) The Baigent apparatus for continuous-flow processing of SEM specimens. *Proc. RMS*, 35/3, 187–194,

Hockley, D., Jackson, M. and Fleck, R.A. (2001) Using the Baigent apparatus for continuous-flow processing of SEM specimens. *Proc. RMS*, 36/2, 110–115.

Hofmann, A.W. (1869) Beiträge zur Kenntnis des Methylaldehyds. *Journal für Praktische Chemie*, 107, 414–424.

Holt, S.J. and Marian Hicks, R. (1961) Studies on formalin fixation for electron microscopy and cytochemical staining purposes. *J. Cell Biol.*, 11, 31–45.

Hopwood, D. (1970) The reactions between formaldehyde, glutaraldehyde and osmium tetroxide, and their fixation effects on bovine serum albumin and on tissue blocks. *Histochemie*, 24, 50–64.

Hopwood, D. (1972) Theoretical and practical aspects of glutaraldehyde fixation. *Histochem. J.*, 4, 267–303.

Hua, Y., Laserstein, P. and Helmstaedter, M. (2015) Large-volume en-bloc staining for electron microscopy-based connectomics. *Nat. Commun.*, 6.

Humbel, B. and Biegelmann, E. (1992) A preparation protocol for postembedding immunoelectron microscopy of *Dictyostelium discoideum* cells with monoclonal antibodies. *Scanning Microsc.*, 6, 817–825.

Humbel, B. and Müller, M. (1986) Freeze substitution and low temperature embedding, in *The Science of Biological Specimen Preparation 1985* (eds M. Müller, R.P. Becker, A. Boyde, J.J. Wolosewick), AMF O'Hare, SEM Inc., IL, pp. 175–183.

Hündgen, M. (1968) Der Einfluß Verschiedener Aldehyde auf die Strukturerhaltung gezüchteter Zellen und auf die Darstellbarkeit von vier Phosphatasen. *Histochemie*, 15, 46–61.

Huxley, H.E. and Zubay, G. (1961) Preferential staining of nucleic acid-containing structures for electron microscopy. *J. Biophys. Biochem. Cytol.*, 11, 273–296.

Ishigaki, Y., Nakamura, Y., Takehara, T., Kurihara, T., Koga, H., Takegami, T., Nakagawa, H., Nemoto, N., Tomosugi, N., Kuwabata, S. and Miyazawa, S. (2011a) Comparative study of hydrophilic and hydrophobic ionic liquids for observing cultured human cells by scanning electron microscopy. *Microsc. Res. Tech.*, 74, 1104–1108.

Ishigaki, Y., Nakamura, Y., Takehara, T., Nemoto, N., Kurihara, T., Koga, H., Nakagawa, H., Takegami, T., Tomosugi, N., Miyazawa, S., and Kuwabata, S. (2011b) Ionic liquid enables simple and rapid sample preparation of human culturing cells for scanning electron microscope analysis. *Microsc. Res. Tech.*, 74, 415–420.

Ito, S., Karnovsky, M.J. (1968) Formaldehyde-glutaraldehyde fixatives containing trinitro compounds. *J. Cell Biol.*, 39:168A.

Jencks, W.P. (1964) Mechanism and catalysis of simple carbonyl group reactions, in *Progress in Physical Organic Chemistry* (eds S.G. Cohen, A. Streitwieser and R.W. Taft), Interscience Publishers, John Wiley & Sons, New York, London, Sydney.

Jiménez, N., Vocking, K., van Donselaar, E., Humbel, B.M., Post, J.A. and Verkleij, A.J. (2009) Tannic acid-mediated osmium impregnation after freeze-substitution: A strategy to enhance membrane contrast for electron tomography. *J. Struct. Biol.*, 166, 103–106.

Johnson, T.J.A. (1985) Aldehyde fixatives: quantification of acid-producing reactions. *J. Electron Microsc. Tech.*, 2, 129–138.

Jones, D. (1973) Reactions of aldehydes with unsaturated fatty acids during histological fixation, in *Fixation in Histochemistry* (ed. P.J. Stoward), Chapman and Hall Ltd, London, pp. 1–45.

Jongebloed, W.L., Stokroos, I., Van der Want, J.J. and Kalicharan, D. (1999) Non-coating fixation techniques or redundancy of conductive coating, low kV FE-SEM operation and combined SEM/TEM of biological tissues. *J. Microsc.*, 193, 158–170.

Jornsanoh, P., Thollet, G., Ferreira, J., Masenelli-Varlot, K., Gauthier, C., Bogner, A. (2011) Electron tomography combining ESEM and STEM: A new 3D imaging technique. *Ultramicroscopy*, 111, 1247–1254.

Jost, P.C., McMillan, D.A. and Griffith, O.H. (1983) Effect of fixation on molecular dynamics in membranes, in *The Science of Biological Specimen Preparation 1983* (eds J.P. Revel, T. Barnard and G.H. Haggis), AMF O'Hare, SEM Inc., IL, pp. 23–30.

Joubert, L.-M., Ferreira, J.A.G., Stevens, D.A., Nazik, H. and Cegelski, L. (2017) Visualization of *Aspergillus fumigatus* biofilms with scanning electron microscopy and variable pressure-scanning electron microscopy: A comparison of processing techniques. *Original Journal of Microbiological Methods*, 132, 46–55.

Karnovsky, M.J. (1965) A formaldhyde-glutaraldehyde fixative of high osmolality for use in electron microscopy. *J. Cell Biol.*, 27, 137A.

Kasthuri, N., Hayworth, K.J., Berger, D.R., Schalek, R.L., Conchello, J.A., Knowles-Barley, S., Lee, D., Vázquez-Reina, A., Kaynig, V., Jones, T.R., Roberts, M., Morgan, J.L., Tapia, J.C., Seung, H.S.,

Roncal, W.G., Vogelstein, J.T., Burns, R., Sussman, D.L., Priebe, C.E., Pfister, H. and Lichtman, J.W. (2015) Saturated reconstruction of a volume of neocortex. *Cell*, 162, 648–661.

Katsen-Globa, A., Puetz, N., Gepp, M.M., Neubauer, J.C. and Zimmermann, H. (2016) Study of SEM preparation artefacts with correlative microscopy: Cell shrinkage of adherent cells by HMDS-drying. *Scanning*, 38, 625–633.

Kiernan, J.A. (2000) Formaldehyde, formalin, paraformaldehyde and glutaraldehyde: What they are and what they do? *Microscopy Today*, 8, 8–12.

Kirk, S.E., Skepper, J.N. and Donald, A.M. (2009) Application of environmental scanning electron microscopy to determine biological surface structure. *J. Microsc.*, 233, 205–224.

Knott, G., Marchman, H., Wall, D. and Lich, B. (2008) Serial section scanning electron microscopy of adult brain tissue using focused ion beam milling. *J. Neurosci.*, 28, 2959–2964.

Kolossow, A. (1892) Ueber eine Neue Methode der Bearbeitung der Gewebe mit Osmiumsäure. *Zeitschrift für wissenschaftliche Mikroskopie und Mikroskopische Technik*, 9, 38–43.

Komai, F., Okada, K., Inoue, Y., Yada, M., Tanaka, O. and Kuwabata, S. (2014) SEM observation of wet lily pollen grains pretreated with ionic liquid. *J. Japan Soc. Hort. Sci.*, 83, 317–321.

Korn, E.D. (1967) A chromatographic and spectrophotometric study of the products of the reaction of osmium tetroxide with unsaturated lipids. *J. Cell Biol.*, 34, 627–638.

Kreshuk, A., Walecki, R., Koethe, U., Gierthmuehlen, M., Plachta, D., Genoud, C., Haastert-Talini, K. and Hamprecht, F.A. (2015) Automated tracing of myelinated axons and detection of the nodes of Ranvier in serial images of peripheral nerves. *J. Microsc.*, 259, 143–154.

Kurzer, F. and Wilkinson, M. (1970) Chemistry of carbohydrazide and thiocarbohydrazide. *Chem. Rev.*, 70, 111–149.

Kushida, H. (1964) Improved methods for embedding with Durcupan. *J, Electron Microsc.*, 13, 139–144.

Kushida, H. (1965) Durcupan as a dehydrating agent for embedding with polyester, styrene and methacrylate resins. *J. Electron Microsc.*, 14, 52–53.

Kushida, T., Nagato, Y. and Kushida, H. (1981) New method of embedding with GMA, Quetol 523 and methyl methacrylate for light and electron microscopic observation of semi-thin sections. *Okajimas Folia Anat. Jpn*, 58, 55–68.

Kuwabata, S., Kongkanand, A., Oyamatsu, D. and Torimoto, T. (2006) Observation of ionic liquid by scanning electron microscope. *Chem. Lett.*, 35, 600–601.

Lee, H. and Neville, K. (1967) *Handbook of Epoxy Resins*, McGraw-Hill Book Company New York.

Leighton, S.B. (1981) SEM images of block faces, cut by a miniature microtome within the SEM – a technical note. *Scanning Electron Microsc.*, 73–76.

Locke, M. (1994) Preservation and contrast without osmication or section staining. *Microsc. Res. Tech.*, 29, 1–10.

Locke, M. and Krishnan, N. (1971) Hot alcoholic phosphotungstic acid and uranyl acetate as routine stains for thick and thin sections. *J. Cell Biol.*, 50, 550–557.

Locke, M., Krishnan, N. and McMahon, J.T. (1971) A routine method for obtaining high contrast without staining sections. *J. Cell Biol.*, 50, 540–544.

Loussert Fonta, C. and Humbel, B.M. (2015) Correlative microscopy. *Arch. Biochem. Biophys.*, 581, 98–110.

Luft, J.H. (1961) Improvements in epoxy resin embedding methods. *J. Biophys. Biochem. Cytol.*, 9, 409–414.

Luft, J.H. (1964) Electron microscopy of cell extraneous coats as revealed by ruthenium red staining. *J. Cell Biol.*, 23, 54A–55A.

Luft, J.H. (1971a) Ruthenium red and violet. I. Chemistry, purification, methods of use for electron microscopy and mechanism of action. *Anat. Rec.*, 171, 347–368.

Luft, J.H. (1971b) Ruthenium red and violet. II. Fine structural localization in animal tissues. *Anat. Rec.*, 171, 369–415.

Lynch, S.P., Webster, G.L. (1975) A new technique of preparing pollen for scanning electron microscopy. *Grana*, 15, 127–136.

MacKenzie, A.P. (1972) Freezing, freeze-drying, and freeze-substitution. *Scanning Electron Microsc.*, II, 273–280.

Matsko, N. and Müller, M. (2005) Epoxy resin as fixative during freeze-substitution. *J. Struct. Biol.*, 152, 92–103.

Maupin-Szamier, P. and Pollard, T.D. (1978) Actin filament destruction by osmium tetroxide. *J. Cell Biol.*, 77, 837–853.

Maupin, P. and Pollard, T.D. (1983) Improved preservation and staining of HeLa cell actin filaments, clatrin-coated membranes, and other cytoplasmic structures by tannic acid-glutaraldehyde-saponin fixation. *J. Cell Biol.*, 96, 51–62.

McDonald, K.L. (1984) Osmium ferricyanide fixation improves microfilament preservation and membrane visualization in a variety of animal cell types. *J. Ultrastruct. Res.*, 86, 107–118.

Meek, K.M. and Chapman, J.A. (1986) Demonstrable fixative interactions in., eds. *Science of Biological Specimen Preparation 1985* (eds M. Müller, R.P. Becker, A. Boyde, and J.J. Wolosewick), AMF O'Hare, SEM Inc., IL, pp. 63–72.

Micheva, K.D. and Smith, S.J. (2007) Array tomography: A new tool for imaging the molecular architecture and ultrastructure of neural circuits. *Neuron*, 55, 25–36.

Mikula, S., Binding, J. and Denk, W. (2012) Staining and embedding the whole mouse brain for electron microscopy. *Nat. Methods*, 9, 1198–1201.

Mikula, S. and Denk, W. (2015) High-resolution whole-brain staining for electron microscopic circuit reconstruction. *Nat. Methods*, 12, 541–546.

Monsan, P., Puzo, G. and Mazarguil, H. (1975) Étude du mécanisme d'établissement des liaisons glutaraldéhyde-protéines. *Biochimie.*, 57, 1281–1292.

Murakami, T., Song, Z.-l., Hinenoya, H., Ohtsuka, A., Taguchi, T., Liu, J.-j. and Sano, T. (1987) Lysine-mediated tissue osmication in combination with a tannin-osmium conductive staining method for non-coated scanning electron microscopy of biological specimens. *Arch. Histol. Jpn*, 50,485–493.

Muscariello, L., Rosso, F., Marino, G., Giordano, A., Barbarisi, M., Cafiero, G., and Barbarisi, A. (2005) A critical overview of ESEM applications in the biological field. *J. Cell Physiol.*, 205, 328–334.

Newman, G.R. and Hobot, J.A. (1987) Modern acrylics for post-embedding immunostaining techniques. *J, Histochem. Cytochem.*, 35, 971–981.

Newman, G.R. and Hobot, J.A. (2001) *Resin Microscopy and On-Section Immunocytochemistry*, II edn, Springer-Verlag, Berlin, Heidelberg, New York.

Newman, G.R. and Jasani, B. (1984) Post-embedding immunoenzyme techniques, in *Immunolabelling for Electron Microscopy* (eds J.M. Polak and I.M. Varndell, Amsterdam, Elsevier Science Publishers, Amsterdam, pp. 53–70.

Nimura, K., Kowai, K., Nakazawa, E., Tomizawa, J. and Ose, Y. (2014) Innovative and fast specimen preparation technique of biological specimens for electron microscopy by the newly developed ionic liquid 'HILEM' IL1000, in *Scientific Instrument News*, Hitachi High-Technologies Corporation, pp. 23–31.

Palade, G.E. (1952) A study of fixation for electron microscopy. *Journal of Experimental Medicine*, 95, 285–298.

Peters, A., Hinds, P.L. and Vaughn, J.E. (1971) Extent of stain penetration in sections prepared for electron microscopy. *J. Ultrastruct. Res.*, 36, 37–45.

Peters, K. and Richards, F.M. (1977) Chemical cross-linking: Reagents and problems in studies of membrane structure. *Annu. Rev. Biochem.*, 46, 523–551.

Plumel, M. (1948) Tampon au cacodylate de sodium. *Bull. Soc. Chim. Biol.*, 30, 129–130.

Porter, K.R., Claude, A. and Fullam, E.F. (1945) A study of tissue culture cells by electron microscopy: Methods and preliminary observations. *Journal of Experimental Medicine*, 81, 233–256.

Porter, K.R. and Kallman, F. (1953) The properties and effects of osmium tetroxide as a tissue fixative with special reference to its use for electron microscopy. *Exp. Cell Res.*, 4, 12–41.

Pourcho, R.G., Bernstein, M.H. and Gould, S.F. (1978) Malachite green: Applications in electron microscopy. *Stain Technol.*, 53, 29–35,

Qin, W., Long, S., Panunzio, M. and Biondi, S. (2013) Schiff bases: A short survey on an evergreen chemistry tool. *Molecules*, 18, 12264–12289.

Reynolds, E.S. (1963) The use of lead citrate at high pH as an electron-opaque stain in electron microscopy. *J. Cell Biol.*, 17, 208–212.

Richards, F.M. and Knowles, J.R. (1968) Glutaraldehyde as a protein cross-linking reagent. *J. Mol. Biol.*, 37, 231–233.

Ripper, D., Schwarz, H. and Stierhof, Y.-D. (2008) Cryo-section immunolabelling of difficult to preserve specimens: advantages of cryofixation, freeze-substitution and rehydration. *Biol. Cell*, 100, 109–123.

Riva, A., Faa, G., Loffredo, F., Piludu, M. and Testa Riva, F. (1999) An improved OsO_4 maceration method for the visualization of internal structures and surfaces in human bioptic specimens by high resolution scanning electron microscopy. *Scanning Microsc.*, 13, 111–122.

Robertson, J.D., Bodenheimer, T.S. and Stage, D.E. (1963) The ultrastructure of Mauthner cell synapses and nodes in goldfish brains. *J. Cell Biol.*, 19, 159–199.

Romeis, B. (1968) *Mikroskopische Technik, 16, Neubearbeitete und Verbesserte Auflage*, R. Oldenbourg Verlag, München, Wien.

Roozemond, R.C. (1969) The effect of fixation with formaldehyde and glutaraldehyde on the composition of phospholipids extractable from rat hypothalamus. *J. Histochem. Cytochem.*, 17, 482–486.

Ryter, A., Kellenberger, E., Birch-Anderson, A. and Maaløe, O. (1958) Etude au microscope électronique de plasmas contenant de l'acide désoxyribonucléique. I. Les nucléoides des bactéries en croissance active. *Zeitschrift für Naturforschung B*, 13, 597–605.

Sabatini. D,D., Bensch, K., Barrnett, R.J. (1962) New means of fixation for electron microscopy and histochemistry. *Anat. Rec.*, 142, 274.

Sabatini, D.D., Bensch, K. and Barrnett, R.J. (1963) Cytochemistry and electron microscopy. The preservation of cellular ultrastructure and enzymatic activity by aldehyde fixation. *J. Cell Biol.*, 17, 19–58.

Sakaue, M., Shiono, M., Tomizawa, J., Nakazawa, E., Kawai, K. and Kuwabata, S. (2014) New preparation method using ionic liquid for fast and reliable SEM observation of biological specimens, in *18th International Microscopy Congress* (ed. P. Hozak), Czechoslovak Microscopy Society, Prague, Czech Republic, pp. 384–385.

Salema, R. and Brandão, I. (1973) The use of PIPES buffer in the fixation of plant cells for electron microscopy. *J. Submicrosc. Cytol.*, 5, 79–96.

Schliwa, M., van Blerkom, J. and Porter, K.R. (1981) Stabilization of the cytoplasmic ground substance in detergent-opened cells and a structural and biochemical analysis of its composition. *Proc. Natl Acad. Sci. USA*, 78, 4329–4333.

Schnepf, E., Hausmann, K. and Herth, W. (1982) The osmium tetroxide-potassium ferrocyanide (OsFeCN) staining technique for electron microscopy: A critical evaluation using ciliates, algae, mosses, and higher plants. *Histochemistry*, 76, 261–271.

Schwarz, H. (1973) Immunelektronenmikroskopische Untersuchungen über Zelloberflächenstrukturen aggregierender Amöben von Dictyostelium discoideum, in *Fachbereich Biologie*, EberhardKarls–Universitaet Tuebingen, Tuebingen.

Schwarz, H. and Humbel, B.M. (2014) Correlative light and electron microscopy using immunolabeled resin sections. *Methods Mol. Biol.*, 1117, 559–592.

Seligman, A.M., Wasserkrug, H.L. and Hanker, J.S. (1966) A new staining method (OTO) for enhancing contrast of lipid-containing membranes and droplets in osmium tetroxide-fixed tissue with osmiophilic thiocarbohydrazide (TCH). *J, Cell Biol.*, 30, 424–432.

Shalla, T.A., Carroll, T.W. and DeZoeten, G.A. (1964) Penetration of stain in ultrathin sections of tobbacco mosaic virus. *Stain Technol.*, 39, 257–265.

Shiono, M., Sakaue, M., Konomi, M., Tomizawa, J., Nakazawa, E., Kawai, K. and Kuwabata, S. (2014) Ionic liquid preparation for SEM observation of minute crustacean, in *18th International Microscopy Congress* (ed. P. Hozak), Prague, Czech Republic, Czechoslovak Microscopy Society, pp. 182–183.

Silva, M.T., Carvalho Guerra, F. and Magalhães, M.M. (1968) The fixative action of uranyl acetate in electron microscopy. *Experientia*, 24, 1074.

Silva, M.T., Santos Mota, J.M., Melo, J.V.C. and Carvalho Guerra, F. (1971) Uranyl salts as fixatives for electron microscopy: Study of the membrane ultrastructure and phospholipid loss in bacilli. *Biochim. Biophys. Acta*, 233,513–520.

Simionescu, N. and Simionescu, M. (1976a) Galloylglucoses of low molecular weight as mordant in electron microscopy. I. Procedure, and evidence for mordanting effect. *J. Cell Biol.*, 70, 608–621.

Simionescu, N. and Simionescu, M. (1976b) Galloylglucoses of low molecular weight as mordant in electron microscopy. II. The moiety and functional groups possibly involved in the mordanting effect. *J. Cell Biol.*, 70, 622–633.

Snowdowne, K.W. and Howell, J.N, (1984) Ruthenium red: Differential effects on excitation and excitation–contraction coupling in frog skeletal muscle. *J. Muscle Res. Cell Motil.*, 5, 399–410.

Spurr, A.R, (1969) A low-viscosity epoxy resin embedding medium for electron microscopy. *J. Ultrastruct Res.*, 26, 31–43.

Stäubli, W. (1963) A new embedding technique for electron microscopy, combining a water-soluble epoxy resin (Durcupan) with water-insoluble Araldite. *J. Cell Biol.*, 16, 197–201.

Stefanini, M., DeMartino, C. and Zamboni, L. (1967) Fixation of ejaculated spermatozoa for electron microscopy. *Nature*, 216, 173–174.

Stierhof, Y.-D., van Donselaar, E., Schwarz, H. and Humbel, B.M. (2008) Cryo-fixation, freeze-substitution, rehydration and Tokuyasu cryo-sectioning, in*Handbook for Cryo-Preparation Methods for Electron Microscopy* (eds A. Cavalier, D. Spehner and B.M. Humbel), CRC Press Inc., Boca Raton, FL, pp. 343–365.

Stoeckenius, W. (1961) Electron microscopy of DNA molecules 'stained' with heavy metal salts. *J. Biophys. Biochem. Cytol.*, 11, 297.

Stokes, D.J. (2003) Recent advances in electron imaging, image interpretation and applications: environmental scanning electron microscopy. *Phil. Trans. R. Soc. Lond. A*, 361, 2771–2787.

Strugger, S. (1956) Die Uranylacetat-Kontrastierung für die Elektronenmikroskopische Untersuchung von Pflanzenzellen. *Die Naturwissenschaften*, 43, 357–358.

Sung, H.-W., Hsu, H.-L., Shih, C.-C. and Lin, D.-S. (1996) Cross-linking characteristics of biological tissues fixed with monofunctional or multifunctional epoxy compounds. *Biomaterials*, 17, 1405–1410.

Takahashi, C., Shirai, T. and Fuji, M. (2013) FE-SEM observation of swelled seaweed using hydrophilic ionic liquid; 1-butyl-3-methylimidazolium tetrafluoroborate. *Microsc. Res. Tech.*, 76, 66–71.

Tanaka, K. and Mitsushima, A. (1984) A preparation method for observing intracellular structures by scanning electron microscopy. *J. Microsc.*, 133, 213–222.

Tapia, J.C., Kasthuri, N., Hayworth, K., Schalek, R., Lichtman, J.W., Smith, S.J. and Buchanan, J. (2012) High contrast en bloc staining of neuronal tissue for field emission scanning electron microscopy. *Nature Protocols*, 7, 193–206.

Teichman, R.J., Fujimoto, M. and Yanagimachi, R. (1972) A previously unrecognized material in mammalian spermatozoa as revealed by malachite green and pyronine. *Biol. Reprod.*, 7, 73–81.

Teichman, R.J., Takei, G.H. and Cummins, J.M. (1974) Detection of fatty acids, fatty aldehydes, phospholipids, glycolipids and cholesterol on thin-layer chromatograms stained with malachite green. *J. Chromatogr.*, 88, 425–427.

Terzakis, J.A. (1968) Uranyl acetate, a stain and a fixative. *J. Ultrastruct Res.*, 22, 168–184.

Thai, T.Q., Nguyen, H.B., Saitoh, S., Wu, B., Saitoh, Y., Shimo, S., Ali Elewa, Y.H., Ichii, O., Kon, Y., Takaki, T., Joh, K., and Ohno, N. (2016) Rapid specimen preparation to improve the throughput of electron microscopic volume imaging for three-dimensional analyses of subcellular ultrastructures with serial block-face scanning electron microscopy. *Med. Mol. Morphol.*, 49, 154–162.

Tidwell, T.T. (2008) Hugo (Ugo) Schiff, Schiff bases, and a century of β-lactam synthesis. *Angew Chem. Int. Ed. Engl.*, 47, 1016–020.

Ting-Beall, H.P. (1980) Interactions of uranyl ions with lipid bilayer membranes. *J. Microsc.*, 118, 221–227.

Tobler, M. and Freiburghaus, A.U. (1990) Occupational risks of (meth)acrylate compounds in embedding media for electron microscopy. *J. Microsc.*, 160, 291–298.

Tokuyasu, K.T. (1973) A technique for ultracryotomy of cell suspensions and tissues. *J. Cell. Biol.*, 57, 551–565.

Torimoto, T., Tsuda, T., Okazaki, K.-i. and Kuwabata, S. (2010) New frontiers in materials science opened by ionic liquids. *Adv. Mater.*, 22, 1196–1221.

Tranfield, E.M., Heiligenstein, X., Peristere, I. and Antony, C. (2014) Correlative light and electron microscopy for a free-floating spindle in Xenopus laevis egg extracts. *Methods Cell Biol.*, 124, 111–128.

Trump, B.F. and Bulger, R.E. (1966) New ultrastructural characteristics of cells fixed in a glutaraldehyde-osmium tetroxide mixture. *Lab. Invest.*, 15, 368–379.

Tsuda, T., Nemoto, N., Kawakami, K., Mochizuki, E., Kishida, S., Tajiri, T., Kushibiki, T. and Kuwabata, S. (2011) SEM observation of wet biological specimens pretreated with room-temperature ionic liquid. *ChemBioChem*, 12, 2547–2550.

Vale-Costa, S., Alenquer, M., Sousa, A.L., Kellen, B., Ramalho, J., Tranfield, E.M. and Amorim, M.J. (2016) Influenza A virus ribonucleoproteins modulate host recycling by competing with Rab11 effectors. *J. Cell Sci.*, 129:, 1697–1710.

Van Donselaar, E., Posthuma, G., Zeuschner, D., Humbel, B.M. and Slot, J.W. (2007) Immunogold labeling of cryo-sections from high-pressure frozen cells. *Traffic*, 8, 471–485.

Venable, J.H. and Coggeshall, R. (1965) A simplified lead citrate stain for use in electron microscopy. *J. Cell Biol.*, 25, 407–408.

Vogt, W. (1972) *Der Wiesbadener Kongress*, Die Arche, Zürich.

Walker, J.F. (1944) *Formaldehyde*, Reinhold Publishing Corporation, New York.

Walther, P. (2003) Recent progress in freeze-fracturing of high-pressure frozen samples. *J. Microsc.*, 212, 34–43.

Walther, P. and Müller, M. (1999) Biological ultrastructure as revealed by high resolution cryo-SEM of block faces after cryo-sectioning. *J. Microsc.*, 196, 279–287.

Walther, P., Wehrli, E., Hermann, R. and Müller, M. (1995) Double-layer coating for high-resolution low-temperature scanning electron microscopy. *J. Microsc.*, 179, 229–237.

Walton, J. (1979) Lead asparate, an en bloc contrast stain particularly useful for ultrastructural enzymology. *J. Histochem. Cytochem.*, 27, 1337–1342.

Wang, L., Humbel, B.M. and Roubos, E.W. (2005) High-pressure freezing followed by cryosubstitution as a tool for preserving high-quality ultrastructure and immunoreactivity in the *Xenopus laevis* pituitary gland. *Brain Res. Protoc.*, 15, 155–163.

Watson, M.L. (1958) Staining of tissue sections for electron microscopy with heavy metals. *J. Biophys. Biochem. Cytol.*, 4, 475–478.

Weakley, B.S. (1977) How dangerous is sodium cacodylate? *J. Microsc.*, 109, 249–251.

White, D.L., Andrews, S.B., Faller, J.W. and Barrnett, R.J. (1976) The chemical nature of osmium tetroxide fixation and staining of membranes by X-ray photoelectron spectroscopy. *Biochim. Biophys. Acta*, 436, 577–592.

White, D.L., Mazurkiewicz, J.E. and Barrnett, R.J. (1979) A chemical mechanism for tissue staining by osmium tetroxide-ferrocyanide mixtures. *J. Histochem. Cytochem.*, 27, 1084–1091

Wigglesworth, V.B. (1957) The use of osmium in the fixation and staining of tissues. *Proc. Roy. Soc. Lond. B Bio.*, 147, 185–199.

Wildhaber, I., Gross, H. and Moor, H. (1982) The control of freeze-drying with deuterium oxide (D_2O). *J. Ultrastruct. Res.*, 80, 367–373.

Wilke, S.A., Antonios, J.K., Bushong, E.A., Badkoobehi, A., Malek, E., Hwang, M., Terada, M., Ellisman, M.H. and Ghosh, A. (2013) Deconstructing complexity: serial block-face electron microscopic analysis of the hippocampal mossy fiber synapse. *J. Neurosci.*, 33, 507–522.

Wilkes, J.S. and Zaworotko, M.J. (1992) Air and water stable 1-ethyl-3-methylimidazolium based ionic liquids *J. Chem. Soc., Chem. Commun.*, 0, 965–967.

Wilson, M.T., Farmer, M.A. and Karwoski, C.J. (1998) Ultrastructure of the frog retina after high-pressure freezing and freeze substitution. *J. Microsc.*, 189, 219–235.

Wollweber, L., Stracke, R., Gothe, U. (1980) The use of a simple method to avoid cell shrinkage during SEM preparation. *J. Microsc.*, 121, 185–189.

Wood, J.G. (1973) The effects of glutaraldehyde and osmium on the proteins and lipids of myelin and mitochondria. *Biochim. Biophys. Acta*, 329, 118–127.

Zacharová, D., Uhrík, B., Henček, M., Lipskaja, E. and Pavelková, J. (1990) Effects of ruthenium red on excitation and contraction in muscle fibres with Ca^{2+} electrogenesis. *Gen. Physiol. Biophys.*, 9, 545–568.

Zamboni, L. and De Martino, C. (1967) Buffered picric acid-formaldehyde: A new rapid fixative for electron microscopy. *J. Cell Biol.*, 35, 148A.

Zhuravlev, O.E., Ivanova, A.I. and Grechishkin, R.M. (2015) Preparation of samples for SEM studies using an ionic liquid. *J. Synch. Investig.*, 9, 904–907.

11

A Brief Review of Cryobiology with Reference to Cryo Field Emission Scanning Electron Microscopy

Roland A. Fleck[1], **Eyal Shimoni**[2] **and Bruno M. Humbel**[3,4]

[1]*Centre for Ultrastructural Imaging, King's College London, UK*
[2]*Department of Chemical Research Support, Weizmann Institute of Science, Rehovot, Israel*
[3]*Electron Microscopy Facility, University of Lausanne, Switzerland*
[4]*Imaging Section, Okinawa Institute of Science and Technology, Onna-son, Okinawa, Japan*

11.1 CRYOPRESERVATION AND BRIEF HISTORY OF LOW TEMPERATURE BIOLOGY

The application of low temperatures in conjunction with rapid freezing to 'fix' biological material has been shown by many authors to provide an excellent approach for the preparation of biological material in a close to native state for investigation by the field-emission-gun scanning electron microscope (FEGSEM). Predating the cryo-FEGSEM, and for that matter cryo-electron microscopy (EM), was the use of low temperature to preserve the viability of biological tissues.

The first well-documented experiment demonstrating that the vivacity of an organism could be reversibly arrested by exposure to low temperatures was carried out by Henry Power in the mid-seventeenth century and reported as observation 30 in Book One, microscopical of his treatise *Experimental Philosophy*, in three books, *Containing New Experiments* (Power, 1664). In the 'resurrection' of eel-worms after several hours frozen in vinegar, the possibility of preserving living material in a state of 'suspended animation' was revealed (Power, 1664). Further interest in low temperature preservation was galvanised when, in 1665, Robert Boyle published *New Experiments and Observations Touching Cold*, a treatise

Biological Field Emission Scanning Electron Microscopy, First Edition.
Edited by Roland A. Fleck and Bruno M. Humbel.

reporting the possibility of using low temperatures to prevent decay in animal and plant tissues (Boyle, 1665). However, following these early reports, little practical success was achieved in the application of low temperatures to the preservation of living material, with the exception of a few, naturally freeze tolerant forms.

From the seventeenth century until the 1940s, understanding of cryobiology remained largely unchanged. *Life and Death at Low Temperatures* (Luyet and Gehenio, 1940) compiled much of the information available before that time, with tentative hypotheses of the mechanisms involved in the preservation of life and causes of death at low temperatures. Among the many works cited by Luyet and Gehenio was that of Mantegazza, who in 1866 reported the ability of human spermatozoa to survive freezing in semen at −17 °C, suggesting that mammalian cells could be recovered following exposure to temperatures below their freezing point (Mantegazza, 1866). Observations by Jahnel that a proportion of human spermatozoa could survive periods of storage in semen frozen to −80 °C (solid CO_2) presented the possibility of using low temperatures for the long-term storage of biological material (Jahnel, 1938; also cited by Luyet and Gehenio, 1940). Parkes revived interest in the cryopreservation of human spermatozoa in 1945 and a few years later, in 1949, one of the most important advances in cryobiology was made by Polge, Smith and Parkes (1949): the discovery that glycerol provided a cryoprotective (antifreeze) benefit in avian spermatozoa, protecting fowl spermatozoa against freezing injury to a temperature of −80 °C (Polge *et al.*, 1949). This hastened the development of a 'general method' for freezing of animal cells (Smith, 1961) and the possibility of routinely preserving material using low temperatures. Major advances in low temperature biology and its applications have followed and continue to this day (e.g. artificial insemination for managed breeding of livestock, *in vitro* fertilisation, conservation of endangered species and preservation of human tissue for medical purposes ranging from transplant and autologous treatments).

The development of cryopreservation techniques has been led by the use of *two-step techniques*, during which, prior to a final step of plunging into liquid nitrogen (LN_2, −196 °C), the sample is immersed in a bath containing a cryoprotectant solution set at an intermediate temperature (e.g. an industrial methylated spirits (IMS) bath at −30 °C). The initial exposure of the sample to a relatively high molarity cryoprotectant solution, usually starting at room temperature, has a double action: (a) concentration of the intracellular material through the removal by osmotic action of intracellular water and/or (b) replacing a proportion of the intracellular water with a cryoprotectant solution. Cryoprotection may also be controlled by cooling in a programmable freezer, at a predetermined rate, to a selected intermediate temperature prior to plunging in LN_2 (Day and McLellan, 1995). Each method exploits the fact that chilling injury is dependent upon the rate of cooling (Grout *et al.*, 1990). Many developments in cryobiology since the formulation of the 'general method' for freezing of animal cells have focused on improving understanding of the modes of freezing injury and damage (Mazur, 1990a, 1990b; Grout and Morris, 1987; Douzou, 1977, 1985; Steponkus, 1992; 1993; 1996; Kartha, 1985a; Ashwood-Smith and Farrant, 1980; Franks, 1985). The rationale behind two-step freezing techniques is explained in following sections.

11.2 HISTORY OF FREEZING FOR ELECTRON MICROSCOPY OBSERVATION

During the early days of electron microscopy frozen samples could not be observed directly at the microscope and freezing techniques were applied mostly in order to immobilise and stabilise biological samples prior to and during their dehydration either by partial or complete sublimation (freeze drying) or by freeze substitution (FS). Freeze drying of suspensions or frozen sections was followed by metal evaporation or shadow casting (Wyckoff, 1946, 1949; Hall, 1950; Williams, 1953), thus enabling their direct observation at the transmission electron microscope (Fernandez-Moran, 1952). The freeze drying and shadow casting techniques evolved later into the freeze fracture-replica technique (Steer, 1957; Moor *et al.*, 1961; Moor and Mühlethaler, 1963) described in detail in subsequent chapters, which in turn was used as the basis for modern cryo-FEGSEM techniques. An alternative approach to remove water from a frozen sample is freeze substitution (FS) which was applied originally for light microscopy (Simpson, 1941a, 1941b). During FS frozen water in the sample is dissolved by organic solvent kept just above its melting temperature. Once dehydrated, the sample is infiltrated with plastic resin and sectioned after the resin is cured (Feder and Sidman, 1958; Fernandez-Moran, 1957; Bullivant, 1960; Van Harreveld and Crowell, 1964; Van Harreveld, Crowell and Malhotra, 1965; Rebhun, 1972; Pease, 1973; Humbel and Müller, 1984).

The stumbling block on the way to making these methods successful was freezing quality. In fact, cryo-EM pioneers encountered the same problems as those encountered by cryobiologists, namely direct damage to samples by ice crystals growth as well as shrinkage of cells due to ice crystals induced osmotic stress. Indeed, during the infancy of cryo-EM there was not much distinction between freezing strategies applied by scientists coming from these two fields: in both, cryoprotectants were often used to suppress ice crystals formation (e.g. Moor, 1964; Steere, 1957; Tokuyasu, 1973) and, in addition, viability tests after thawing were used even by cryo-EM specialists as criteria for the success of freezing (Moor, 1964). However, there were two forces driving cryo-EM towards branching out from cryobiology and to favour one rapid cooling step over slow cooling processes, which first brought the system close to thermodynamic equilibrium prior to its final cooling down to cryogenic temperatures (i.e. two-step techniques): (1) the understanding that ice crystals growth as well as the introduction of cryoprotectants distort the native structure of the sample and should, therefore, be avoided as much as possible; (2) specimens prepared for electron microscopy observation are mostly smaller than those used for cryobiology, with the thickness of the former lying between about 100 nm to a couple of hundreds of μm. These small dimensions, favourable for rapid cooling in terms of heat transfer, raised the hope that given the right set of conditions samples could be either vitrified or at least be 'adequately frozen' (Studer, Michel and Müller, 1989), that is frozen with ice crystals that are smaller than the detection limit of the microscope.

The challenges involved in achieving the vitrification holy grail were understood but problems lay either in engineering or in physical limitations (e.g. the use of improper coolants such as LN and liquid helium II or the use of too-thick samples). Many brilliant scientists, technicians and engineers contributed to the evolution of cryo-EM by constructing and developing ingenious freezing devices and techniques. Some of these contributions are described in Section 11.8, Vitrification the 'key' to cryo-FEGSEM, in this chapter, and the reader is also referred to other sources, such as Benedetti and Favard, 1973; Robards and Slytr, 1985; Steinbrecht and Zierold, 1987. Modern cryo-EM owes its success to the work of two groups that revolutionised the field by devising techniques that resulted in vitrification: the group of Jacques Dubochet at EMBL and later at the University of Lausanne and the group of Hans Moor at the ETH in Zurich.

Water was vitrified by vapour deposition on a cold surface in 1935 and by fast cooling of microdroplets in 1980 (Burton and Oliver, 1935; Brüggeller and Mayer, 1980), but these were either not useful or not amenable for direct observation at the electron microscope. It was Jacques Dubochet's group that managed for the first time to vitrify thin films of water supported by the surface tension of water itself between the bars of EM grids

(Jacques Dubochet, Nobel Prize for Chemistry 2017 stemming from his pioneering work in vitrification of thin aqueous films; Dubochet and McDowall, 1981). The water films could be vitrified because they were plunged into liquid ethane rather than into liquid nitrogen or helium, which are not good cryogens due to the Leidenfrost effect (a film of insulating vapour that slows down cooling). The frozen samples could then be observed directly at their frozen hydrated state in the cryo-electron microscope and vitrification could be verified by electron diffraction. The plunging technique developed by this group is the basis of sample preparation used for modern single particle reconstruction, reaching near atomic resolution. The success of the method lies not only in choosing the right cryogen but also in the high cooling rates that can be achieved during the freezing of extremely thin aqueous films (~100 nm), which are possibly in the range of 10^7 °C/s (Steinbrecht and Zierold, 1987). As Jacques Dubochet wrote: '*It is the good fortune of the electron microscopist that, reducing sample size in order to increase the cooling speed, vitrification becomes easy just when the dimensions of the specimen are those suitable for electron microscopy*' (Dubochet and Lepault, 1984). But what about samples that could not be thinned down to such small dimensions, for example plant or animal tissue or even cell monolayers that are a few µm in thickness? Without the addition of cryoprotectants such samples cannot be vitrified by plunging or, for that matter, by any other freezing method applied at ambient pressure: due to the poor thermal conductivity of water, cooling of such samples would be too slow to allow for vitrification and samples would thus be badly damaged by ice crystals. If samples contained high enough concentrations of native cryoprotectants they could be, at best, adequately frozen at least at their periphery by the application of techniques such as plunging (Feder and Sidman, 1958; Handley, Alexander and Chien, 1981; Altmann, 1894), propane jet freezing (Müller, Meister and Moor, 1980; Plattner and Knoll, 1984), slam freezing (Van Harreveld and Crowell, 1964; Heuser *et al.*, 1979), etc.

It was the invention of the high pressure freezing technique by Hans Moor and his colleagues (Moor and Riehle, 1968; Riehle, 1968) that enabled the vitrification of samples that were up to about 200 µm thick. Moor's group, which had already developed the first commercial freeze fracture device (Moor *et al.*, 1961), used the technique to directly visualise freezing-related damage to cells (Moor, 1964) and as a result could use it to investigate the improvements brought by high pressure freezing (Riehle and Hoechli, 1973). The high pressure freezing (HPF) technique is based on the physics of water: pressure of 2100 bars applied during freezing acts as a physical cryoprotectant; high pressure counteracts the expansion of water upon crystallisation, thus slows down ice crystal formation and reduces the critical cooling-rate required for vitrification. The technique was further developed and became commercial mainly thanks to the efforts of Martin Müller from Moor's group and to his collaboration with BAL-TEC AG in Liechtenstein, which developed the first commercial high pressure freezing machine. During his attempts to improve the method Müller also used his military experience and applied explosives in order to produce a wave of highly pressurised liquid nitrogen (needless to say, no HPF machine ended up using explosives). Martin Müller summed up and formulated a series of principles that led the field up until today. According to Müller, the preparation of samples for EM observation requires the application of *consecutive steps*, with each step a potential source for loss of information. Thus, the *information density* in the resulting image may be much degraded compared to that of the native sample. Once lost, information cannot be recreated in the final image, no matter how sophisticated the post-processing strategies are. By rapid lowering of thermal energy, leading to vitrification, the sample is immobilised. This process freeze-frames the sample throughout all levels of organisation down to the molecular level at a particular point in time, which enhances the microscopist's capacity to resolve details at high spatiotemporal resolution, theoretically with no loss of information density. The distinction between adequate freezing and vitrification was also elaborated by Müller, as mentioned above (Studer, Michel and Müller, 1989), which allowed microscopists to obtain excellent and reliable results even under conditions that were suboptimal for full vitrification. The benefits of cryo-immobilisation, compared to conventional chemical fixation techniques, stem almost as a direct consequence of this philosophy.

In the coming sections we will expand on states of water at low temperatures, damage related to ice nucleation and growth and strategies to avoid such damage.

11.3 TEMPERATURE AND THE CONDENSED PHASES OF WATER

Water, the major component of biological systems and prerequisite for life, remains the subject of current research in chemical physics and physical chemistry (Gallo *et al.*, 2016). Water has a number of anomalies in its properties and considering its comparatively simple structure and behaviour as a liquid, it has a complex phase diagram (Figure 11.1). The phase diagram of water is reviewed in depth by Gallo *et al.* (2016) and consists of different types of ices, amorphous phases and anomalies. Water has unique thermodynamics in its supercooled liquid state, high and low density amorphous phases that are still poorly understood (Gallo *et al.*, 2016). These anomalous states and phase changes are often encountered at the temperatures and pressures used in cryopreservation, cryo-EM sample preparation and cryo-EM.

11.3.1 Supercooling

Observers in the early nineteenth century recognised that on cooling a liquid below its thermodynamic freezing point the onset of crystallisation was found to be highly unpredictable

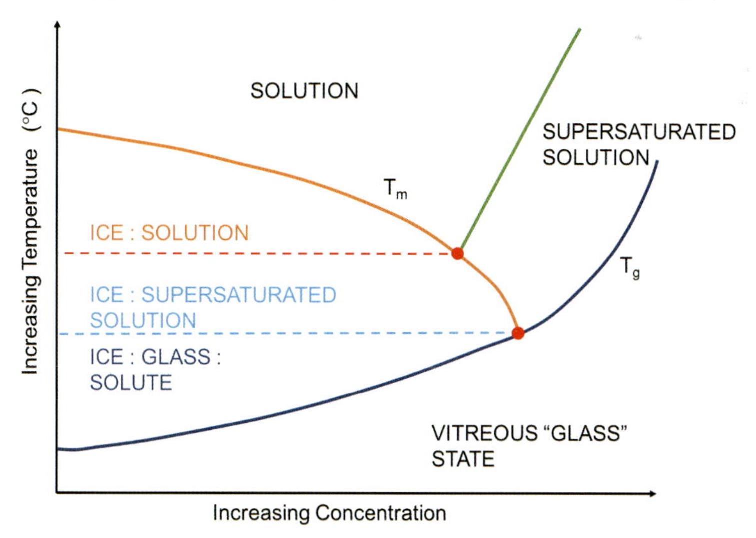

Figure 11.1 Simple phase diagram for water showing that an increase in temperature will cause water to change from one state to another (solid to liquid). The area where stable vitreous solutions exist resides below the curve representing the glass transition temperature (T_g). The eutectic point (T_m) denotes the melting/freezing point of the liquid. Below the orange line and above T_g a complex multiphase condition exists. These are reviewed in depth by Gallo *et al.* (2016).

and sample-dependent (Taylor, 1987). Although the freezing point of cytoplasm is usually above −1 °C, cells can, and generally do, remain unfrozen and therefore supercool to −10 or −15 °C, even when ice is present in the external medium (Mazur, 1970). The ability of cells to supercool, even when external ice is present, indicates the absence of effective ice nucleators within the interior of the cells and the ability of the cell membrane to block nucleation (Mazur, 1970).

The ability of solutions to cool beyond the equilibrium freezing point (T_m) is governed by the fact that before crystallisation takes place it must first be preceded by a nucleation event in which a solid-to-liquid interface is created (Taylor, 1987). The 'nucleus' is a cluster of water molecules with a configuration that can be identified by other water molecules as 'ice like', around which other molecules in the solution then condense (Taylor, 1987). In cooled solutions these 'ice nuclei' formed by random motion occur spontaneously, but they must exist for sufficient time to allow crystal growth through condensation of further molecules. In a solution at 0 °C the number of molecules necessary for nucleus formation is high, with the effect that the number of such nucleating clusters in a given sample of liquid is very low and their lifetime is very short. However, with further cooling the number of molecules required to act as a critical nucleus decreases, with the result that the number of nucleation points in a given sample increases. The progression of intracellular ice formation has been studied extensively and directly visualised by high speed video microscopy, providing insight into the progression from nucleation to a frozen intracellular space (Pegg, 2010; Yang, Zhang and Xu, 2009; Stott and Karlsson, 2009; Mazur *et al.*, 2005).

11.3.2 Homogeneous Ice Nucleation

The spontaneous nucleation of ice by the above events and the absolute lowest temperature at which the liquid state exists is the homogeneous ice nucleation temperature (T_h). To form the first particle of the new phase a surface must be formed between the particle and the surrounding liquid. The creation of the surface requires energy, supplied by the release of energy resulting from the crystallisation process (MacFarlane, 1987). At, or just below, the normal equilibrium melting point, the free energy associated with crystallisation is very small and hence the surface energy problem can only be solved by forming a very large particle (MacFarlane, 1987; Mehl, 1996a). The spontaneous occurrence of such a particle is highly improbable and hence crystallisation is unlikely to occur until a much lower temperature where the free energy of crystallisation is larger (MacFarlane, 1987). The temperature at which the liquid begins to spontaneously form crystallites at a significant rate is termed the homogeneous nucleation temperature (T_h) (Mazur, 1970; Toner, 1993; Stillinger, 1995). However, if suitable foreign surfaces are present, such that the interfacial tension between the new crystallite and the surface is lower, then crystal formation will occur preferentially on those sites and this is termed heterogeneous nucleation (Mehl, 1996a).

11.3.3 Heterogeneous Ice Nucleation

The presence of particulate matter and/or temperature gradients within a solution can provide interference and/or a template for ice formation, allowing heterogeneous ice nucleation at temperatures above T_h. Ice formation in aqueous solutions is usually by heterogeneous nucleation, at temperatures well above the homogeneous ice nucleation temperature (−38.5 °C) of water (Hobbs, 1974).

Furthermore, the apparent inability of membranes to block intracellular ice nucleation below −15 °C has prompted the hypothesis that membranes contain water filled pores similar to those proposed by Solomon and others (Mazur, 1970). The barrier properties of membranes arise because at higher subzero temperatures (−10 to −20 °C) ice crystals small enough to pass through such pores do not exist (Mazur, 1970; MacFarlane, 1987). Alternatively, it has been hypothesised that intracellular ice formation may occur as a result of nucleation within the cell catalysed by either the plasma membrane or intracellular particles (Toner, Karel and Cravalho, 1990). Muldrew and McGann (1994) hypothesised that the agent inducing intracellular ice formation may in fact be the osmotically driven water efflux, which occurs during freezing (cryodehydration). This efflux of water is hypothesised to produce a rupture of the plasma membrane, permitting the propagation of extracellular ice into the cytoplasm (Muldrew and McGann, 1994).

This recognition that biological membranes are not static but have a liquid nature in their interior is important to cryopreservation (Franks, 1985; Mazur, 1970; Toner, 1993; Taylor, 1987). The multicomponent nature of cell membranes allows more than one phase to exist at any one time, that is the system will not show a single sharp melting temperature since it is possible to vary the temperature of the system and still have both phases present (solid and liquid) (Taylor, 1987). This means that a membrane can gradually change its fluidity over a wide temperature range, allowing fine control of cell function determined by the temperature of its surroundings (Taylor, 1987).

11.3.4 Post-Nucleation

The liquid–solid phase transition is completed by the growth of ice nuclei into crystals of variable size and shape influenced by the extent of supercooling, rate of cooling and the nature and concentration of dissolved solutes (Franks, 1985). The rate of nucleation rapidly increases with increasing cooling rate, with the opposite being true of ice crystal growth, with slow cooling producing a small number of large crystals and rapid cooling, producing a multitude of very small crystals (Taylor, 1987). Small ice crystals have high surface energies, which render them thermodynamically less stable. Ultimately, unless a solution is frozen very slowly with a minimum of supercooling, metastable small ice crystals will form, which will attempt to fuse, forming larger, more stable crystals with lower surface energies (Franks, 1985). This process is encountered in the food industry where, even at constant storage temperatures, small ice crystals decrease in size while larger crystals grow (Donhowe and Hartel, 1996; Taylor, 1987). The number of crystals therefore decreases with time while their average size increases (Reid, 1983).

In the food industry the mechanism of recrystallisation and crystal growth have attracted a great deal of interest, particularly in the maintenance of ice cream quality through a series of freeze–thaw cycles where recrystallisation will reduce product quality and shelf life (Donhowe and Hartel, 1996). Refrigerated ice cream is subjected to several recrystallisation mechanisms: melt–refreeze recrystallisation under oscillating-temperature conditions, rounding occurring at constant temperatures and migratory recrystallisation rarely occurring (Donhowe and Hartel, 1996). Recrystallisation has been reported to occur even with very small temperature fluctuations, with the crystallisation rate dependent upon the size of temperature fluctuation and storage temperature (Donhowe and Hartel, 1996). Storage temperature, amplitude of temperature fluctuation and period of fluctuation have all been demonstrated to influence ice crystal size and distribution (Donhowe and Hartel, 1996). These observations are applicable in the storage of cryopreserved material, which may be stored for extended periods of time, presenting the possibility that a number of temperature fluctuations may be encountered. Although studies on ice cream were carried out at high subzero temperatures (−20 °C), ice crystal growth and recrystallisation can be detected as low as −130°C (Taylor, 1987).

11.3.5 Vitrification

Vitrification occurs where cells are immobilised in an amorphous solid, presenting the possibility of circumventing crystallisation. This is a potentially desirable attribute in the exploitation of low temperature systems and techniques (Franks, 1985). If T_h (the homogeneous nucleation temperature) is sufficiently low then it is possible that ice nucleation can be avoided; instead, as the liquid cools it becomes generally more viscous until it reaches a point where it can no longer flow on a measurable time scale (MacFarlane, 1987). The liquid now possesses the structural properties of a liquid but the mechanical properties of a solid and has become a glass (MacFarlane, 1987). The glass itself is commonly described in terms of viscosity (approximately 10^{14} N s/m^2 or $>10^{13}$ poises) (Franks, 1985; Grout, Morris and McLellan, 1990) and the glass transition viscosity corresponds to a state in which there is effectively no diffusive movement of molecules on a measurable time scale (Franks, 1985). The phenomenon of glass formation presents itself in two important ways in cryobiology. Firstly, in cryopreservation protocols that utilise moderate levels of cryoprotectants and moderate cooling rates, where ice formation is inevitable, the 'unfrozen' water fraction of these solutions vitrifies, resulting in a partially frozen solution (MacFarlane, 1987). Secondly, high concentrations of solutes may be employed, to attempt to avoid ice formation altogether by promoting the vitrification of the entire solution (MacFarlane, 1987; MacFarlane, Forsyth and Barton, 1992; Steponkus, Langus and Fujikawa, 1992; Mehl, 1996a, 1996b).

While in the glassy state, the vitrified material will relax towards its equilibrium state with time, dependent upon temperature, with the possibility of ice nucleation below the glass transition state (Mehl, 1996b). Fracturing in vitreous samples has been observed in many systems and has been suggested to be a cause of mortality (Hunt *et al.*, 1994; Pegg, Wusteman and Boylan, 1997). Fracture events are believed to be points of ice nucleation and crystal growth that cause irreversible mechanical damage within the cell. Fracture formation is believed to be intrinsic to the glassy state and is related to the ability of the glass to overcome thermal stresses within it. In practical vitrification solutions, ice nucleation may occur during cooling through heterogeneous processes, but it is limited in highly concentrated vitrification solutions. During cooling, stable and unstable ice nuclei form and, during storage, stabilisation of the unstable ice nuclei is possible.

As a glass relaxes, densification of the sample limits ice crystal growth, promoting storage at or just below the glass transition temperature (T_g) rather than at lower temperatures where fracture events are more likely (Mehl, 1996b). At, or close to, the glass transition temperature, the diffusion of water molecules will stabilise the ice nuclei, but prevent their growth. The homogeneous distribution of ice nuclei may also deplete the glassy matrix of excess water molecules limiting the crystal growth during subsequent warming (Mehl, 1996b). This may contribute to the stability of the vitreous state whilst a sample is being imaged in the cryo-FEGSEM.

11.3.6 Thawing

Thawing can also have complex effects on cryopreserved material. The influence of warming rates and ice crystals on freeze fracture and devitrification (e.g. recrystallisation) events and the theories of vitrification/devitrification have been discussed in detail by Mehl (1996a, 1996b), Franks (1985), MacFarlane *et al.* (1992) and Stillinger (1995). By reducing the warming rate post-thaw function can be improved (Pegg *et al.*, 1997). In systems where fractures have been encountered and have probably resulted from thermal stresses created by rapid warming, fractures may be avoided by employing comparatively slow initial warming rates which allow the vitreous material to soften, reducing stresses and avoiding fractures, followed by rapid warming to ambient temperatures avoiding/limiting ice crystal nucleation/growth (Pegg, Wusteman and Boylan, 1996, 1997).

In specimens cooled under extreme non-equilibrium conditions, recrystallisation will occur during warming, as the very small ice crystals formed on cooling, which are thermodynamically metastable, minimise their surface-to-volume ratios by fusing or growing into larger, more stable ice forms. Although recrystallisation generally occurs rapidly at high subzero temperatures (T_r) it has been detected as low as −130 °C (Fernández-Morán, 1960; Taylor, 1987; Mayer, 1985).

Changes in X-ray diffraction patterns in aqueous solutions correspond to the transformation of cubic ice to hexagonal ice, $I_c \rightarrow I_h$. It has been proposed that recrystallisation will occur along preferred axes, so presenting diffraction's ability to relate to different crystal shapes. In pure water, crystal growth rates along different axes have been demonstrated to vary substantially, dependent upon conditions such as supercooling and the presence of fish antifreeze glycoproteins (MacFarlane *et al.*, 1992). There have been several types of recrystallisation identified, each of which have been discussed in detail by Luyet (1965) (see below). Maintaining a vitreous sample is considered essential for cryo-EM; however, Cyrklaff and Kühlbrandt (1994) reported a stabilising effect from the presence of cubic ice. In studies with Tobacco Mosaic Virus (TMV), cubic ice was determined to provide a rigid stabilising support during transmission electron microscopy to ~1Å with no disruption to the virus structure evident from external crystallisation of vitreous water (Cyrklaff and Kühlbrandt, 1994).

11.3.7 Irruptive Recrystallisation

Ephemeral spherulites formed during rapid cooling resume their inhibited crystalline growth when the temperature reaches a specific narrow range (T_r) during slow rewarming. The event is easily visualised by the abrupt shift from a transparent preparation to an intensely opaque one (Luyet, 1965; Franks, 1985; Taylor, 1987).

11.3.8 Migratory Recrystallisation

This is a gradual growth of large crystals in a population at the expense of the smaller ones. The migration of molecules from small to large crystals occurs as the specimen is warmed gradually from T_r to T_m, with the rate increasing with rising temperature (Luyet, 1965; Franks, 1985; Taylor, 1987).

11.3.9 Spontaneous recrystallisation

This can occur during rapid cooling, where the latent heat released during freezing is able to cause localised temperature rises, which can give rise to recrystallisation (Luyet, 1965; Franks, 1985; Taylor, 1987).

All recrystallisation occurs due to the highly unstable (metastable) nature of solutions that have undergone rapid cooling (Rasmussen and MacKenzie, 1971). In these solutions the system will attempt, at a given temperature, to reduce its total surface area in order to reach its equilibrium state (the state achieved on slow cooling with initial freezing at T_r).

11.4 FREEZE DRYING

Once frozen, water may be 'removed' from a sample by freeze drying, leaving a dry sample with structures and elements largely in their native state (MacKenzie, 1972). Freeze drying of a frozen biological sample involves the sublimation of ice (transition from the solid state to the gaseous state without melting), and the simultaneous and/or subsequent removal of some or all of the water that was not converted to ice during the initial freezing (Strong and MacKenzie, 1993). Freeze drying reduces the distortion (due to forces from water changing from liquid to vapour phase) and shrinkage when hydrated, unfrozen specimens are dried by evaporation. Freeze drying overcomes this problem by sublimation under vacuum – a process that avoids the liquid phase and thereby reduces distortion. The rate of sublimation is a function of temperature and vacuum, with drying taking several hours. Freeze drying only occurs if the partial pressure of the vapour in the drying chamber is lower than the water vapour pressure above the specimen. A vacuum chamber is not essential, as sublimation can take place at atmospheric pressures, as seen when snow evaporates without melting in cold, dry weather (where dry air has low partial pressure of water vapour) (Heldman and Hohner, 1974; Karel, 1975).

In practice, however, freeze-drying is carried out under vacuum at low pressures (for commercial freeze driers: 10 Pa, 0.0001 bar). Under these conditions water vapour has a large specific volume and a vacuum pump with large displacement capacity is required. To support drying and reduce load on the vacuum system, vapours are condensed as ice on the surface of condensers kept at low temperature (for commercial freeze driers: <–40 °C or less). This principal is also employed in the SEM where an LN cooled surface acts as an anticontaminator, attracting water vapour to condense on its surface. Freeze drying occurs in two stages: the first stage is sublimation drying, in which sublimation of the frozen water occurs, and the second stage is desorption drying, during which most of the unfrozen water adsorbed on the solid matrix is removed (MacKenzie, 1965). Typically, freeze drying is carried to a final moisture content of ~1–3% (Berk, 2013).

During sublimation the loss of water leaves a porous structure, which promotes further evaporation of unfrozen water. The process is kinetic and in materials cooled to an ice phase (leaving water molecules at a lowest available energy state), a proportion of water remains unfrozen and is not converted to ice by further or slower cooling. In this case the force driving freezing tends to zero. A generic description of a biological sample would be a water-soluble/protein-rich system and a completely frozen sample, cooled under non-vitrifying conditions that leaves a non-frozen water component of as much as 25% (as a % of dry weight) (Strong and MacKenzie, 1993). Rapid cooling with liquid cryogen (plunge freezing) can increase this unfrozen component to 30%. In the case of high pressure frozen or truly vitreous material the component would be 100%. For ultrastructural preservation, when the hydration shell is removed a collapse of protein fine structure may occur (Gross, 1987; Wildhaber, Gross and Moor, 1982).

Sublimation is driven by a vapour pressure difference, which is determined by temperature. At −30 °C ice in a tissue exerts a vapour pressure of ~0.28 mmHg (0.38 mbar/38 Pa), at −50 °C ~0.029 mmHg (0.039 mbar/3.8 Pa), at −90 °C 0.00007 mmHg (0.0001 mbar/0.009 Pa) and at −100 °C 0.00001 mmHg (0.000013 mbar), equating to approximately a threefold reduction in vapour pressure per 10 °C reduction in temperature. Thus, in a freeze drying apparatus the net driving force for sublimation is the difference between the sample and the condensing surface. For example, a biological sample at −30 °C has a vapour pressure of 0.38 mbar and with a condenser set at −60 °C able to exert a vapour pressure of 0.01 mbar, the sublimation force is 0.37 mbar. In the case of a vitrified tissue where ultrastructural preservation is critical, the sample must be maintained below its glass transition temperature (T_g), which is likely to be significantly lower than −30 °C (T_g for pure water is −135 °C), requiring the condenser surface to be colder still. However, at much lower temperatures (i.e. sample temperatures at or below −100 °C and condenser lower still), the achieved vapour pressure will be very low (<0.000013 mbar). Predicted drying times by Perry (1976) for a 10 mm block of tissue freeze dried at −60 °C is 1 year and by Urist, Mikolski and Boyd (1975) for bone powder at −75 °C, for a block of 22 mm depth >10 years, for a block 2 mm in depth 4 months and 3 to 4 days for a block of 2 mm^3 or less. Stephenson, (1953) performed extensive tests freeze drying pigs liver to develop a general theory of freeze drying, in doing so he showed that for a given temperature of sublimation the minimum drying time for a tissue was significantly longer than for an equivalently sized piece of ice (due to resistance in the drying layer to the passage of water vapour). A paradox in the role of vacuum exists whereby at too low a pressure the vacuum can act as a barrier to effective heat transfer and reduce the sublimation rate.

The temperature of recrystalisation (T_r), reported by Luyet over 50 years ago, is determined by differential scanning calorimetry (DSC) for a glass forming aqueous system to correspond with the glass transition temperature (T_g) of the solution. Under suboptimal freeze drying conditions, collapse of the freeze dried system occurs. This process, the T_g and thermomechanical properties of sugar-water vitreous systems, has been reviewed by Levine and Slade (1988), which can be highly informative in determining the freeze drying limitations of a biological sample for ultrastructural imaging. For electron microscopy, freeze drying should, ideally, be carried out at temperatures below the recrystallisation point of vitreous water (>−135 °C), but this would require very long drying times (see Table 11.1). In practice temperatures of −60 °C give reasonable results (Figure 11.2). However, for delicate specimens temperatures of −80 °C and lower are required (Gross, 1987).

If applied to successfully vitrified material (vitrification is discussed in detail below), a sample may be freeze dried for elemental analysis (see Chapter 28 by Warley and Skepper in Volume II) or coated for direct observation in the SEM exploiting the reduced shrinkage (~20%) when compared with critical point dried (~65%) materials (Boyde and Maconnachie, 1979). For high resolution imaging in the FEGSEM, freeze drying followed by heavy metal staining permits high resolution imaging without the requirement for complex cryo-stages. These high resolution techniques were developed for the TEM (see work by Nermut and Frank, 1971; Moor *et al.*, 1961, Branton *et al.*, 1975; Moor, 1970, 1971, 1973; Moor and Mühlethaler, 1963; Mühlethaler, Hauenstein and Moor, 1973) and are extensively reviewed by Gross (1987). However, with the modern FEGSEM, comparable studies can be readily undertaken in the SEM and gain from the topographical information generated. For structural resolution, high quality specimen preparation (cryofixation by vitrification) is essential and discussed above (Moor, 1986). Once frozen, freeze drying (lyophilisation) is performed under conditions optimised to minimise collapse, devitrifcation and shrinkage and then coated at high resolution (Wildhaber *et al.*, 1982; see also Chapter 12 by Tacke, Lucas, Woodward, Gross and Wepf).

Table 11.1 Sublimation rates with temperatures of ice under a vacuum condition at least 100× > saturation vapour pressure of the specimen. The table shows the rapidly diminishing rate of sublimation with reduced vapour pressure and corresponding reduction in the drying volume. Adapted from the BAL-TEC BAF060 User Manual (BAL-TEC AG, Liechtenstein, after Umrath, 1983)

Temperature (°C)	Saturation pressure (mbar)	Sublimation rate (g/cm² s)	Etch rate (1/s)
−10	2.60	2.98×10^{-2}	324 μm
−20	1.03	1.21×10^{-2}	131 μm
−30	3.81×10^{-1}	4.54×10^{-3}	49.3 μm
−40	1.29×10^{-1}	1.57×10^{-3}	17.0 μm
−50	3.93×10^{-2}	4.90×10^{-4}	5.30 μm
−60	1.08×10^{-2}	1.37×10^{-4}	1.48 μm
−70	2.59×10^{-3}	3.37×10^{-5}	364 nm
−80	5.36×10^{-4}	7.16×10^{-6}	77.0 nm
−85	2.29×10^{-4}	3.10×10^{-6}	33.3 nm
−90	9.35×10^{-5}	1.28×10^{-6}	13.7 nm
−95	3.61×10^{-5}	5.02×10^{-7}	5.39 nm
−100	1.32×10^{-5}	1.70×10^{-7}	2.00 nm
−105	4.57×10^{-6}	6.55×10^{-8}	0.70 nm
−110	1.48×10^{-6}	2.15×10^{-8}	0.23 nm
−115	4.45×10^{-7}	6.58×10^{-9}	70.4 pm
−120	1.24×10^{-7}	1.86×10^{-9}	19.9 pm
−130	7.38×10^{-9}	1.15×10^{-10}	1.22 pm
−140	2.88×10^{-10}	4.64×10^{-12}	49.5 fm
−150	6.68×10^{-12}	1.12×10^{-13}	1.19 fm
−160	8.02×10^{-14}	1.40×10^{-15}	0.01 fm

11.5 MECHANISMS OF LOW TEMPERATURE DAMAGE AND INJURY

The application of low temperature to biological systems has a number of extraordinary complex effects, even in the non-frozen state.

The responses of plants to environmental stresses, encountered due to perturbations in their natural environment, particularly wide extremes of temperature, have been of interest to humans since early agriculturists began identifying 'hardy' plants that survive environmental stresses and 'tender' plants that do not (Levitt, 1980). As early as 1778, Bierkander reported that some plants were killed between 1 °C and 2 °C, and similar findings were reported by Göppert (1830) and Kunisch (1880), prompting Molish to suggest in 1897 that low temperature damage in the absence of freezing should be called chilling injury as opposed to freezing injury (Levitt, 1980). However, some plants can survive prolonged exposure to low temperatures (Hirsh, 1987); in addition, Desmids have been reported to be able to survive in frozen sediment for in excess of 70 days (Nizam, 1960). Algae are also found associated with ice in the Antarctic and the unicellular planktonic algae *Dunaliella* spp., *Chlamydomonas* spp. and *Pyramimonas* spp. remain motile to temperatures as low as −10 °C (Watanabe, 1988; Burch and Marchant, 1983).

These effects are primarily due to the temperature dependence of so many structures and processes. For example, cooling alters, to varying degrees, dissociation constants, retards

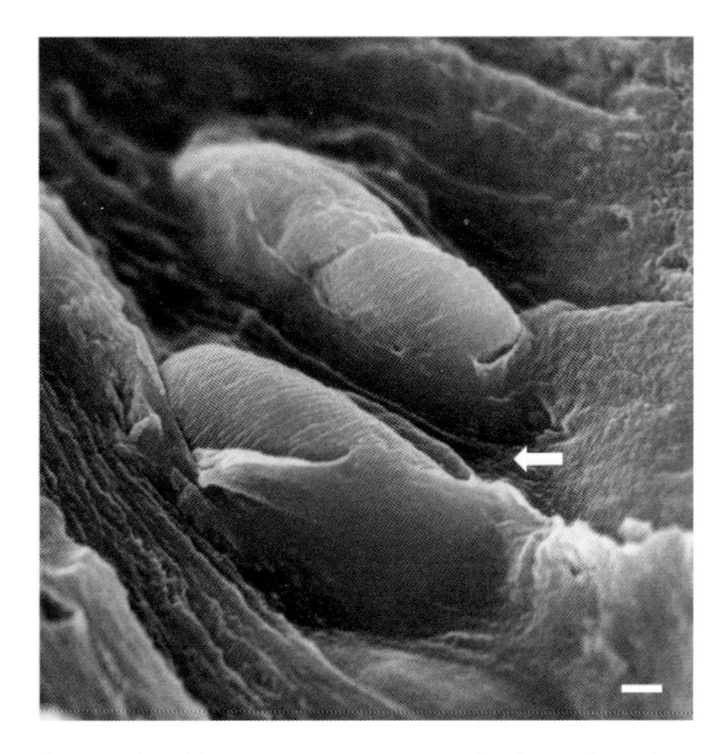

Figure 11.2 *Euglena gracilis* Klebs CCAP 1224/5Z encapsulated in sodium alginate, cryoprotected and cryopreserved. Fractured from the cryopreserved state, without thawing, in nitrogen slush and freeze-dried. Samples were imaged using a JEOL JEM 100CX-STEM using the Secondary Electron Detector. Arrow points to two *Euglena* protists encapsulated in alginate, scale bar represents 1 μm.

chemical reactions and induces phase changes (notably in water and lipids), which can result in complex secondary effects (Franks, 1985).

At temperatures used for cryopreservation (−196 °C) normal cellular chemical reactions cease, as kinetic energy levels are too low to allow necessary molecular motion (Grout *et al.*, 1990). At −196 °C the cell is effectively in an "arrested" state and theoretically the possibility exists for the long-term maintenance of material with little or no loss of viability. Indeed, studies on long-term cryopreserved algal cultures demonstrated no significant reduction in viability on up to 22 years of cryostorage (Day *et al.*, 1997), with no further reduction in viability to this day (J.G. Day, personal communication).

The earliest theories of freezing injury proposed that ice crystals formed in the extracellular solution pierced or prised cells and intracellular structures apart, that is via direct mechanical damage. It is generally true that intracellular freezing is lethal; however, the same cannot be said of extracellular freezing, as extracellular solutions are commonly seen to freeze during a two-step cryopreservation procedure. Osmotic stress (caused by ice forming and concentrating the solution around the cells), rather than ice crystal themselves, is attributed to freezing injury. The incorporation of cryoprotectant solutions, for example containing glycerol, act to protect cells by modulating the rise in salt concentration during freezing, possibly by moderating the osmotic gradient between intracellular and extracellular space (Polge *et al.*, 1949; Mazur, 1970; Pegg, 1987). Cryopreservation protocol development has in recent years included vitrification via single-step cooling, with cells being rapidly cooled to allow the cells and cell suspension to achieve a 'glassy' vitreous state

(Holt, 1997; Massip, 1996; Fahy, Levy and Ali, 1987) and the use of alginate encapsulation techniques (Figure 11.2), which ultimately permit the vitrification of the cells (Fabre and Dereuddre, 1990; Dereuddre *et al.*, 1992; Dereuddre, 1991; Janeiro, Vieitez and Ballester, 1996; Phunchindawan *et al.*, 1997; Mari *et al.*, 1995; Matsumoto and Sakai, 1995).

Membranes, as sites of freezing injury, have been observed as lysis of the cell or protoplast, leakage of electrolytes and other cell constituents and the breakdown of fine structure (Singh and Miller, 1985). The concept of membrane reduction during cell contraction following cryodehydration has been proposed as a mechanism of plasmolysis and extracellular freezing injury and was based on direct observations of reductions in cell volumes during plasmolysis (Williams, Willemont and Hope, 1981; Williams and Hope, 1981). Ultrastructural analysis of frozen-fixed rye cells during extracellular freezing, to lethal temperatures, highlighted membrane responses in cells susceptible to injury, where membranes have been reported to roll up and fuse, forming multilayered vesicles (Steponkus, Uemura and Webb, 1993). Model liposome studies demonstrated that the water of hydration around the phospholipid head groups is not only critical in the maintenance of the phospholipid lamellar lattice but is also instrumental in the prevention of fusion of adjacent bilayers (Singh and Miller, 1985). Perturbation of the water status in the vicinity of phospholipid bilayers by either dehydration or increased concentrations of intracellular cations (e.g., Ca^{2+}), as encountered during extracellular freezing, may cause destabilisation of the bilayers leading to fusion (Singh and Miller, 1985).

Free radicals may be generated and released without appropriate regulation by antioxidant species in response to freezing or cold shock. Free radicals themselves are a natural and intrinsic part of metabolic reactions in all living cells and one of their primary sources is oxygen. Although ubiquitous in photosynthetic organisms, oxygen at elevated levels may thus be deleterious promoting generation of toxic free radicals derived from oxygen (Fleck *et al.*, 2000). One of the primary reasons for the toxicity of free radicals is their highly reactive nature, which promotes the production of cascades of damaging chain reactions in living tissue (Benson and Withers, 1987; Benson and Noronha-Dutra, 1988; Benson, Lynch and Jones, 1992; Benson and Roubelakis-Angelakis, 1994). Reactive singlet oxygen can target membranes, nucleic acids and proteins, promoting the production of radical species. In many biological reactions there is a close relationship between free radicals and transition metal ions, which are involved in both radical generation and scavenging, for example the dismutation of superoxide. In the presence of a metal catalyst (copper or iron) the Haber-Weiss and Fenton reactions play an important role producing the (highly toxic) hydroxyl radical $OH^{\cdot}$ from hydrogen peroxide which may attack lipids, DNA, sugars, etc., causing a wide variety of damage (Esterbauer, Zollner and Schaur, 1988; Esterbauer, Eckl and Ortner, 1990). Free radicals, and in particular those derived from oxygen, have been associated with the responses of plants to many external stresses including dehydration, temperature extremes, heavy metals, herbicides, ionising radiation disease and senescence (Benson, 1990; Field, 1981, 1984). Although these free radical responses are associated with stresses, which have been encountered throughout their evolution in response/adaptation to their environment, for example drought, flooding, frost, etc., plants also encounter stresses during cryopreservation, for example cryodehydration, the use of high molarity cryoprotectant solutions and temperature extremes (−196 °C).

Free radical damage also occurs in tissues exposed to low and ultralow temperature storage and radical injury is exacerbated when suboptimal storage protocols are applied. Oxidative stress still occurs in mammalian transplant organs during low temperature storage (Fuller and Green, 1986; Cotterill *et al.*, 1989) and free radical-mediated loss of organ function and tissue damage can be ameliorated by applying antioxidants and free radical scavenging agents (Whiteley, Fuller and Hobbs, 1992a, 1992b; Pickford *et al.*, 1989; McAnulty and Huang, 1996, 1997; Green *et al.*, 1986). Free radical damage has been implicated in the low temperature storage recalcitrance of germplasm from a wide range of higher plant species (Benson, 1990). Sensitivity to seed storage treatments has been associated with free radical processes (Hendry *et al.*, 1992; Magill *et al.*, 1994) and a diverse range of antioxidant treatments have been used to enhance the longevity of seeds stored at subzero temperatures (Benson, 1990; Basu and Dasgupta, 1978; Gorecki and Harman, 1987). Similarly, cryopreserved cells and tissues of higher plant species can undergo free radical mediated oxidative stress, as evidenced by singlet oxygen formation, lipid peroxidation and free radical generation (Magill *et al.*, 1994; Benson and Withers, 1987; Benson and Noronha-Dutra, 1988; Benson and Roubelakis-Angelakis, 1994). Importantly, environmentally induced acclimation responses of higher plants to cold stress have been associated with the enhancement of antioxidant enzyme activities (Anderson, Prasad and Stewart, 1995). Knowledge of these natural protective mechanisms may be used to advantage by adapting them as cryoprotective strategies (Luo and Reed, 1997). These events are, however, less likely to influence ultrastructure in EM due to lower operating temperatures (<−130 °C).

Changes in temperature can alter membrane electrical properties independently of cryoprotectants (Chekurova, Kislov and Veprintsev, 1990) and must be considered in the development of any cryopreservation protocol. Exposure to solutions at lower temperatures (exposure temperature of 34 °C and 0 °C) have been demonstrated to reduce, by 18-fold, the lethality of solutions of epsilon-toxin (Lindsay, 1996). Reduced toxicity at lower temperatures may be due to the toxin interacting with cell surfaces via a temperature sensitive mechanism, for example permeability. At low temperatures, below the phase transition temperature, the freezing of lipids results in segregation of proteins (Haest *et al.*, 1974). The possibility of reducing the toxicity of a solution by exposing cells to the solution at lower temperatures clearly has important implications in cryobiology. This has been demonstrated in the cryopreservation of human skin with propane-1,2-diol, where the use of low permeation temperatures (4 °C) yielded higher post-treatment viability using a controlled two-step cooling protocol (Villalba *et al.*, 1996). Further studies investigating the effect of temperature and incubation time on the toxicity of propane-1,2-diol to mouse zygotes demonstrated that toxicity acted in a temperature and time dependent fashion (Mahadevan and Miller, 1997).

After the cryoprotectant exposure step, during which cells may be exposed to potentially toxic solutions (at or above 0 °C), the material for cryopreservation is further cooled. Cooling will commonly permit the intracellular and extracellular solutions to undergo some degree of supercooling with a further reduction in temperature inducing phase changes. Cryopreserved biological materials are likely to encounter three distinct condensed solute phases: liquid, glassy (vitreous) or crystalline. The condensed (i.e., water below 100 °C) phases of the solution owe their existence to the interactions between the constituent particles: atoms, ions or molecules. Each state has important consequences for the successful preservation of viable material.

11.6 CRYOPROTECTANTS

Despite the considerable advantages of cryofixation, the propensity for water to readily form crystals during freezing makes the fixation of material at low temperatures without damage to ultrastructure the most difficult and limiting step in EM cryo-specimen preparation. Thus, in much the same way as in cryopreservation, cryoprotectants (e.g. sugars and alcohols) can be introduced to the system to aid vitrification and mitigate against ice crystal formation (they are believed to attract/bind water molecules and reduce their mobility; during freezing this reduces their ability to integrate into a crystal). Tokuyasu cryosectioning employs sucrose as a cryoprotectant to allow vitrification at comparatively slow cooling rates (~15–50 °C/s). Sucrose, a cryoprotectant commonly used in plant tissue cryopreservation and also associated with low temperature acclimation (the process whereby an organism changes its physiology to allow it to survive low temperatures, e.g. during winter) is added to the chemically prefixed samples to prevent ice crystal formation (Tokuyasu, 1973).

The cryoprotectant solutions may be nominally divided into penetrating and non-penetrating compounds. Larger non-penetrating cryoprotectants are believed to act by protective dehydration of the cell, reducing the amount of cellular water available for intracellular ice formation (Kartha, 1985a, 1985b; Farrant, 1980; Benson, 1990). External cryoprotectants may also modify freezing, not only via dehydration but also by changing the medium surrounding the cells to provide a cryoprotective benefit by inhibiting ice crystals that would be formed in a more dilute medium from propagating into the cell (Karlsson 2010). The smaller penetrating, low molecular weight compounds are thought to protect the cell by lowering the temperature at which the intracellular water freezes (Kartha, 1985b; Franks, 1985; Benson, 1990). The solutes can, however, confer membrane protection by mechanisms other than by colligative action (an action dependent on the concentration of chemical and not on its nature) by affecting membrane stability through specific solute membrane interactions, for example reduction of the solute permeability of the membranes (Santarius, 1996). In systems with lower freezing temperatures the initial concentration of cryoprotectant must be higher to prevent the concentration of the potentially membrane-toxic solute in the unfrozen liquid from exceeding its critical limit (Santarius, 1996; Mazur, 1970). This non-specific behaviour of the solutes explains the additive benefits of cryoprotectants (Santarius, 1996).

At low temperatures, insufficient concentrations of rapidly permeating cryoprotectants may be cryosensitisers, accelerating membrane damage (Santarius, 1996). Under strongly hypertonic conditions, encountered during freezing, a gradual influx of non-penetrating solutes, including electrolytes, into the intrathylakoid space can take place. This may lead to membrane injury, and the reports of increasing thylakoid damage with reductions in temperature in the presence of limited concentrations of permeating cryoprotectants may be explained by an increase in the membrane permeability to electrolytes (Santarius, 1996). During rapid freezing regimes, non-penetrating cryoprotectants have been demonstrated to be efficient in protecting thylakoid membranes (Santarius, 1996). The protective action of sugars is considered to be due to their ability to reduce the permeability of thylakoid membranes to solutes (Santarius, 1996; Hincha *et al.*, 1996). Sugars and carbohydrates are also important in the overwinter survival of insects (Ushatinskaya, 1993). The cryoprotective efficiency of different solutes varies dramatically between different biological membrane systems and may indicate a complementary biochemical basis for cryoprotection. Cellular membranes have been identified as one of the primary sites of injury in frozen tissues and may therefore be protected through chemical stabilisation (Franks, 1985; Kartha, 1985b; Douzou, 1977; Benson, 1990). The naturally occurring cryoprotectants, trehalose and proline, have found success in the cryopreservation of plant germplasm and mammalian tissue (Dalimata and Graham, 1997; Iwasaki *et al.*, 1995; Saha *et al.*, 1996; Benson, 1990) and are believed to conserve membrane integrity during dehydration by the substitution of water molecules in membranes with sugar (Rudolf and Crowe, 1985).

In studies employing the cryoprotectants DMSO, glycerol and proline, maintenance of ATPase activity, coupling, unidirectional transport and membrane integrity were demonstrated in frozen and thawed endoplasmic reticulum (as reviewed by Benson, 1990). It has also been demonstrated that DMSO, ethylene glycol and formamide decrease both the cell membrane permeability for ions and the membrane potential (Chekurova *et al.*, 1990). Furthermore, cryoprotectants have also been demonstrated to act as free radical scavengers (Benson, 1990; Benson and Withers, 1987). In studies where DMSO, a water miscible solvent, was used as the cryoprotectant, evolution of methane has been detected, indicating that DMSO was acting as a free radical scavenger (Benson and Withers, 1987). DMSO is also known to be able to induce cellular differentiation and act as a radioprotectant (as reviewed by Yu and Quinn, 1994). These studies support non-colligative protective mechanisms of cryoprotectants (sugars, DMSO, glycol, etc.) in plants and animals (Benson, 1990; Piironen, 1993).

Cryoprotectants, which are essential for minimising cryoinjury during freezing, may also be toxic to biological systems. Monohydric alcohols, dimethyl sulfoxide (DMSO) and ethylene glycol (EG) are known to denature enzymes at room temperature (Adam, Rana and McAndrew, 1995). DMSO destabilises proteins at high temperatures; however, it has also been reported that DMSO protects isolated enzymes during freezing (Adam *et al.*, 1995; Anchordoguy *et al.*, 1992). This apparent paradox was attributed to temperature dependent, hydrophobic interactions between DMSO and non-polar moieties of proteins (Arakawa *et al.*, 1990). In studies on the interaction of DMSO with phospholipid bilayers, leakage from phospholipid vesicles has been attributed to the destabilisation of phospholipid membranes through a hydrophobic association between DMSO and the bilayer (Anchordoguy *et al.*, 1992). DMSO's action on thc stability of the liquid matrix of cell membranes appears to be responsible; these actions also appear to cause related effects on membrane permeability and fusion (Yu and Quinn, 1994). In addition to the above, cryoprotectants such as DMSO, ethylene glycol and formamide can decrease both cell membrane permeability for ions and the membrane potential with changes in membrane electric parameters depending upon the cryoprotectant type and concentration (Chekurova *et al.*, 1990). A similar strategy using dimethylformamide (DMF) has also been employed in electron microscopy to improve preservation of photoreceptors by cryofixation followed by freeze substitution (Meissner and Schwarz, 1990).

11.7 CRYOPRESERVATION OF BIOLOGICAL SYSTEMS

Animals from polar seas exhibit numerous adaptations to low temperatures, which prevent lethal freezing injury. Unique adaptations for freezing avoidance include the synthesis of low molecular mass ice-nucleating proteins, glycoproteins or peptides, which control and induce extracellular ice formation (Johnston, 1990). Marine poikilotherms also exhibit a range of adaptations that increase the rate of some physiological processes, partially compensating for the effects of low temperature (Johnston, 1990). Some aspects of the physiology of polar marine species, such as low metabolism and slow growth rates, probably result from low temperature adaptations and the highly seasonal nature of food supplies (Johnston, 1990). In addition, wood frogs can endure freezing for at least 2 weeks with up to 65% of their total body water in the form of ice (Storey *et al.*, 1996). Their adaptations to freezing include the capability to exhibit control over ice crystal growth in plasma with ice-nucleating proteins, the accumulation of low molecular weight cryoprotectants to minimise intracellular dehydration and the stabilisation of macromolecular components, and good ischemia tolerance by all organs (possibly including metabolic arrest mechanisms to reduce organ energy requirements whilst frozen) (Storey *et al.*, 1996; Storey, 1990).

Accumulation of the disaccharide trehalose has been observed in yeast, fungal spores, brine shrimp cysts and soil dwelling nematodes, and the accumulation the amino acid proline in haploid plants and marine invertebrates are examples of natural cryoprotectants (Rudolph and Crowe, 1985; Behm, 1997). The strategy of accumulating compounds that prevent deleterious alterations in membranes during reduced water states (a common point of cryoinjury) have also been successfully employed in the cryopreservation of tissues (Rudolph and Crowe, 1985; Madden *et al.*, 1993; Crowe *et al.*, 1985; Bhandal *et al*, 1985).

The mechanisms employed in nature to protect against freezing injury explain the rationale behind two-step cryopreservation protocols. These protocols are dependent upon cell dehydration and the concentration of the cytosol, in order to preclude or minimise ice crystallisation during freezing (Ashwood-Smith and Farrant, 1980; Farrant and Ashwood-Smith, 1980). The observation that intracellular ice formation was influenced by the cooling rate (and the importance of supercooling the tissue below its freezing point for ice formation to occur) was first made by Müller-Thurgau whilst studying frozen potatoes. Only certain parts of the tissue formed ice, which corresponded with dead tissue on thawing (Müller-Thurgau, 1886). More recently, cryopreservation protocols that control concentration of the cytosol, thus minimising injury allowing viable cells to be recovered post-thaw, have been developed (Day and McLellan, 1995). During freezing, as ice forms in the extracellular solution, a gradient of water potential is established between the unfrozen intracellular space and the partially frozen extracellular compartment. The consequence is that the chemical potential of the water in the partially frozen extracellular solution decreases. If the cooling process is slow enough the cell will respond by losing water across the semipermeable plasma membrane until the osmotic potential across the plasma membrane is returned to a state of equilibrium. In cells that are cooled too rapidly the chemical potential of the extracellular solution decreases much faster than the rate at which water can diffuse from the cell. The end result is intracellular ice formation in the cytosol (Figure 11.3C) (Franks, 1985; Farrant, 1980; Toner, 1993). However, freeze-induced concentration of solutes is not a continuous function of temperature. The concentration of the remaining solution will eventually reach a point at which it will form a glass, thus preventing any further concentration of the now supercooled cytosol (Figure 11.1). If sufficient concentration of the cytosol also occurs the cytosol will also vitrify. This will occur if sufficient time is allowed for osmotic equilibrium of the cytosol to occur, that is in slowly cooled specimens (Franks, 1985; Farrant, 1980; Steponkus *et al.*, 1992). An optimal cooling rate exists (Figure 11.3B) because at too-rapid cooling rates the probability of intracellular ice formation is increased (Figure 11.3C), whereas at excessively slow cooling rates (Figure 11.3A) the time the specimen is exposed to freeze concentrated solution is increased and cells suffer from dehydration (Moor, 1964; Franks, 1985; Farrant, 1980; Steponkus *et al.*, 1992). Cryoprotectant solutions are commonly introduced to mitigate injury encountered during freeze-induced dehydration (Farrant, 1980; Steponkus *et al.*, 1992). Once the cryoprotected system is brought to a state close to thermodynamic equilibrium during cooling to subzero temperatures (−30 °C), it is safe to rapidly cool down the sample to cryogenic temperatures since the sample is now cryoprotected due to partial dehydration and elevated solutes concentration.

The optimal cooling rate required for viability cryopreservation depends on the specific characteristics of the sample and needs to be found out empirically. Different tissues have different concentrations of native cryoprotectants and membranes differ in their permeability to water molecules, and hence to dehydration rates upon freezing. Figure 11.4 describes the optimal cooling rate as an inverted U-shaped function with a peak located at different positions on the cooling rate axis depending on tissue or organism type.

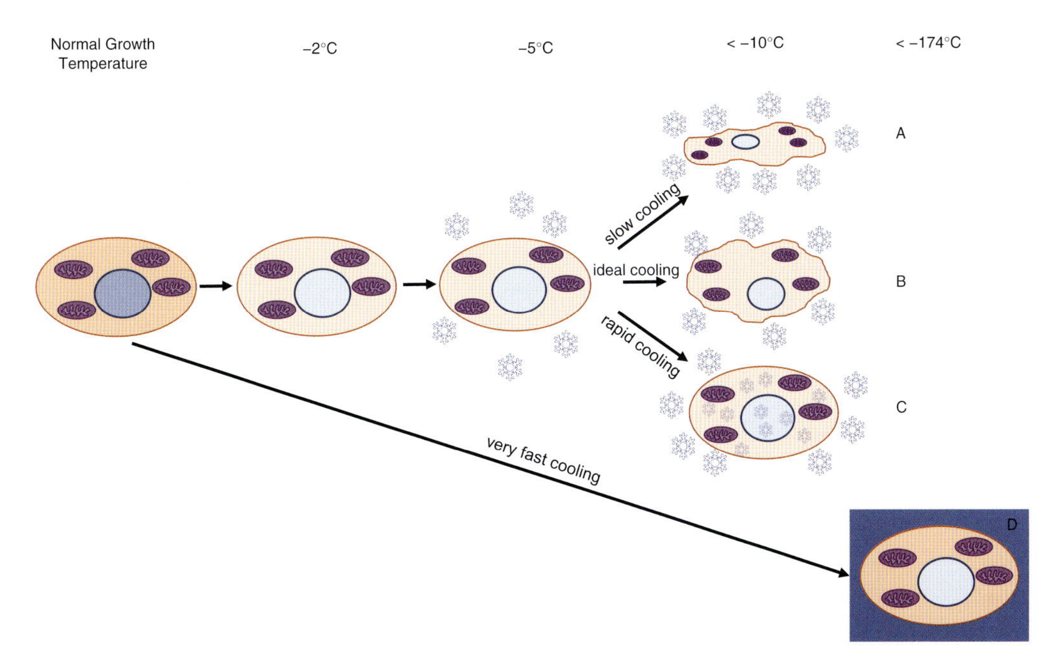

Figure 11.3 Preservation of structure and viability upon freezing depends on the cooling rate of the sample. Preservation of viability is obtained when cooling rates allow for partial dehydration of cells by the action of growing extracellular ice crystals, as well as by the addition of cryoprotectants (B). If cooling is too slow cells may die due to excessive dehydration (A). If cooling is too fast cells may die due to the formation of intracellular ice crystals (C). For structural preservation the samples should be free of ice crystals and cryoprotectants and vitrification can be achieved only when cooling is extremely fast (D). Adapted from Mazur (1977) and Mullen and Critser (2007).

The combination of knowledge of the process of cryodehydration and its capacity to concentrate intracellular space to a point where vitrification can be achieved at comparatively slow cooling rates allowed processes associated with plant cold acclimation (tolerance to low temperature and freezing injury) to be addressed through direct observation of membrane injury by freeze fracture (Fleck, 2001) (Figures 11.5 and 11.6). In this example the leaf was subjected to extracellular ice nucleation, simulating the effect of freezing of surface water on a leaf (frost) (Figure 11.5). The extracellular freezing induced cryodehydration of the tissue (the principal cause of cell damage) and supported manual plunge freezing into liquid propane from subzero temperatures as high as −4 °C after only a few minutes of cryodehydration (<5 min) (Figure 11.5). This allowed the role of COR genes and changes in soluble sugar concentrations in cold acclimation to be deciphered (Steponkus and Webb, 1992; Webb *et al.*, 1996; Artus *et al.*, 1996).

Further adaptations are in encapsulation methodology, which may limit the effects of exposure to the frozen crystalline environment. In addition, frozen materials of relatively low solubility, for example alginate, have been reported to exhibit inhibition of spherulite ice crystal formation (Franks, 1985; Luyet and Rapatz, 1958). The benefits of encapsulation have been applied to the two-step preservation of gametophytes of Laminaria (Vigneron, Arbault and Kaas, 1997). In addition, encapsulation has been employed as a procedure to avoid fracture damage during freezing and thawing of rabbit embryos (Kojima *et al.*, 1990). Studies comparing encapsulated and non-encapsulated embryos have reported that encapsulated embryos, cooled by conventional controlled two-step cooling techniques, have

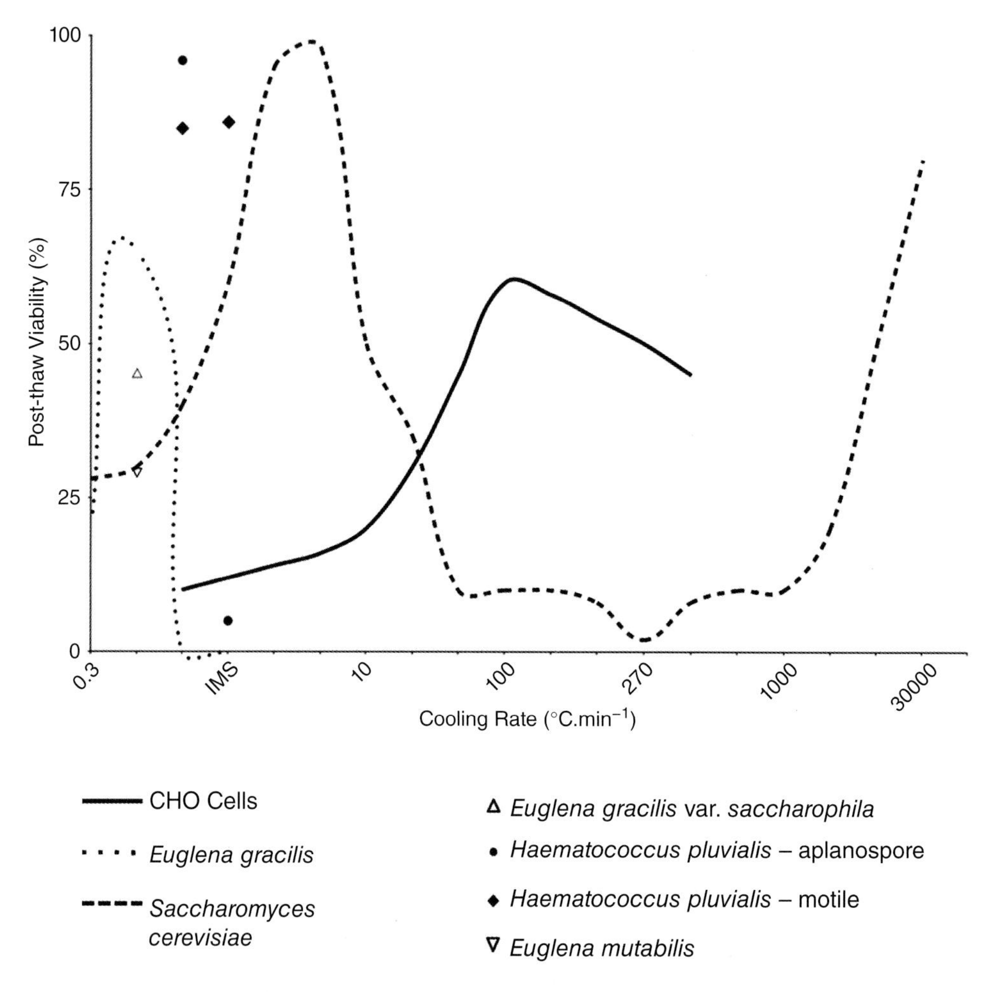

Figure 11.4 Effect of cooling rate on post-thaw viability of a range of organisms. Inverted 'U' cooling curves are evident with slower cooling rates, resulting in loss of viability through excessive cryodehydration, while higher water permeability allows faster cooling rates without cell death due to intracellular ice nucleation and faster cooling rates, reducing viability through intracellular ice nucleation: *Euglena gracilis* (CCAP 1224/5Z) (dotted line), mammalian CHO cells (solid line), *Saccharomyces cerevisiae* (dashed line). In the data shown for *Saccharomyces cerevisiae* very fast cooling rates >1000 °C/min circumvent both cryodehydration and intracellular ice nucleation injury by vitrification. Two related euglenoids cryopreserved under identical conditions highlight variations within species to cryopreservation tolerance *Euglena gracilis* var. *saccharophila* (CCAP 1224/7B) (upward pointing triangle) and *Euglena mutabilis* (CCAP 1224/40) (downward pointing triangle). For the two distinct life stages of *Haematococcus pluvialis* (CCAP 34/8) a significant variation in tolerance to cooling rate is shown (solid diamonds: motile stage and solid dots: aplanospore stage) likely to be due to differences in permeability between the dormant resting stage and motile stage. Adapted and redrawn from Mazur (1970), Dumont, Marechal and Gervais (2003) and Fleck (1998).

enhanced function and survival after thawing, which ultimately permits increased numbers of offspring from frozen and thawed embryos (Kojima *et al.*, 1990). An alternative to the controlled cooling of encapsulated cells involves the dehydration of encapsulated cells with a highly concentrated vitrification solution (PVS2) followed by rapid cooling by plunging in LN (Phunchindawan *et al.*, 1997).

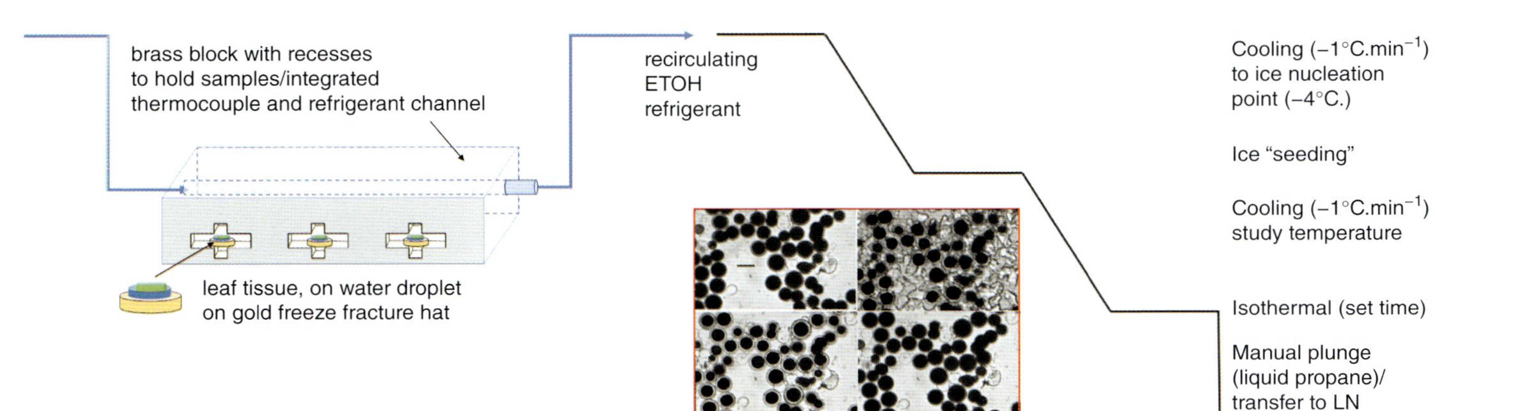

Figure 11.5 Experimental system to allow controlled modelling of the effect of extracellular ice formation on leaf tissues, simulating the effect of winter frost forming on a leaf. The leaf section is placed on a water droplet on a standard 'gold hat' freeze fracture carrier. The carrier is then placed in small chambers machined in a brass block with flow through refrigeration channels. The whole system is progressively cooled following a predetermined cooling regime. At a predetermined subzero temperature extracellular ice nucleation is triggered by carefully brining a lightly frosted tip into contact with the water droplet. Further cooling to a temperature nadir, a fixed isothermal period and direct plunging into liquid propane are applied. In a controlled cooling experiment simulating the process of frost forming on the outer surface of a leaf, cryodehydration is modelled and mechanisms of cold-induced injury deduced. The inserted image (boxed red) shows *Haematococcus pluvialis* Flotow CCAP 34/8 undergoing cryodehydration (top right plate) and rehydration post-thaw (bottom two plates).

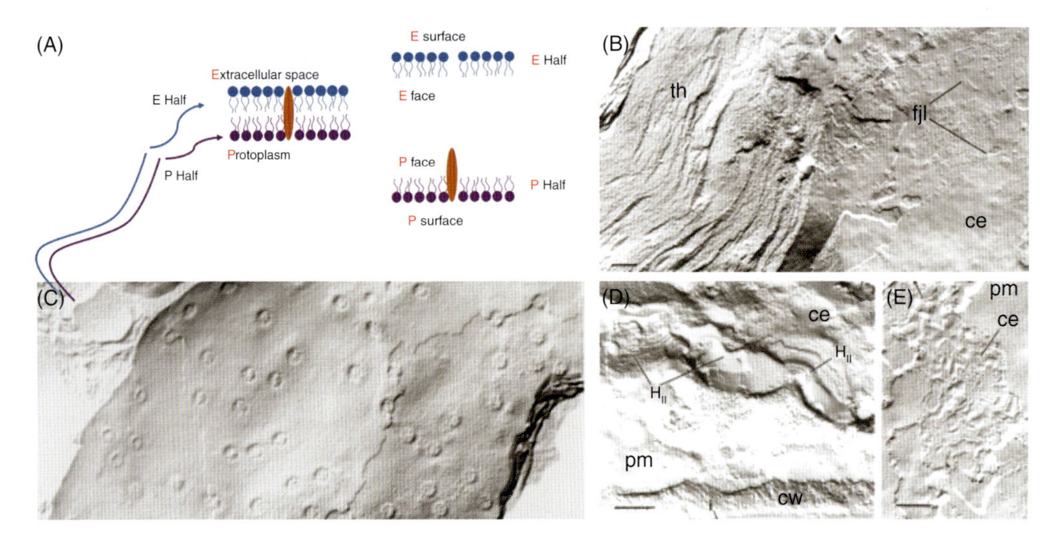

Figure 11.6 Freeze fracture/freeze etch replica viewed by transmission electron microscopy to investigate mechanisms of freeze induced dehydrative injury and cold acclimation in plants. (A) shows the theory behind freeze fracture in biological tissues where a fracture plane extends through membrane bilayers to reveal external and internal faces and surfaces. (C) shows a fracture replica through the nuclear envelope showing p-face and e-surface planes with nuclear pores linking the two. (B) and (D) show commonly observed freeze induced dehydrative injury, where (B) is the fracture jump lesion (fjl) in the chloroplast envelope (ce) identified by the presence of thylakoid membrane in the replica (th) and (D) is hexagonal II phase transformations of a lipid bilayer and (E) the preceding whorl like changes to the bilayer. Specific 'notation' is employed to describe the surface being viewed. The interior views revealed when the membrane is split by freeze fracture are termed faces to distinguish them from the true surfaces (potentially exposed by etching). 'Face' refers to the fracture face within a membrane exposed by the fracture, 'Surface' refers to the natural surface of the membrane exposed by etching. Biological membranes can be considered as being a lipid bilayer with intercalated proteins (Singer and Nicolson, 1972; Kopp, 1972, 1973). The half-membrane leaflet adjacent to the extracellular space is termed the E half and that adjacent to the protoplasm is termed the P half. The P face is the fracture face of the P half of the membrane and will always have protoplasm beneath it. The E face is the fracture face of the E half of the membrane and has extracellular space beneath it (Branton *et al.*, 1975). Scale bars represent 200 nm.

11.8 VITRIFICATION THE 'KEY' TO CRYO-FEGSEM

The avoidance of crystalline ice upon freezing is also fundamental to unlocking the resolving potential of cryo-electron microscopy. Vitrification allows for structures to be viewed in their native state, avoiding the artefacts commonly associated with chemical fixation and dehydration preparation techniques as well as the ultrastructural disruption attributed to intracellular ice formation (Dubochet *et al.*, 1983; Ebersold, Cordier and Lüthy, 1981; Murk *et al.*, 2003, with protocols reviewed by Fleck, 2015). Cryoimmobilisation, if correctly applied to a biological tissue, proffers avoidance or minimisation of the following deleterious effects: loss and redistribution of ions and molecules, extraction of lipids, depolymerisation of proteins, denaturation of enzymes, disruption of membrane permeability, osmotic effects, shrinkage and loss of antigenicity. Thus, vitrification techniques have been extensively adopted for the preparation of biological material for ultrastructural investigation by electron microscopy (Murk *et al.*, 2003; Studer, Hennecke and Müller, 1992; Studer, Chiquet and Hunziker, 1996; Verkleij *et al.*, 1985; Hunziker & Herrmann, 1987).

Direct single-step cryopreservation protocols aim to prevent intracellular ice growth and limit or exclude the exposure of the material to the dehydration effects of extracellular ice by adopting rapid cooling rates and vitrifying the system (Armitage, 1989). Vitrification strives to avoid/reduce the complex set of interacting variables that must be simultaneously optimised during two-step protocols, circumventing problems of chilling sensitivity/injury and removing the requirement for specialised expensive equipment.

Vitrification (discussed above) is readily achieved by cooling a viscous liquid below its thermodynamic freezing point, through its metastable supercooled regime and finally below its 'glass transition' temperature T_g (Figure 11.3) (Stillinger, 1995). The formation of a vitreous state is an intrinsic property of all liquids, and pure water and aqueous solutions will readily vitrify given a sufficiently high cooling rate (MacFarlane, 1987). Characteristically, vitrification solutions employ molecular liquids (e.g. propylene glycol) at concentrations in excess of 40% (w/w) (MacFarlane, 1987). Vitrification solutions successfully developed for plant and animal systems are commonly comprised of high concentrations of conventional cryoprotectants (Armitage, 1986; Armitage and Rich, 1991) and/or more novel sugar constituents (sucrose) (Ishikawa *et al.*, 1996, 1997; Matsumoto *et al.*, 1997; Takagi *et al.*, 1997). As such, a sample prepared in this way should be referred to as being in a glass or a vitreous state and not a glassy or vitreous ice!

The principal of vitrification for the preservation of tissue relies on the concept of rapid cooling to allow liquid water to reach a metastable amorphous (vitreous state) and thus 'fix' a tissue without the changes in volume and osmotic challenges associated with freezing of water (Figure 11.7). The most common approach in cryopreservation of tissue by vitrification is to introduce a cryoprotectant as demonstrated by Sakai to preserve a range of plant tissues using a PVS2 vitrification solution containing PVS2 (w/v) 30% glycerol, 15% ethylene glycol and 15% DMSO and 0.15 M sucrose (Sakai, Kobayashi and Oiyama, 1990). Others adopted vitrification procedures that precluded intracellular ice formation in Drosophila embryos by employing ultrarapid cooling and warming rates to minimise chilling injury (Steponkus *et al.*, 1990). The materials and methods section describes rapid

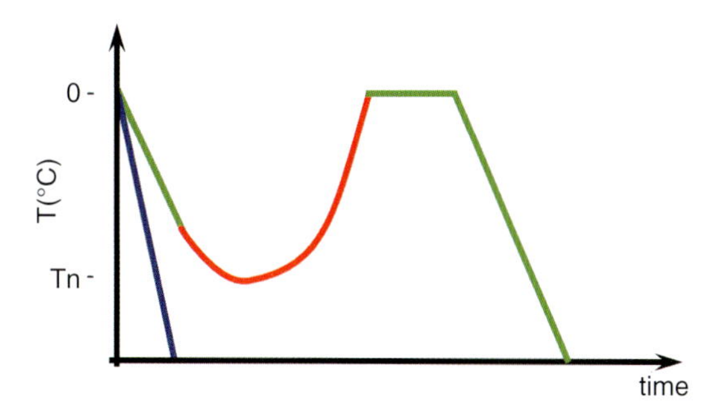

Figure 11.7 For vitrification very fast rates of cooling are required (blue line), allowing the sample to proceed to a temperature below the nucleation point and achieve a vitreous amorphous state. This avoids both the volume changes and exothermic events (red line) associated with ice nucleation when samples are cooled at slower rates (green line). There can be a large temperature difference between the temperatures of nucleation (T_n) where heterogeneous nucleation may occur at −2 °C and homogeneous nucleation (in the absence of nucleation sites) at −38 °C or lower. Graph adapted from Robards and Sleytr (1985).

Table 11.2 Differences between cryogen solid and boiling points (°C)

Cryogen	T solid (°C)	T boiling (°C)	T difference (°C)
Nitrogen	−210.05	−195.85	−14.2
Ethane	−183.45	−88.75	−94.7
Propane	−189.85	−42.25	−147.6

cooling in a manner not dissimilar to that currently employed to produce vitreous thin films for cryotransmission electron microscopy (Dubochet and McDowall, 1981) or material for fracture/cryo-SEM. Five strategies were described:

1. Plunging directly into liquid nitrogen with the volume of material reduced and surface area available for cooling increased by loading the embryos in straws. The authors describe the use of polypropylene straws as a specimen container and liquid nitrogen as the cryogenic fluid. They measure a relatively slow cooling rate, attributed to the Leidenfrost effect (Leidenfrost occurs because LN is used at or just below its boiling temperature and when a warm object is placed in the LN the temperature of the LN is easily and rapidly raised above its boiling point creating an insulating layer of vapour phase nitrogen; see Table 11.2) (Leidenfrost, 1756; Wares and Bell, 1966). In this case vaporised liquid nitrogen around the straw impedes heat transfer and was determined as limiting the cooling rate between 0 and −60 °C to 15 °C/s and between −60 and −120 °C to 50 °C/s (Steponkus *et al.*, 1990).
2. Plunging of straws into liquid propane liquefied with liquid nitrogen attained a faster initial cooling rate (0 to −60 °C) determined as >55 °C/s (Steponkus *et al.*, 1990). In this case the increased separation between the solid and boiling point of the cryogen allows the liquid propane cryogen to be cooled to close to the temperature at which it solidifies. This allows the cryogen to conduct heat away from the sample without warming sufficiently to form a vapour phase (Table 11.2).
3. A cooling procedure where the volume was reduced to 20 μl drops expelled directly from a syringe into liquid propane allowed the group to record the first successful recovery of viable eggs from storage in liquid nitrogen (Steponkus *et al.*, 1990). Although they were not able to determine the exact cooling rate, it is likely to have been faster than with straws. The 20 μl volume corresponds to the observations that plunge freezing (slam freezing, plunge freezing, propane jet freezing and spray freezing) techniques employed in electron microscopy can achieve – a maximum of 10–20 μm of vitrification and the spray freezing techniques described by Bachmann and Schmitt (1971) (e.g. high surface area/volume ratio, which was also used for the first time for vitrifying water from the liquid state) (Figure 11.8).
4. The first successful cryopreservation of Drosophila embryos was further refined by placing eggs on a copper grid and forcibly propelling them into liquid propane in much the same way as a guillotine plunge freezing instrument works (e.g. the commercially available Vitrobot (Thermo Fisher Scientific, USA), EM GP2 (Leica Microsystems, Austria), CP3 (Gatan, USA)) to provide faster streaming of cryogen on both surfaces (Frederik and Storms, 2005; Handley *et al.*, 1981, Bellare *et al.*, 1988; Dubochet *et al.*, 1981; Fernández-Morán, 1960). The measured cooling rate (0 to −60 °C) was determined to be 700 °C/s (Steponkus *et al.*, 1990).
5. To mitigate toxicity to the embryos from the propane itself a final optimised protocol was described that utilised liquid nitrogen slush as the cryogenic fluid. This minimised the generation of an insulating vapour phase between the embryos and the cryogen by cooling

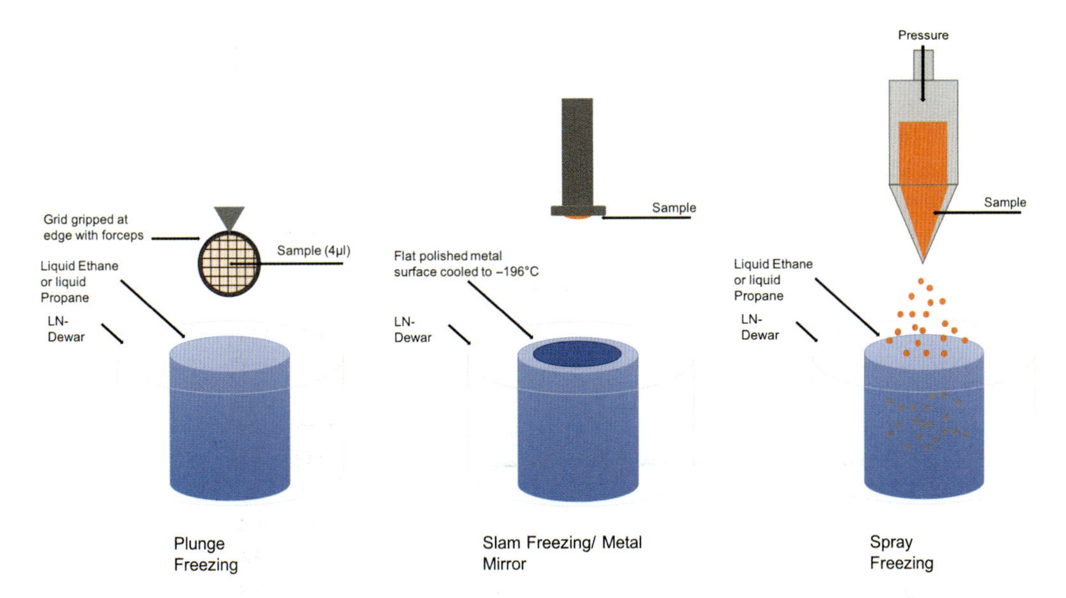

Figure 11.8 Examples of three rapid freezing systems commonly employed in electron microscopy able to achieve modest vitrification depths of between 10 and 20 µm (Van Harreveld and Crowell, 1964; Heuser, Reese and Landis, 1976; Bachmann and Schmitt, 1971; Knoll, Verkleij and Plattner, 1987; Dubochet and McDowall, 1981). Adaptations of plunge freezing and spray freezing were used to successfully vitrify and recover viable Drosophila embryos.

the LN to an average temperature of −207 °C (Sansinena et al., 2012), by evaporation, to close to its solidification temperature and was designed to provide a cooling rate (0 to −60 °C) of 400 °C/s (Steponkus *et al.*, 1990).

Plunge freezing methods for vitrifying samples for cryo-EM have been widely adopted and aim to rapidly immerse a sample in a liquid cryogen with the intention of rapidly cooling the specimen and achieving a vitreous state (Dubochet and McDowall, 1981). Today plunge freezing is routinely used to prepare small sample volumes mounted on TEM grids for cryo-TEM. Plunging cells grown directly on grids can also be effectively employed to prepare material for cryo-FEGSEM, where cell surfaces are of interest (Figure 11.9). There have been many descriptions of plunge freezing equipment since first postulated by Altmann (1894) and Fernández-Morán (1960), ranging from simple mechanical plunging systems (Handley *et al.*, 1981) to the more elaborate (Bellare *et al.*, 1988; Frederik and Sommerdijk, 2005; Frederik and Storms, 2005). Ballistic cryofixation used a rifle to fire a hypodermic needle (~400 m/s) through muscle into liquid propane (Monroe *et al.*, 1968). A rubber-band 'powered' injector for TEM grids containing specimens was described by Rebhun (1972), with the precursor of the modern plunge freezer described by Glover and Garvitch (1974), which used a spring powered piston that also acted as a shock absorber. Attempts to improve freezing by super cooling liquid nitrogen to its freezing point (−210 °C) were described by Umrath (1974), where the liquid nitrogen was evacuated to allow it to cool to its freezing point. This approach is seen in the liquid nitrogen 'slush' systems used to freeze tissues for the Gatan and Quorum Technologies cryo-SEM preparation systems. The modern guillotine plunge freezing devices were first named by Luyet and Gonzales (1951) and described by Costello and Corless (1978), whereby a gantry supported a plunge-rod mechanism.

Less common approaches to vitrification of tissues include spray freezing (where the small microdroplet are effectively plunge cooled exploiting the least surface area-to-volume ratio of the specimen to provide efficient cooling). Early applications used a nebuliser and spray

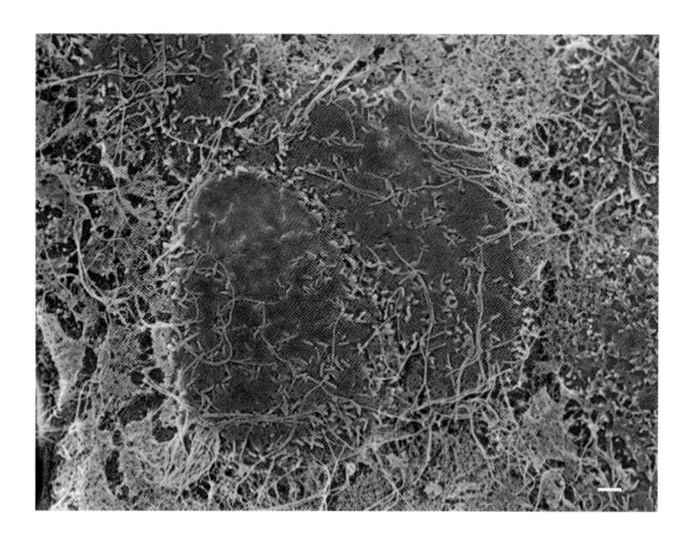

Figure 11.9 Mammalian cells infected with influenza, imaged with a JEOL 7401F Cold Field Emission SEM at 3 keV. Infected cells were grown on TEM grids and plunge frozen in liquid ethane (CP3, Gatan) and transferred (VCT100, BAL-TEC AG, Liechtenstein) to the FEGSEM after multiangle coating with 4 nm Pt/C (BAF060, BAL-TEC AG, Liechtenstein). Scale bar represents 1 μm.

gun for the freeze drying of virus (Williams, 1954) and then refined for freeze fracture by Bachmann and Schmitt (1971), Bachmann and Schmitt-Fumian (1973a, 1973b) and Plattner, Schmitt-Fumian and Bachmann (1973) using artist airbrushes. In fact, it was the freezing of microdroplets that led for the first time to true vitrification of liquid water (Brüggeller and Mayer, 1980). Attempts to improve thermal conductivity and improve freezing still further have employed cold metal surfaces (Gersch, 1932), which led to the development of compression 'plier' types with copper plates cooled in liquid nitrogen (Eránkö, 1954; Hagler and Buja, 1984; Tvedt *et al.*, 1988). Slam freezing against silver or copper blocks has been effectively described by Escaig (1982, 1984), Heuser *et al.* (1979), Van Harreveld and Crowell (1964) and Van Harreveld and Trubatch (1979). Refinements and variations on mechanical slam freezing techniques are descried by Akahori *et al.* (1980), Boyne (1979), Sjöström, Johansson and Thornell (1974), Verna (1983), Heath (1984), Allison, Daw and Rorvik (1987), Edelmann (1989), Livesey *et al.* (1989), Escaig (1982) and Escaig, Géraud and Nicolas (1977).

An alternative strategy to increase the conductivity of cooling is jet cooling, whereby a liquid cryogen (liquid propane) is jetted across the specimen. First introduced by Moor, Müller and Kistler (1976), it was later refined by Müller, Meister and Moor (1980) and Haggis (1986). Variants include the 'single sided' jet (Pscheid, Schudt and Plattner, 1981; Knoll, Oebel and Plattner, 1982) and 'double sided' jet (Müller *et al.*, 1980). These technologies have largely been superseded by commercial high pressure freezers, which is the other principle the vitrification technique employed today when the vitrification depth desired exceeds that achievable with plunge freezing (Studer *et al.*, 1989). High pressure freezing is where a high pressure double jet of pressurised liquid nitrogen is used to maximise thermal efficiency and conductive cooling to achieve vitrification under the application of high pressure (Riehle, 1968; Riehle and Hoechli, 1973).

Thus, improved vitrification may be achieved with high pressure freezing (Moor *et al.*, 1980; Müller and Moor, 1984; Riehle, 1968; Studer *et al.*, 1995, 2001; Studer, Humbel and Chiquet, 2008; Galway *et al.*, 1995). High pressure freezing (HPF) can extend the depth of vitrification of the sample to 200 μm by employing high pressure with rapid cooling (2100 bar, 10 000 °C/s), which restricts the expansion of water during freezing, thus inhibiting crystallisation whilst cooling the system below the T_g (~−140 °C) and allowing vitrification of complex tissues (Kanno, Speedy and Angell, 1975; Studer et al., 1995, 2008). At sufficiently high pressures (210 MPa) the cooling rate required for the vitrification of water is reduced from several 100 000 °C/s to a few 1000 °C/s, making vitrification of relatively thick samples practicable (Studer *et al.*, 2008; Shimoni and Müller, 1998). Good heat conduction through the sample and between the sample and the carrier upon which it is mounted is vital for effective freezing. Air-filled spaces between the tissue and the sample holder used for freezing can reduce the freezing efficiency by acting as insulators. Furthermore, air-filled spaces are likely to collapse when high pressures are applied, deforming the tissue. To solve these problems, sample holders are often loaded with cryoprotectants, acting as both a cryoprotectant to improve freezing and protecting the sample against pressure induced shearing, or only fillers to protect against pressure artefacts (Hohenberg, Mannweiler and Müller, 1994; Studer *et al.*, 1992, 1989; Galway *et al.*, 1995).

Employing vitrification techniques (with or without the aid of a cryoprotectant) to cryo-immobilise tissues has allowed substantial investigations of membranes to be performed by a technique referred to as freeze fracture, or when the sample is also subjected to a degree of water sublimation following fracture – freeze fracture freeze etch (Steere, 1957, 1973; Moor *et al.*, 1961; Moor, 1970; Moor and Mühlethaler, 1963). Freeze fracturing of a frozen specimen cracks the material to reveal a plane through the tissue. The fracture occurs along weak hydrophobic planes of the tissue such as membranes or surfaces of organelles and, once cleaved, both surfaces are available for viewing. This is particularly powerful when applied to study membrane structure and organisation (Figures 11.6B and 11.10) (Kopp, 1972; Branton, 1973).

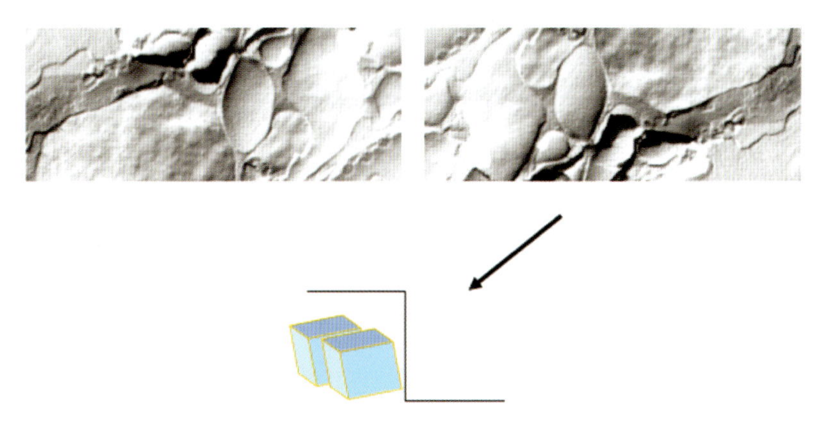

Figure 11.10 Correct orientation of an image is critical in freeze fracture. The correctly orientated image is on the top left. However, by rotating the image through 180° the topography of the image is inverted (top right). This is due to the unidirectional fixed angle shadowing (45°) with Pt/C applied to the fractured membrane to generate the surface replica. This is illustrated by the lower cartoon with the surface coating applied at 45° (arrow), only covering the sample surfaces (blue boxes) 'visible' to the coating source (black line). Surfaces orientated towards the shadowing source have a thicker coating of Pt/C than surfaces obscured from the shadowing source. When photographed in the TEM, differences in coating thickness contribute to the different levels of contrast and give the impression of 3D topography.

Traditionally, the samples were prepared for imaging in the transmission electron microscope (TEM) as the SEMs available at the time lacked the spatial resolution at operating conditions suitable for directly imaging the fracture face (for more information refer to Chapter 12 by Tacke, Lucas, Woodward, Gross and Wepf). To view fractured surfaces and membranes in the TEM a protracted and complex workflow (which may take many days in the case of plant material) had to be completed:

1. Cells are quickly frozen (by HPF, directly in liquid nitrogen or intermediate cryogen) to immobilise cell components.
2. A block of frozen cells is fractured under vacuum. This fracture is irregular and occurs along lines of weakness like the plasma membrane or surfaces of organelles (Kopp, 1972; Branton, 1973; Pinto da Silva and Branton, 1970).
3. Surface ice and unbound water from tissues can be removed by specific temperature/vacuum combinations (freeze etching).
4. The surface is shadowed, possibly from a single direction, with platinum/carbon (Moor, 1970).
5. A layer of carbon is evaporated vertically on to the surface to stabilise the surface.
6. The organic material is digested away by acid, leaving a replica.
7. The platinum–carbon (Pt/C) replica is collected on a grid and examined by a transmission electron microscope.

In addition to being complex, freeze fracture requires the scientist to have specialist knowledge of how the fracture is created and how the image is generated in the TEM. Knowing the direction of shadowing is essential to understand whether a structure is elevated or depressed. The TEM image may appear in 3D to the viewer and the impression of topography may be implied. However, as the image is a 2D projection of the 3D replica with differences in contrast primarily due to differences in the thickness of the Pt/C replica, correct orientation of the image is necessary to reveal the true topography of the sample (see Figure 11.10).

In a FEGSEM with good resolution at low accelerating voltage operating conditions, similar or higher resolution fracture images can be generated directly from the fractured frozen surface without the need to generate a replica. Images generated from the secondary electron detector provide correctly orientated topography without the need to orientate the image with respect to the shadowing source. Coating may be applied by sputter or electron beam (ebeam) approaches (refer to Chapter 12 by Tacke, Lucas, Woodward, Gross and Wepf and Chapter 9 by Resch for discussions of coating technologies and their applications). Electron-beam coating with Pt/C, Cr or Ta can be conducted by double axis rotational shadowing (DARS) (Shibata, Arima and Yamamoto, 1984; Hermann *et al.*, 1988) where shadowing/coating involves initial shadowing with Pt/C at a constant angle (e.g. 1.5 nm 45°), followed by Pt/C (1.5 nm) coating during rotation of the sample (one rotation axis) as well as of the electron beam gun itself used for Pt/C evaporation (second rotation axis) (Shibata *et al.*, 1984; Hermann *et al.*, 1988). Alternatively, multiangle rotational shadowing (MARS) can be used as a single evaporated layer of Pt/C or Cr (4 nm or 3 nm), which is applied with continuous rotation and tilting of the sample or evaporation source to provide an even coating of complex surfaces (Figures 11.11 and 11.12). Other high resolution shadowing conditions have been described that employ a double layer coating (Pt/C, 2 nm, 45° followed by C, 6 nm, 90°) to allow both high resolution and excellent charge dispersion for imaging (Walther and Müller, 1997).

By combining high resolution 3D imaging techniques possible with cryo-TEM with the capacity to view large surface areas in the FEGSEM at high resolution (comparable to those

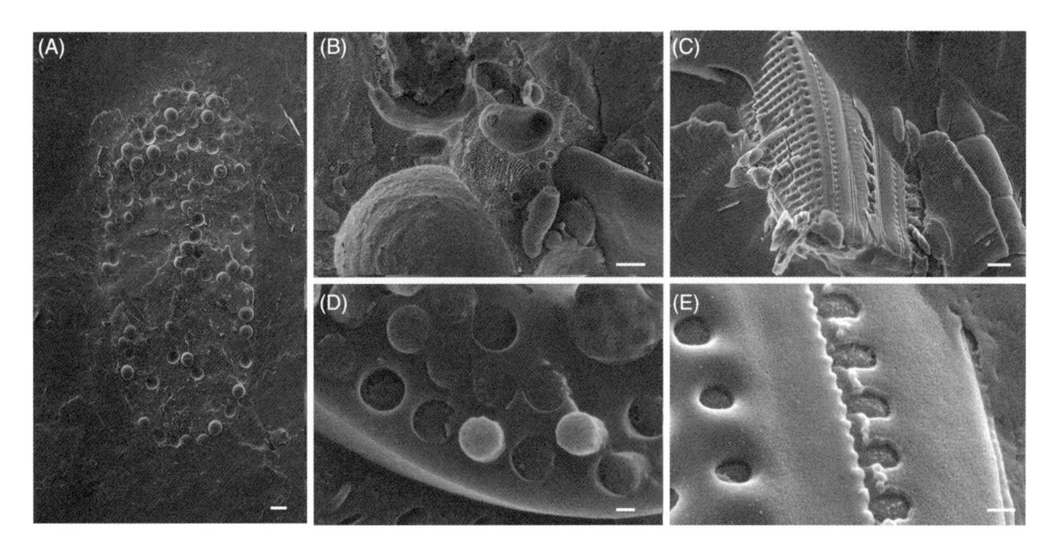

Figure 11.11 The versatility of the cryo-FEGSEM when combined with freeze fracture. (A) Transverse fracture of the *Paramecium bursaria* Focke CCAP 1660/3B packed full of symbiotic Chlorella; scale bar represents 10 μm. (B) Cross fracture of *Euplotes vannus* Minkjewicz CCAP 162/12 highlighting various intracellular organelle; nucleus with nucleopore structures (bottom centre) and Golgi apparatus (central), scale bar represents 1 μm. (C and E) Mixed diatom culture containing both filamentous and pennate forms; scale bars represent 1 μm and 100 nm in (C) and (E) respectively. (D) Transverse fracture through the flagella of *Paramecium bursaria* clearly showing the microtubule doublet organisation and central Axoneme; scale bar represents 100 nm. Samples for FEGSEM were prepared by high pressure freezing (HPM010, BAL-TEC AG, Liechtenstein), fractured (BAF060, BAL-TEC AG, Liechtenstein) and coated with 4 nm Pt/C using a MARS shadowing technique. Imaging was at 3 keV on a JEOL 7401F equipped with a VCT100 cryostage (BAL-TEC AG, Liechtenstein).

of the TEM tomography studies) it has been possible to elucidate the complex structure, organisation and distribution of knob structures presented on the surface of erythrocytes infected with Plasmodium (Watermeyer *et al.*, 2016) (Figure 11.12). Findings demonstrated that the knob membrane has a distinct distribution of membrane proteins different from the surrounding erythrocyte membrane and included a discrete structure at the apex (Watermeyer *et al.*, 2016). The same group were also able to combine soft X-ray tomography and FEGSEM to investigate loss of mechanical integrity in the erythrocyte cell membrane during the final stages of egress (Hale et al., 2017). In this case material for the FEGSEM was prepared using conventional chemical fixation and critical point drying as the study did not require the high resolution membrane structural information of the previous study, but rather a quantitative assessment of the frequency of loss of mechanical integrity in a synchronised infection of human blood cells.

The versatility of the FEGSEM has also allowed Khalifa *et al.* (2016) to correlate fluorescence with ultrastructural and elemental data acquired by energy dispersive X-ray spectroscopy (EDS). During this study into the biomineralisation pathway that occurs linking the marine habitat of the unicellular protist foraminifera shells (*Amphistegina lobifera* and *Amphistegina lessoni*), the correlative workflow required the application of a few consecutive steps: samples were first high pressure frozen, then fractured and viewed in the FEGSEM with EDS and/or subsequently viewed with an epifluorescent microscope cryo-adapted with a Linkam CMS196 cryo-stage (Linkam Scientific, United Kingdom). Thus, by combining information derived from a few imaging and analysis modalities, this

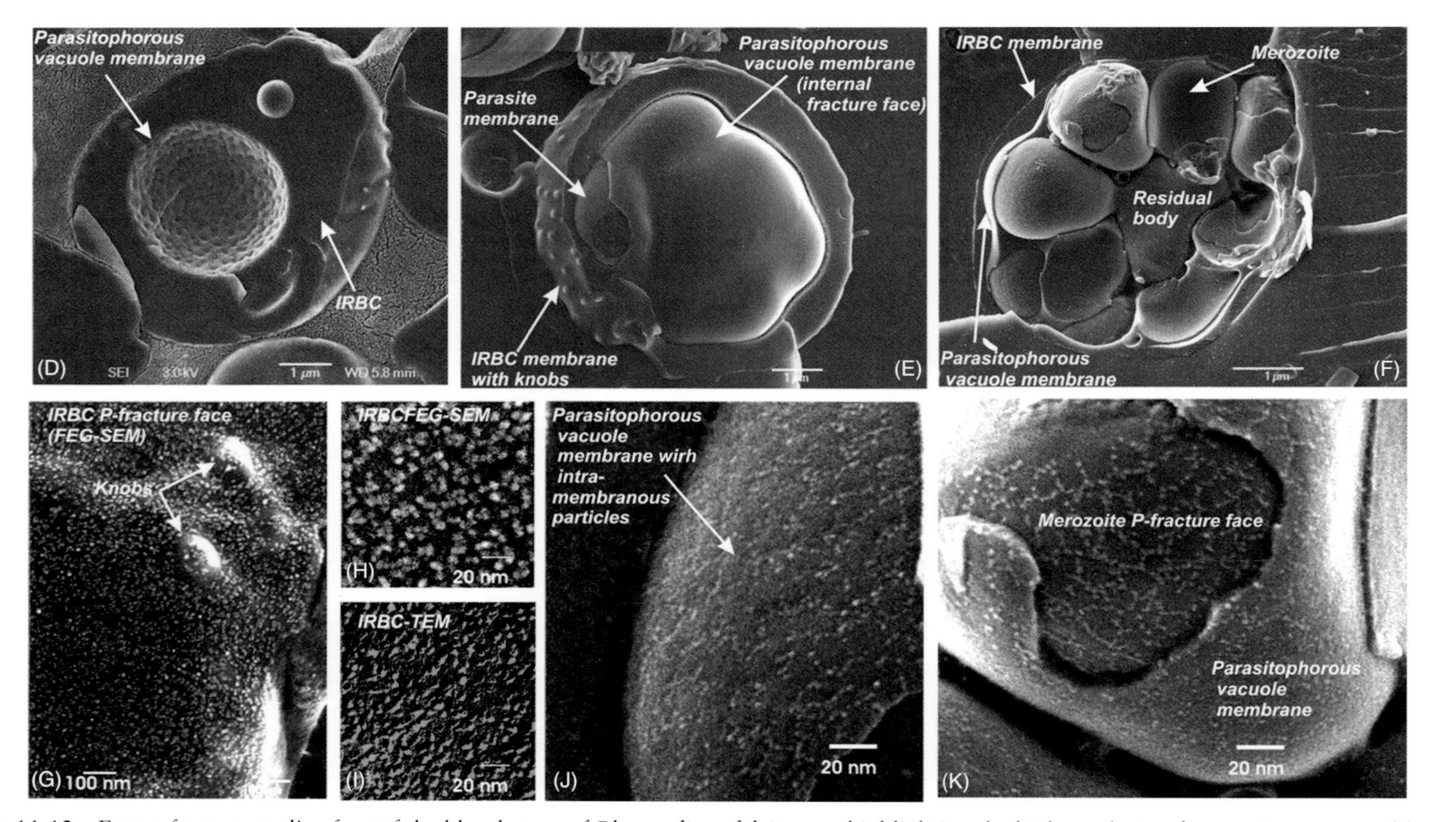

Figure 11.12 Freeze fracture replica face of the blood stage of *Plasmodium falciparum* highlighting the high resolution fracture images possible with a modern FEGSEM. Plates h and i compare the information level attainable with FEGSEM and freeze fracture-TEM. Samples for FEGSEM were prepared by high pressure freezing (HPM010, BAL-TEC AG, Liechtenstein), fractured (BAF060, BAL-TEC AG, Liechtenstein) and coated with 4 nm Pt/C using a MARS shadowing technique. Imaging was at 3 keV with a gentle beam enabled on a JEOL 7401F equipped with a VCT100 cryo-stage (BAL-TEC AG, Liechtenstein). Plate kindly provided by M.J. Blackman, H.R. Saibil and R.A. Fleck.

approach was successfully employed to reveal novel insights into the biomineralisation pathway of foraminifera (Khalifa *et al.*, 2016).

The FEGSEM can also be combined with an ion beam (FIB) for milling of samples. The FIB milling of plunge frozen vitreous samples is rapidly becoming a key preparative step in the work-flow for cellular structural biology (refer to Chapter 19 by Fukuka, Leis and Rigort). The ability for FIB to precisely target and prepare electron-transparent volumes that harbour molecular machines within their functional context is important for cell biology (refer to Engel et al., 2015, Mahamid et al., 2015, Arnold et al., 2016, Mahamid et al., 2016 for results which demonstrate FIB milling as a preparative step in the work-flow for cellular structural biology). However, the vitrification depth achieved by plunge freezing is limiting (~20 nm). The HPF can be used to prepare complex tissues for cryo FIB milling of lamella but these lamella must subsequently be recovered for cryo electron tomography by cryo lift out. Cryo lift out is a process where vitreous lamella are transfered to a TEM grid for electron tomography (without devitrification or contamination) allowing the potential for quantitative investigation of 3D cellular structures to be truly realised.

11.9 CONCLUSION

Understanding the many complex processes that can occur during the cryopreperation of biological systems is important if one is to be aware of potential artefacts and pitfalls in cryo-FEGSEM imaging, to be able to troubleshoot unsuccessful protocols and/or design effective robust strategies for cryofixing and preparing specimens for cryo-FEGSEM. It is hoped this general discussion of low temperature biology will provide a useful background to cryo techniques. Examples of cryo-FEGSEM have been shown above; these are by no means exhaustive or sufficiently comprehensive to fully answer the question regarding the power, versatility and capacity of a FEGSEM. Instead, the remaining chapters in Volume I and those in Volume II set out to answer these questions.

ACKNOWLEDGEMENTS

The support of the team at the Cornell Integrated Microscopy Unit who introduced RAF to freeze fracture/freeze etch TEM is appreciated, as is the assistance of Dr John Day, Curator Culture Collection of Algae and Protozoa for the provision of samples for testing in the cryo-FEGSEM. ES and BMH are in dept to Dr Martin Müller and Dr Hans Moor at ETHZ for introducing them to the great world of cryopreparation for electron microscopy. The ETH lab was one of the great places dedicated to image cell biology as close to the natural habitat as possible.

REFERENCES

Adam, M.M., Rana, K.J. and McAndrew, B.J. (1995) Effect of cryoprotectants on activity of selected enzymes in fish embryos. *Cryobiology*, 32, 92–104.

Akahori, H., Tamada, E., Usukura, J. and Takahashi, H. (1980) Development of a rapid freezing device. *J. Clin. Electron Microscopy*, 13, 576–577.

Allison, D.P., Daw, C.S. and Rorvik, M.C. (1987) The construction and operation of a simple inexpensive slam freezing device for electron microscopy. *J. Microscopy*, 147, 103–108.

Altmann, R. (1894) *Die Elementarorganismen und ihre Beziehungen zu den Zellen*, von Veit & Company.

Anchordoguy, T.J., Carpenter, J.F., Crowe, J.H. and Crowe, L.M. (1992) Temperature-dependent perturbation of phospholipid bilayers by dimethylsulfoxide. *Biochim. Biophys. Acta*, 1104 , 117–122.

Anderson, M.D., Prasad, T.K. and Stewart, C.R. (1995) Changes in isozyme profiles of catalase, peroxidase, and glutathione reductase during acclimation to chilling in mesocotyls of maize seedlings. *Plant Physiol.*, 109, 1247–1257.

Arakawa, T., Carpenter, J.F., Kita, Y.A. and Crowe, J.H. (1990) The basis for toxicity of certain cryoprotectants, a hypothesis. *Cryobiology*, 27, 401–415.

Armitage, W.J. (1986) Feasibility of corneal vitrification. *Cryobiology*, 23, 566.

Armitage, W.J. (1989) Survival of corneal endothelium following exposure to a vitrification solution. *Cryobiology*, 26, 318–327.

Arnold, J., Mahamid, J., Lucic, V., de Marco, A., Fernandez, JJ., Laugks, T., Mayer, T., Hyman, AA., Baumeister, W., Plitzko, JM. (2016) Site-specific cryo-focused ion beam sample preparation guided by 3D correlative microscopy. *Biophys. J.* 110, 860–869.

Artus, N.N., Uemura, M., Steponkus, P.L., Gilmour, S.J., Lin, C. and Thomashow, M.F. (1996) Constitutive expression of the cold-regulated *Arabidopsis thaliana* COR15a gene affects both chloroplast and protoplast freezing tolerance. *Proc. Natl Acad. Sci. USA*, 93, 13404–13409.

Ashwood-Smith, M.J. and Farrant, J. (1980) *Low Temperature Preservation in Medicine and Biology*, Pitman Medical Ltd, Kent, UK.

Bachmann, L. and Schmitt, W.W. (1971) Improved cryofixation applicable to freeze etching. *Proc. Natl Acad. Sci. USA*, 68, 2149–2152.

Bachmann, L. and Schmitt-Fumian, W.W. (1973a) Spray-freeze-etching of dissolved macromolecules, emulsions and subcellular components, in *Freeze-Etching: Techniques and Applications* (eds E.L. Benedetti and P. Favard), Société Fançaise de Microsopie Électronique, Paris, pp. 63–71.

Bachmann, L. and Schmitt-Fumian, W.W. (1973b) Spray-freezing and freeze-etching, in *Freeze-Etching: Techniques and Applications* (eds E.L. Benedetti and P. Favard), Société Fançaise de Microsopie Électronique, Paris, pp. 73–79.

Basu, R.N. and Dasgupta, M. (1978) Control of seed deterioration by free radical controlling agents. *Indian J. Exp. Bot.*, 16, 1070–1073.

Bellare, J.R., Davis, H.T., Scriven, L.E. and Talmon, Y. (1988) Controlled environment vitrification system: an improved sample preparation technique. *J. Electron Microsc. Tech.*, 10, 87–111.

Benedetti, E.L. and Favard, P. (eds) (1973) *Freeze-Etching: Techniques and Applications*, Société Française de Microscopie Électronique, Paris.

Benson, E.E. (1990) *Free Radical Damage in Stored Plant Germplasm*, IBPGR, Rome.

Benson, E.E., Lynch, P.T. and Jones, J. (1992) The detection of lipid peroxidation products in cryoprotected and frozen rice cells consequences for post-thaw survival. *Plant Sci. (Limerick)*, 85, 107–114.

Benson, E.E. and Noronha-Dutra, A.A. (1988) Chemiluminescence in cryopreserved plant tissue cultures the possible involvement of singlet oxygen in cryoinjury. *Cryo-Letters*, 7, 120–131.

Benson, E.E. and Roubelakis-Angelakis, K.A. (1994) Oxidative stress in recalcitrant tissue cultures of grapevine. *Free Radical Biology and Medicine*, 16, 355–362.

Benson, E.E. and Withers, L.A. (1987) Gas chromatographic analysis of volatile hydrocarbon production by cryopreserved plant tissue cultures, a non-destructive method for assessing stability. *Cryo-Letters*, 8, 35–46.

Berk, Z. (2013) Freeze drying (lyophilization) and freeze concentration, in *Food Process Engineering and Technology*, 2nd edn (ed. S.L. Taylor), Elsevier, New York, pp. 567–581.

Bierkander (1778) De l'Action et de l'Effet du froid sur les Végétaux, published in the Memoirs of the Stockholm Academy.

Boyde, A. and Maconnachie, E. (1979) Volume changes during preparation of mouse embryonic tissue for scanning electron microscopy. *Scanning*, 2, 149–163.

Boyle, R. (1665) *New Experiments and Observations Touching Cold, or an Experimental History of Cold, Begun to Which are Added an Examen of Antiperistasis, and an Examen of Mr Hobs's Doctrine about Cold, by the Honorable Robert Boyle*, J. Crook, London.

Boyne, A.F. (1979) A gentle, bounce-free assembly for quick-freezing tissues for electron microscopy: application to isolated torpedine ray electrocyte stacks. *J. Neurosci. Meth.*, I, 353–364.

Branton, D. (1973) The fracture process of freeze-etching, in *Freeze-Etching Techniques and Applications* (eds E.L. Benedetti and P. Favard), Société Française de Microscopie Électronique, Paris, pp. 107–112.

Branton, D., Bullivant, S., Gilula, N.B., Karnovsky, M.J., Moor, H., Mühlethaler, K., Northcote, D.H., Packer, L., Satir, B., Satir, P., Speth, V., Staehelin, L.A., Steere, R.L. and Weinstein, R.S. (1975) Freeze-etching nomenclature. *Science*, 190, 54–56.

Brüggeller, P. and Mayer, E. (1980) Complete vitrification in pure liquid water and dilute aqueous solutions. *Nature*, 288, 569–571.

Bullivant, S. (1960) The staining of thin sections of mouse pancreas prepared by the Fernandez-Moran helium II freeze substitution method. *J. Biophys. Biochem. Cytol.*, 8(3), 639–647.

Burch, M.D. and Marchant, H.J. (1983) Motility and microtubule stability of antarctic algae at sub-zero temperatures. *Protoplasma*, 115, 240–242.

Burton, E.F. and Oliver, W.F. (1935) The crystal structure of ice at low temperatures. *Proceedings of the Royal Society of London*, Series A, 153, 166–172.

Chekurova, N.R., Kislov, A.N. and Veprintsev, B.N. (1990) The effect of cryoprotectants on the electrical characteristics of mouse embryo cell membranes. *Kriobiologiya*, 1, 25–30.

Costello, J.M. and Corless, M.J. (1978) Direct measurement of temperature changes within freeze-fracture specimens during quenching in liquid coolants. *J. Microsc.*, 112, 17–37.

Cotterill, L.A., Gower, J.D., Fuller, B.J. and Green, C.J. (1989) Oxidative stress during hypothermic storage of rabbit kidneys possible mechanisms by which calcium mediates free radical damage. *Cryo-Letters*, 10, 119–126.

Cyrklaff, M. and Kühlbrandt, W. (1994) High-resolution electron microscopy of biological specimens in cubic ice. *Ultramicroscopy*, August, 55 (2), 141–153.

Dalimata, A.M. and Graham, J.K. (1997) Cryopreservation of rabbit spermatozoa using acetamide in combination with trehalose and methyl cellulose. *Theriogenology*, 48, 831–841.

Day, J.G. and McLellan, M.R. (1995) *Methods in Molecular Biology*, vol. 38, *Cryopreservation and Freeze-Drying Protocols*, Humana Press Inc., Totowa, New Jersey, USA.

Day, J.G., Watanabe, M.M., Morris, G.J., Fleck, R.A. and Mclellan, M.R. (1997) Long-term viability of preserved eukaryotic algae. *Journal of Applied Physiology*, 9, 121–127.

Dereuddre, J. (1991) Encapsulation/dehydration – a new approach to cryopreservation of plant cells. *Cryobiology*, 28, 539.

Dereuddre, J., Tannoury, M., Hassen, N., Kaminski, M. and Vintejoux, C. (1992) Resistance of coated somatic embryos *Daucus carota* L. to dehydration and cooling in liquid nitrogen cytological study. *Bull. Soc. Bot. Fr. Lett. Bot.*, 139, 15–33.

Donhowe, D.P. and Hartel, R.W. (1996) Recrystallization of ice in ice cream during controlled accelerated storage. *International Dairy Journal*, 6, 1191–1208.

Douzou, P. (1977) *Cryobiochemistry, An Introduction*, Academic Press, New York, USA.

Douzou, P. (1985) Interactive effects of cryosolvents and ionic and macromolecular solvents on protein structures and functions. *Cryobiology*, 22, 607.

Dubochet, J. and Lepault, J. (1984) Cryo-electron microscopy of vitrified water. *Journal de Physique Colloques*, 45 (C7), C7-85–C7-94.

Dubochet, J. and McDowall, A.W. (1981) Vitrification of pure water for electron microscopy. *J. Microsc.*, 124, RP3–RP4.

Dubochet, J., Lepault, J., Freeman, R., Berriman, J.A. and Homo, J.-Ci. (1982) Electron microscopy of frozen water and aqueous solutions. *J. Microsc.*, 128, 219–237.

Dubochet, J., McDowall, A.W., Menge, B., Schmid, E.N. and Lickfeld, K.G. (1983) Electron microscopy of frozen-hydrated bacteria. *J. Bacteriol.*, 155, 381–390.

Dumont, F., Marechal, P.A. and Gervais, P. (2003) Influence of cooling rate on *Saccharomyces cerevisiae* destruction during freezing: unexpected viability at ultra-rapid cooling rates. *Cryobiology*, 46 (1), 33–42.

Ebersold, H.R., Cordier, J.-L., Lüthy, P. (1981) Bacterial mesosomes: method dependent artifacts. *Arch. Microbiol.*, 130, 19–22.

Edelmann, L. (1989) The contracting muscle: A challenge for freeze-substitution and low temperature embedding. *Scanning Microscopy, Supplement* 3, 241–252.

Engel, B.D., Schaffer, M, Kuhn Cuellar, L., Villa, E., Plitzko, J.M., Baumeister, W. (2015) Native architecture of the Chlamydomonas chloroplast revealed by in situ cryo-electron tomography. *Elife* 4: e04889.

Eränkö, O. (1954) Quenching of tissues for freeze-drying. *Acxta Anat.*, 22, 331–336.

Escaig J. (1982) New instruments which facilitate rapid freezing at 83 K and 6 K. *J. Microsc.*, 126, 221–229.

Escaig, J. (1984) Control of different parameters for optimal freezing conditions, in *Science of Biological Specimen Preparation, 1983* (eds J.P. Revel, T. Barnard and G.H. Haggis), SEM Inc., AMF O'Hare, IL, USA, pp. 117–122.

Escaig, J., Géraud G. and Nicolas, G. (1977) Congélation rapide de tissus biologiques. Mesures des temperatures et des vitesses de congelation par thermocouple en couche mince. *C.R. Acad. Sci. Paris D*, 284, 2289–2292.

Esterbauer, H., Eckl, P. and Ortner, A. (1990) Possible mutagens derived from lipids and lipid precursors. *Mutat. Res.*, 238, 223–234.

Esterbauer, H., Zollner, H. and Schaur, R.J. (1988) Hydroxyalkenals: cytotoxic products of lipid peroxidation. *ISI Atlas of Science: Biochemistry*, 1, 311–317.

Fabre, J. and Dereuddre, J. (1990) Encapsulation/dehydration: A new approach to cryopreservation of Solanum shoot-tips. *Cryo-Letters*, 11, 413–426.

Fahy, G.M., Levy, D.I. and Ali, S.E. (1987) Some emerging principles underlying the physical properties biological actions and utility of vitrification solutions. *Cryobiology*, 24, 196–213.

Farrant, J. (1980) General observations on cell preservation, in *Low Temperature Preservation in Medicine and Biology* (eds M.J. Ashwood-Smith and J. Farrant), Pitman Medical Ltd, Kent, UK, pp. 1–18.

Farrant, J. and Ashwood-Smith, M.J. (1980) Practical aspects, in *Low Temperature Preservation in Medicine and Biology* (eds M.J. Ashwood-Smith and J. Farrant), Pitman Medical Ltd, Kent, UK, pp. 285–310.

Feder, N. and Sidman, R.L. (1958) Methods and principles of fixation by freeze-substitution. *J. Biophys. Biochem. Cytol.*, 4 (5), 593–600.

Fernández-Moran, H. (1952) Application of the ultrathin freezing-sectioning technique to the study of cell structures with the electron microscope. *Arkiv. Fysik.*, 4, 471.

Fernández-Moran, H. (1957) Electron microscopy of nervous tissue, in *Metabolism of the Nervous System* (ed. D. Richter), Pergamon Press, New York.

Fernández-Morán, H. (1960) Low-temperaure preparation techniques for electron microscope of biological specimens based on rapid freezing with liquid helium II. *Ann. N.Y. Acad. Sci.*, 85, 689–713.

Field, R.J. (1981) The role of 1-aminocyclopropane-1-carboxylic acid in the control of low-temperature-induced ethylene production in leaf tissue of *Phaseolus vulgaris* L. *Ann. Bot.*, 54, 61–67.

Field, R.J. (1984) The effect of low temperature on ethylene production by leaf tissue of *Phaseolus vulgaris* L. *Ann. Bot.*, 47, 215–223.

Fleck, R.A. (1998) Mechanisms of cell damage and recovery in cryopreserved freshwater protists. PhD Thesis, University of Abertay Dundee.

Fleck, R.A. (2001) A guide to freeze fractured/freeze-etched *Arabidopsis thaliana*. *The Quekett Journal of Microscopy*, 39 (2), 163–177.

Fleck, R.A. (2015) Low temperature electron microscopy, in, *Methods in Cryopreservation and Freeze-Drying* (eds W. Wolkers and H. Oldenhof), Lab Protocol Series, *Methods in Molecular Biology*, Springer, Berlin, pp. 243–274.

Fleck, R.A., Benson, E.E., Bremner, D.H. and Day, J.G. (2000) Studies of free radical mediated cryoinjury in the unicellular alga *Euglena gracilis* using a novel nondestructive hydroxyl radical assay: A novel approach for developing protistian cryopreservation strategies. *Free Radical Research*, 32 (2), 57–170.

Franks, F. (1985) *Biophysics and Biochemistry at Low Temperatures*, Cambridge University Press, Cambridge, UK.

Frederik, P.M. and Sommerdijk, N.A.J.M. (2005) Spatial and temporal resolution in cryo-electron microscopy – A scope for nano-chemistry. *Curr. Opin. Colloid. Interface Sci.*, 10, 245–249.

Frederik, P.M. and Storms M.M.H. (2005) Automated, robotic preparation of vitrified samples for 2D and 3D cryo electron microscopy. *Microscopy Today, November* 2005, 32–38.

Fuller, B.J. and Green, C.J. (1986) Oxidative stress in organs stored at low temperatures for transplantation, in *Free Radicals Cell Damage and Disease* (ed. C. Rice-Evans), Richelieu Press, London.

Gallo, P., Amann-Winkel, K., Angell, C.A., Anisimov, M.A, Caupin, F., Chakravarty, C., Lascaris, E., Loerting, T., Panagiotopoulos, A.Z., Russo, J., Sellberg, J.A., Stanley, H.E., Tanaka, H., Vega, C., Xu, L. and Pettersson, L.G.M. (2016) Water: A tale of two liquids. *Chemical Reviews*, 116, 7463–7500.

Galway, M.E., Heckman, J.W., Hyde, G.J. and Fowke, L.C. (1995) Chapter 1: Advances in high-pressure and plunge-freeze fixation, in *Methods in Cell Biology* (eds L.D.W. Galbraith, H.J. Bohnert and D.P. Bourque), vol. 49, Elsevier, Philadelphia, PA, USA, pp. 3–19.

Gersch, I. (1932) The Altman technique for fixation by drying while freezing. *Anat. Rec.*, 53, 309–337.

Glover, A.J. and Garvitch, Z.S. (1974) The freezing rate of freeze-etch specimens for electron microscopy. *Cryobiology*, 11, 248–254.

Göppert, H.R. (1830) *Ueber die Wärme-Entwicklung in den Pflanze, deren Gefieren und die Schutzmittel gegen Dasselbe*, Breslau.

Gorecki, R.J. and Harman, G.E. (1987) Effects of antioxidants on viability and vigor of ageing pea seeds. *Seed Sci. Technol.*, 15, 109–118.

Green, C.J., Healing, G., Simpkin, S., Fuller, B.J. and Lunec, J. (1986) Reduced susceptibility to lipid peroxidation in cold ischemic rabbit kidneys after addition of desferrioxamine mannitol or uric-acid to the flush solution. *Cryobiology*, 23, 358–365.

Gross, H. (1987) High resolution metal replication of freeze-dried specimens, in *Cryotechniques in Biological Electron Micoscopy* (eds R.A. Steinbrecht and K. Zierold), Springer-Verlag: Berlin, Heidelberg, pp. 203–215.

Grout, B.W.W. and Morris, G.J. (1987) *The Effects of Low Temperatures on Biological Systems*, Edward Arnold, Suffolk, UK.

Grout, B., Morris, J. and McLellan, M. (1990) Cryopreservation and the maintenance of cell lines. *Tibtech*, 293–297.

Haest, C.W.M., Verkleij, A.J., De Gier, J., Scheek, R., Ververgaert, P.H.J.T. and Van Deenen, L.L.M. (1974) The effect of lipid phase transitions on the architecture of bacterial membranes. *Biochim. Biophys. Acta*, 356, 17–26.

Haggis, G.H. (1986) Study of the conditions necessary for propane-jet freezing of fresh biological tissues without detectable ice formation. *J. Microsc.*, 143, 275–282.

Hagler, H.K. and Buja, L.M. (1984) New techniques for the preparation of thin freeze-dried cryosections for X-ray microanalysis, in *The Science of Biological Specimen Preperation* (eds J.-P. Revel, T. Barnard and G.H. Haggis), SEM Inc., AMF O'Hare, IL, USA, pp. 161–166.

Hale, V.L., Watermeyer, J.M., Hackett, F., Vizcay-Barrena, G., van Ooij, C., Thomas, J.A., Spink, M.C., Harkiolaki, M., Duke, E., Fleck, R.A., Blackman, M.J. and Saibil, H.R. (2017) Parasitophorous vacuole poration precedes its rupture and rapid host erythrocyte cytoskeleton collapse in *Plasmodium falciparum* egress. *PNAS*, 114 (13), 3439–3444.

Hall, C.E. (1950) A low temperature replica method for electron microscopy. *Journal of Applied Physics*, 21 (1), 61–62.

Handley, D.A., Alexander, J.T. and Chien, S. (1981) The design and use of a simple device for rapid quench-freezing of biological samples. *J. Microsc.*, 121, 273–282.

Heath, B. (1984) A simple and inexpensive liquid helium cooled 'slam freezing' device. *J. Microsc.*, 134, 75–82.

Heldman, D.R. and Hohner, G.A. (1974) An analysis of atmospheric freeze drying. *J. Food Sci.*, 39 (1), 147–155.

Hendry, G.A.F., Finch-Savage, W.E., Thorpe, P.C., Atherton, N.M., Buckland, S.M., Nilsson, K.A. and Steel, W.E. (1992) Free radical processes and loss of seed viability during desiccation in the recalcitrant species *Quercus robur* L. *New Phytol.*, 122, 273–279.

Hermann, R., Pawley, J., Nagatani, T. and Müller, M. (1988) Double-axis rotary shadowing for high-resolution scanning electron-microscopy. *Scanning Microscopy*, 2, 1215–1230.

Heuser, J.E., Reese, T.S. and Landis, D.M.D. (1976) *Preservation of synaptic structure by rapid freezing. Cold Spring Harbor Symp. Quant. Biol.*, XL, 17–24.

Heuser, J.E., Reese, T.S., Dennis, M.J., Jan, Y., Jan, L. and Evans, L. (1979) Synaptric vesicle exocytosis captured by quick freezing and correlated with quantal transmitter release. *J. Cell Biol.*, 81, 257–300.

Hincha, D.K., Sieg, F., Bakaltcheva, I., Köth, H. and Schmitt, M. J. (1996) Freeze-thaw damage to thylakoid membranes: Specific protection by sugars and proteins, in *Advances in Low-Temperature Biology* (ed. P.L. Steponkus), Jai Press Ltd, USA, pp. 141–184.

Hirsh, A.G. (1987) Vitrification in plants as a natural form of cryoprotection. *Cryobiology*, 24, 214–228.

Hobbs, P.V. (1974) *Ice Physics*, Oxford University Press, Oxford.

Hohenberg, H., Mannweiler, K. and Müller, M. (1994) High-pressure freezing of cell suspensions in cellulose capillary tubes. *J. Microsc.*, 175, 34–43.

Holt, W.V. (1997) Alternative strategies for the long-term preservation of spermatozoa. *Reproduction Fertility and Development*, 9, 309–319.

Humbel, B. and Müller M. (1984) In Proceedings of 8th European Congress on Electron Microscopy (eds P. Csanady, P. Röhlichand D. Szabo D.), pp. 1789–1798, Budapest.

Hunt, C.J., Song, Y.C., Bateson, E.A. and Pegg, D.E. (1994) Fractures in cryopreserved arteries. *Cryobiology*, 31, 506–515.

Hunziker, E.B. and Herrmann, W. (1987) In situ localization of cartilage extracellular matrix components by immunoelectron microscopy after cryotechnical tissue processing. *J. Histochem. Cytochem.*, 35, 647–655.

Ishikawa, M., Tandon, K., Suzuki, M. and Yamaguishi-Ciampi, A. (1996) Cryopreservation of bromegrass (*Bromus inermis* Leyss) suspension cultured cells using slow prefreezing and vitrification procedures. *Plant Sci. (Shannon)*, 120, 81–88.

Ishikawa, K., Harata, K., Mii, M., Sakai, A., Yoshimatsu, K. and Shimomura, K. (1997) Cryopreservation of zygotic embryos of a Japanese terrestrial orchid (*Bletilla striata*) by vitrification. *Plant Cell Rep.*, 16, 754–757.

Iwasaki, T., Kojima, T., Zeniya, Y. and Totsukawa, K. (1995) Viability of porcine embryos after thawing at various temperatures with freezing slowly. *Animal Sci. Technol.*, 66, 705–712.

Jahnel, F. (1938) Uber die Wiederstands Fahigkeit von Menschlichen Spermatozoen Gegeneber Starker Kalte. *Klin. Wochenschr.*, 17, 1273.

Janeiro, L.V., Vieitez, A.M. and Ballester, A. (1996) Cryopreservation of somatic embryos and embryonic axes of *Camellia japonica* L. *Plant Cell Rep.*, 15, 699–703.

Kanno, H., Speedy, R.J. and Angell, C.A. (1975) Supercooling of water to –92 degrees C under pressure. *Science*, 189 (4206), 880–881.

Karel, M. (1975) Heat and mass transfer in freeze drying, in *Freeze Drying and Advanced Food Technology* (eds S.A. Goldblith, L. Rey and W.W. Rothmayr), Academic Press, London.

Karlsson, J.O. (2010) Effects of solution composition on the theoretical prediction of ice nucleation kinetics and thermodynamics. *Cryobiology*, 60 (1), 43–51.

Kartha, K.K. (1985a) *Cryopreservation of Plant Cells and Organs*, CRC Press Ltd, Florida, USA.

Kartha, K.K. (1985b) Meristem culture and germplasm preservation, in *Cryopreservation of Plant Cells and Organs* (ed. K.K. Kartha), CRC Press Ltd, Florida, USA, pp. 115–134.

Khalifa, G.M., Kirchenbuechler, D., Koifman, N., Kleinerman, O., Talmon, Y., Elbaum, M., Addadi, L., Weiner, S. and Erez, J. (2016) Biomineralization pathways in a foraminifer revealed using a novel correlative cryo-fluorescence-SEM-EDS technique. *Journal of Structural Biology*, 196 (2), 155–163.

Knoll, G., Oebel, G. and Plattner, H. (1982) A simple sandwich-cryogen-jet procedure with high cooling rates for cryofixation of biological materials in the native state. *Protoplasma*, 111, 161–176.

Knoll, G., Verkleij, A.J. and Plattner, H. (1987) Cryofixation of dynamic processes in cells and organelles, in *Cryotechniques in Biological Electron Micoscopy* (eds R.A. Steinbrecht and K. Zierold), Springer-Verlag, Berlin, Heidelberg, pp. 256–271.

Kopp, F. (1972) *Zur Membranstruktur: Lokalisation von Membranlipiden im Hefeplasmalemma*, Cell Biology, Federal Institute of Technology (ETH).

Kopp, F. (1973) Morphology of the plasmalemma of baker's yeast *Saccharomyces cerevisiae*, in *Freeze-Etching Techniques and Applications* (eds E.L. Benedetti and P. Favard), Societé Française de Microscopie Électronique, Paris, pp. 181–185.

Kunisch, H. (1880) Über die tödliche Einwirkung niederer Temperaturen auf die Pflanzen, Inaugural Dissertation, Breslau, 55 pp.

Leidenfrost, J.G. (1756) *De Aquae Communis Nonnullis Qualitatibus Tractatus*, Duisburg.

Levine, H. and Slade, L. (1988) Thermomechanical properties of small-carbohydrate-water glasses and 'rubbers'. Kinetically metastable systems at sub-zero temperatures. *J. Chem. Soc., Faraday Trans. 1*, 84 (8), 2619–2633.

Levitt, J. (1980) *Responses of Plants to Environmental Stress*, Academic Press, New York, USA.

Lindsay, C.D. (1996) Assessment of aspects of the toxicity of *Clostridium perfringens* epsilon-toxin using the MDCK cell line. *Human and Experimental Toxicology*, 15, 904–908.

Livesey, S.A., Buescher, E.S., Krannig, G.L., Harrison, D.S., Linner, J.G. and Chiovetti, R. (1989) Human neutrophil granule heterogeneity: Immunolocalisation studies using cryofixed, dried and embedded specimens. *Scanning Microscopy, Suppl.* 3, 231–240.

Luo, J. and Reed, B.M. (1997) Abscisic acid-responsive protein, bovine serum albumin, and proline pretreatments improve recovery of *in vitro* currant shoot-tip meristems and callus cryopreserved by vitrification. *Cryobiology*, 34, 240–250.

Luyet, B.J. (1965) Phase transitions encountered in the rapid freezing of aqueous solutions. *Annals of the New York Academy of Science*, 125, 502–521.

Luyet, B.J. and Gehenio, P.M. (1940) *Life and Death at Low Temperatures*, St Louis University, Missouri, USA.

Luyet, B. and Gonzales, F. (1951) Recording ultrarapid changes in temperature. *Refrig. Engng*, 59, 1191–1193, 1236.

Luyet, B. and Rapatz, G. (1958) Patterns of ice formation in some aqueous solutions. *Biodynamica*, 8, 1–68.

MacFarlane, D.R. (1987) Physical aspects of vitrification in aqueous solutions. *Cryobiology*, 24, 181–195.

MacFarlane, D.R., Forsyth, M. and Barton, C.A. (1992) Vitrification and devitrification in cryopreservation in *Advances in Low-Temperature Biology* (ed. P.L. Steponkus), Jai Press Ltd, USA, pp. 221–278.

MacKenzie, A.P. (1965) Factors affecting the mechanism of transformation of ice into water vapor in the freeze drying process. *Annals of the New York Academy of Sciences*, 125, 522–547.

MacKenzie, A.P. (1972) Freezing, freeze-drying, and freeze-substitution. *Scanning Electron Microsc.*, II, 273–280.

Magill, W., Deighton, N., Pritchard, H.W., Benson, E.E. and Goodman, B.A. (1994) Physiological and biochemical studies of seed storage parameters in Carica papaya. *Proceedings of the Royal Society of Edinburgh Section B (Biological Sciences)*, 102B, 439–442.

Mahadevan, M.M. and Miller, M.M. (1997) Deleterious effect of equilibration temperature on the toxicity of propanediol during cryopreservation of mouse zygotes. *Journal of Assisted Reproduction and Genetics*, 14, 51–54.

Mahamid, J., Schampers, R., Persoon, H., Hyman, A.A., Baumeister, W., Plitzko, J.M. (2015) A focused ion beam milling and lift-out approach for site-specific preparation of frozen-hydrated lamellas from multicellular organisms *J. Struct. Biol.* 192, 262–269.

Mahamid, J., Pfeffer, S., Schaffer, M., Villa, E., Danev, R., Cuellar, L.K., Förster, F., Hyman, A.A., Plitzko, J.M., Baumeister, W. (2016) Visualizing the molecular sociology at the HeLa cell nuclear periphery Science 351, 969–972.

Mantegazza P. (1866) Sullo sperma umano. *Rendiconti Reale Istituto Lombardo di Scienze e Lettere*, 3, 183–196.

Mari, S., Engelmann, F., Chabrillange, N., Huet, C. and Michaux-Ferriere, N. (1995) Histo-cytological study of apices of coffee (*Coffea racemosa* and *C. sessiliflora*) *in vitro* plantlets during their cryopreservation using the encapsulation/dehydration technique. *Cryo-Letters*, 16, 289–298.

Massip, A. (1996) Vitrification: An interesting method for cryopreservation of mammalian embryos. *Contraception Fertilite Sexualite*, 24, 665–673.

Matsumoto, T. and Sakai, A. (1995) An approach to enhance dehydration tolerance of alginate-coated dried meristems cooled to –196 °C. *Cryo-Letters*, 16, 299–306.

Mayer, E. (1985) Vitrification of pure liquid water. *Journal of Microscopy*, 140 (1), 3–15.

Mazur, P. (1970) Cryobiology: The freezing of biological systems. *Science*, 168, 939–949.

Mazur, P. (1977) The role of intracellular freezing in the death of cells cooled at supraoptimal rates. *Cryobiology*, 14 (3), 251–272.

Mazur, P. (1990a) Equilibrium and nonequilibrium freezing of mammalian embryos. *Cryobiology*, 27, 648–649.

Mazur, P. (1990b) Kinetics of the killing of unfrozen intact Drosophila embryos at low temperatures. *Cryobiology*, 27, 650–651.

Mazur, P., Seki, S., Pinn, I.L., Kleinhans, F.W. and Edashige, K. (2005) Extra- and intracellular ice formation in mouse oocytes. *Cryobiology*, 51, 29–53.

McAnulty, J.F. and Huang, X.Q. (1996) The effect of simple hypothermic preservation with trolox and ascorbate on lipid peroxidation in dog kidneys. *Cryobiology*, 33, 217–225.

McAnulty, J.F. and Haung, X.Q. (1997) The efficacy of antioxidants administered during low temperature storage of warm ischemic kidney tissue slices. *Cryobiology*, 34, 406–415.

Mehl, P.M. (1996a) Thermodynamics and dynamics of the glass transition with application to biology. *Cryobiology*, 33, 623.

Mehl, P.M. (1996b) Crystallisation and vitrification in aqueous glass-forming solutions, in *Advances in Low-Tempertaure Biology* (ed. P.L. Steponkus), Jai Press Ltd, USA, pp. 185–256.

Meissner, D.H. and Schwarz, H. (1990) Improved cryofixation and freeze-substitution of embryonic quail retina: A TEM study on ultrastructural preservation. *J. Electron Microsc. Tech.*, 14, 348–356.

Molish, H. (1897) Untersuchungen über das Erfieren der Pflanzen, Fisher, Jena, p. 1.

Monroe, R.G., Gamble, W.J., La Farge, C.G., Gamboa, R., Morgan, C.L., Rosenthal, A. and Bullivant, S. (1968) Myocardial ultrastructure in systole and diastole using ballistic cryofixation. *J. Ultrastruct. Res.*, 22, 22–36.

Moor, H. (1964) Die Gefrier-Fixation Lebender Zellen und Ihre Anwendung in der Elektronenmikroskopie. *Zeitschrift fiir Zellforschung*, 62, 546–580.

Moor, H. (1970) High resolution shadow casting by the use of an electron gun. *Proceedings of the 7th International Congress on Electron Microscopy*, vol. 1, pp. 413–414.

Moor, H. (1971) Recent progress in the freeze-etching technique. *Phil. Trans. Roy. Soc. Lond. B*, 261, 121–131.

Moor, H. (1973) Evaporation and electron guns, in *Freeze-Etching Techniques and Applications* (eds E.L. Benedetti and P. Favard), Société Française de Microscopie Électronique, Paris, pp. 27–30.

Moor, H. (1986) Recent progress in high pressure-freezing. *J. Electron Microsc.*, 35, 1961–1964.

Moor, H. and Mühlethaler, K. (1963) Fine structure in frozen-etched yeast cells. *J. Cell Biol.*, 17, 609–628.

Moor, H., Müller, M. and Kistler, J. (1976) Freezing in a propane jet. *Experientia*, 32, 805.

Moor, H. and Riehle, U. (1968) Snap-freezing under high pressure: A new fixation technique for freeze-etching. *Proc. Fourth Europ. Reg. Conf. Elect. Microsc.*, 2, 33–34.

Moor, H., Mühlethaler, K., Waldner, H., Frey-Wyssling, A. (1961) A new freezing-ultramicrotome. *J. Biophys. Biochem. Cytol*, 10, 1–13.

Moor, H., Bellin, G., Sandri, C., Akert, K. (1980) The influence of high pressure freezing on mammalian nerve tissue. *Cell Tissue Res.*, 209, 201–216.

Morris, G.J. (1981) *Cryopreservation: An Introduction to Cryopreservation in Culture Collections*, Institute of Terrestrial Ecology, Cambridge. UK.

Mühlethaler, K., Hauenstein, W. and Moor, H. (1973) Double fracturing method for freeze-etching, in *Freeze-Etching Techniques and Applications* (eds E.L. Benedetti and P. Favard), Société Française de Microscopie Électronique, Paris, pp. 101–106.

Muldrew, K. and McGann, L.E. (1994) The osmotic rupture hypothesis of intracellular freezing injury. *Biophys. J.*, 66, 532–541.

Mullen, S.F. and Critser, J.K. (2007) The science of cryobiology. *Cancer Treat. Res.*, 138, 83–109.

Müller-Thurgau, H. (1886) Über das Gefrieren und Erfrieren der Pflanzen. 2 Teil. *Landw. Jahrb.*, 15, 453–610.

Müller, M., Meister, N. and Moor, H. (1980) Freezing in a propane jet and its application in freeze-fracturing. *Mikroskopie (Wien)*, 36, 129–140.

Müller, M. and Moor, H. (1984) Cryofixation of thick specimens by high pressure freezing, in: Eds. Science of Biological Specimen Preparation (eds J.P. Revel, T. Barnard and G.H. Haggis), SEM Inc., AMF O'Hare, Chicago, IL, USA, pp. 131–138.

Murk, J.L.A.N., Posthuma, G., Koster, A.J., Geuze, H.J., Verkleij, A.J., Kleijmeer, M.J. and Humbel, B.M. (2003) Influence of aldehyde fixation on the morphology of endosomes and lysosomes: Quatitative analysis and electron tomography. *J. Microsc.*, 212, 81–90.

Nermut, M.V. and Frank, H. (1971) Fine structures of influenza A2 (Singapore) as revealed by negative staining, freez-drying and freeze-etching. *J. Gen. Virol.*, 10, 37–51.

Nizam, J. (1960) Nuclear Cytology, Culture Conditions and Radiation Effects on Certain Species of Desmids: Radiation Effects on Chlorella, PhD Thesis, University of London.

Pease, D.C. (1973) Glycol methacrylate copolymerized with glutaraldehyde and urea as an embedment retaining lipids. *J. Ultrastruct. Res.*, October, 45 (1), 124–148.

Pegg, D.E. (1987) Mechanisms of freezing damage, in *Symposia of the Society for Experimental Biology, No. XXXXI. Temperature and Animal Cells Meeting, Durham, England, UK, September 10–12, 1986* (eds K. Bowler and B.J. Fuller), The Biochemical Society Book Depot, Colchester, pp. 363–378.

Pegg, D.E. (2010) The relevance of ice crystal formation for the cryopreservation of tissues and organs. *Cryobiology*, 60, S36–S44.

Pegg, D.E., Wusteman, M.C. and Boylan, S. (1996) Fractures in cryopreserved elastic arteries: Mechanism and prevention. *Cryobiology*, 33, 658–659.

Pegg, D.E., Wusteman, M.C. and Boylan, S. (1997) Fractures in cryopreserved elastic arteries. *Cryobiology*, 34, 183–192.

Perry, V.P. (1976) Freeze-drying for the preservation of human tissues, in Tissue banking for transplantqtion (eds K.W. Sell and G.E. Friedlaender), Gruneand Stratton, New York, pp. 189–193.

Phunchindawan, M., Hirata, K., Sakai, A. and Miyamoto, K. (1997) Cryopreservation of encapsulated shoot primordia induced in horseradish (*Armoracia rusticana*) hairy root cultures. *Plant Cell Rep.*, 16, 469–473.

Pickford, M.A., Simpkin, S., Fryer, P. and Green, C.J. (1989) Lipid peroxidation after 48-hour flush preservation of single left lung isografts in rats: a comparison of hypertonic citrate racemic verapamil hydrochloride or desferrioxamine with isotonic saline. *Cryobiology*, 26, 578–579.

Piironen, J. (1993) Cryopreservation of sperm from brown trout (*Salmo trutta* m. *lacustris* L.) and Artic charr (*Isalvelinus alpinus* L.). *Aquaculture*, 116 (2–3), 275–285 (1993).

Pinto da Silva, P. and Branton, D. (1970) Membrane splitting in freeze-etching. *J. Cell Biol.*, 45, 598–605.

Plattner, H. and Knoll, G. (1984) Cryofixation of biological materials for electron microscopy by the methods of spray-, sandwich-, cryogen-jet-, and sandwich-cryogen-jet-freezing: a comparison of techniques, in *Science of Biological Specimen Preparation 1983* (eds J.P. Revel, T. Barnard and G.H. Haggis, SEM Inc., AMF O'Hare, Chcago, IL, USA, pp. 139–146.

Plattner, H., Schmitt-Fumian, W.W. and Bachmann, L. (1973) Cryofixation of single cells by spray-freezing, in *Freeze-Etching: Techniques and Applications* (eds E.L. Benedetti and P. Favard),. Société Française de Microsopie Électronique, Paris, pp. 63–71.

Polge, C., Smith, A.U. and Parkes, A.S. (1949) Revival of spermatazoa after vitrification and dehydration at low temperatures. *Nature*, 164, 666.

Power, H. (1664) *Experimental Philosophy*, Book One, New Experiments Microscopical, London, 83 pp. Observat XXX – Of the Little White Eels or Snigs, in Vineger or Aleger.

Pscheid, P., Schudt, C. and Plattner, H. (1981) Cryofixation of monolayer cell cultuers for freeze-fracture without chemical pre-treatments. *J. Microsc.*, 121, 149–167.

Rasmussen, D.H. and MacKenzie, A.P. (1971) The glass transition in amorphous water. Application of the measurements to problems arising in cryobiology. *J. Phys. Chem.*, 75, 967–973.

Rebhun, L.I. (1972) Freeze-substitution and freeze-drying, in *Principals and Techniques of Electron Microscopy, Biological Applications*, vol. 2 (ed. M.A. Hayat), Van Nostrand Reinhold Co., New York, pp. 3–52.

Reid, D.S. (1983) Fundamental physicochemical aspects of freezing. *Food Technology*, 37, 110–115.

Riehle, U. (1968) *Über die Vitrifizierung Verdünnter Wässriger Lösungen*, Federal Institute of Technology (ETH).

Riehle, U. and Hoechli, M. (1973) The theory and technique of high pressure freezing, in *Freeze-Etching Techniques and Applications* (E.L. Benedetti and P. Favard), Société Française de Microscopie Électronique, Paris, pp. 31–61.

Robards, A.W. and Sleytr, U.B. (1985) Low temperature methods in biological electron microscopy, in *Practical Methods in Electron Microscopy* (ed. A.M. Glauert), vol. 10, Elsevier, Amsterdam.

Rudolph, A.S. and Crowe, J.H. (1985) Membrane stabilization during freezing: The role of 2 natural cryoprotectants, trehalose and proline. *Cryobiology*, 22, 367–377.

Saha, S., Otoi, T., Takagi, M., Boediono, A., Sumantri, C. and Suzuki, T. (1996) Normal calves obtained after direct transfer of vitrified bovine embryos using ethylene glycol, trehalose, and polyvinylpyrrolidone. *Cryobiology*, 33, 291–299.

Sakai, A., Kobayashi, S. and Oiyama, I. (1990) Cryopreservation of nucellar cells of navel orange (*Citrus sinensis* Osb. var. brasiliensis Tanaka) by vitrification. *Plant Cell Rep.*, 9 (1), 30–33.

Sansinena, M., Santos, M.V., Zaritzky, N. and Chirife, J. (2012) Comparison of heat transfer in liquid and slush nitrogen by numerical simulation of cooling rates for French straws used for sperm cryopreservation. *Theriogenology*, 77(8), 1717–21.

Santarius, K.A. (1996) Freezing of isolated thylakoid membranes in complex media. X. Interactions among various low molecular weight cryoprotectants. *Cryobiology*, 33, 118–126.

Shibata, Y., Arima, T. and Yamamoto, T. (1984) Double-axis rotary replication for deep-etching. *J. Microsc.*, 136, 121–123.

Shimoni, E. and Müller, M. (1998) On optimizing high-pressure freezing: from heat transfer theory to a new microbiopsy device. *J. Microsc.*, 192 (Pt 3), 236–247.

Simpson, W.L. (1941a) An experimental analysis of the Altmann technic of freeze-drying. *Anat. Rec.*, 80, 173–189.

Simpson, W.L. (1941b) The application of the Altmann method to the study of the Golgi apparatus. *Anat. Rec.*, 80, 329–345.

Singer, S.J. and Nicolson, G.L. (1972) The fluid mosaic model of the structure of cell membranes. *Science*, 175, 720–731.

Singh, J. and Miller, R.W. (1985) Biophysical and ultrastructural studies of membrane alterations in plant cells during extracellular freezing: molecular mechanisms of membrane injury, in *Cryopreservation of Plant Cells and Organs* (rd. K.K. Kartha), CRC Press Ltd, Florida, USA, pp. 61–74.

Sjöström, M., Johansson, R. and Thornell, L.E. (1974) Cryoultramictrotomy of muscles in defined state. Methodological aspects, in *Electron Microscopy and Cytochemistry* (eds E, Wise, W.Th. Daens, I. Molenaar and P. van Duijin, North-Holland, Amsterdam, 387 pp.

Smith, A.U. (1961) *Biological Effects of Freezing and Supercooling*, Edward Arnold Ltd, London.

Steere, R.L. (1957) Electron microscopy of structural detail in frozen biological specimen. *J. Biophys. Biochem. Cytol.*, 3, 45–63.

Steere, R.L. (1973) Preparation of high-resolution freeze-etch, freeze-fracture, frozen-surface, and freeze-dried replicas in a single freeze-etch module, and the use of stereo electron microscopy to obtain maximum information from them, in *Freeze-Etching Techniques and Applications* (eds E.L. Benedetti and P. Favard), Société Française de Microscopie Électronique, Paris, pp. 223–255.

Steinbrecht, R.A. and Zierold, K. (eds) (1987) *Cryotechniques in Biological Electron Microscopy*, Springer-Verlag, Berlin, Heidelberg, New York, London, Paris, Tokyo.

Stephenson, J.L. (1953) Theory of vacuum drying of frozen tissues. *Bull. Math. Biophy.*, 15, 411–429.

Steponkus, P.L. (1992) *Advances in Low-Temperature Biology*, Jai Press Ltd, USA.

Steponkus, P.L. (1993) *Advances in Low-Temperature Biology*, Jai Press Ltd, USA.

Steponkus, P.L. (1996) *Advances in Low-Temperature Biology*, Jai Press Ltd, USA.

Steponkus, P.L., Langis, R. and Fujikawa, S. (1992) Cryopreservation of plant tissues by vitrification, in *Advances in Low-Temperature Biology* (ed. P.L. Steponkus), Jai Press Ltd, USA, pp. 1–61.

Steponkus, P.L. and Webb, M.S. (1992) Freeze-induced dehydration and membrane destabilization in plants, in *Water and Life: Comparative Analysis of Water Relationships at the Organismic, Cellular and Molecular Level* (ed. G.N. Somero, C.B. Osmond and C.L. Bolis), Springer-Verlag, Berlin, pp. 338–362.

Steponkus, P.L., Uemura, M. and Webb, M.S. (1993) A contrast of the cryostability of the plasma membrane of winter rye and spring oat: Two species that widely differ in their freezing tolerance and plasma membrane lipid composition, in *Advances in Low-Temperature Biology* (ed. P.L. Steponkus), Jai Press Ltd, USA, pp. 211–312.

Steponkus, P.L., Myers, S.P., Lynch, D.V., Gardner, L., Bronshteyn, V., Leibo, S.P., Rall, W.F., Pitt, R.E., Lin, T.-T. and Macintyre, R.J. (1990) Cryopreservation of *Drosophila melanogaster* embryos. *Nature*, 345, 170–172.

Stillinger, F. (1995) A topographic view of supercooled liquids and glass formation. *Science*, 267, 1935–1939.

Stott, S.L. and Karlsson, J.O.M. (2009) Visualization of intracellular ice formation using high-speed video cryomicroscopy. *Cryobiology*, 58, 84–95.

Strong, D.M. and MacKenzie, A.P. (1993) Freeze-drying of tissues, in *Musculoskeletal Tissue Banking* (ed. W.W. Tomford), Raven Press, New York, pp. 181–208.

Studer, D., Chiquet, M., Hunziker, E.B. (1996) Evidence for a distinct water-rich layer surrounding collagen fibrils in articular cartilage extracellular matrix. *J. Struct. Biol*, 117, 81–85.

Studer, D., Hennecke, H. and Müller, M. (1992) High-pressure freezing of soybean nodules leads to an improved preservation of ultrastructure. *Planta*, 188, 155–163.

Studer, D., Humbel, B.M. and Chiquet, M. (2008) Electron microscopy of high pressure frozen samples: bridging the gap between cellular ultrastructure and atomic resolution. *Histochem. Cell Biol.*, 130, 877–889.

Studer, D., Michel, M. and Müller, M. (1989) High pressure freezing comes of age. *Scanning Microsc. Suppl.*, 3, 253–268.

Studer, D., Graber, W., Al-Amoudi, A. and Eggli, P. (2001) A new approach for cryofixation by high-pressure freezing. *J. Microsc.*, 203, 285.

Studer, D., Michel, M., Wohlwend, M., Hunziker, E.B. and Buschmann M.D. (1995) Vitrification of articular-cartilage by high-pressure freezing. *J. Microsc.-Oxford*, 179, 321–332.

Takagi, H., Thinh, N.T., Islam, O.M., Senboku, T. and Sakai, A. (1997) Cryopreservation of *in vitro* grown shoot tips of taro (*Colocasia esculenta* (L) Schott) by vitrification. 1. Investigation of basic conditions of the vitrification procedure. *Plant Cell Rep.*, 16, 594–599.

Taylor, M.J. (1987) Physico-chemical principles in low temperature biology, in *The Effects of Low Temperatures on Biological Systems* (eds B.W.W. Grout and G.J. Morris), Edward Arnold, London, UK, pp. 3–71.

Tokuyasu, K.T. (1973) A technique for ultracryotomy of cell suspensions and tissues. *J. Cell Biol.*, 57 (2), 551–565.

Toner, M. (1993) Nucleation of ice crystals inside biological cells, in *Advances in Low-Temperature Biology* (ed. P.L. Steponkus), Jai Press Ltd, USA, pp. 53–100.

Toner, M., Karel, M. and Cravalho, E. (1990) Thermodynamics and kinetics of intracellular ice formation during freezing of biological cells. *J. Appl. Phys.*, 67, 1582–1593.

Tvedt, K.E., Halgunset, J., Kopstad, G. and Haugen, O.A. (1988) Quick sampling and perpendicular cryosectioning of cell monolayers for the X-ray microanalysis of diffusible substances. *J. Microsc.*, 151, 49–59.

Umrath, W. (1974) Cooling bath for rapid freezing in electron microscopy. *J. Microsc.*, 101, 103–105.

Umrath, W. (1983) Calculation of freeze drying times for electron microscopic preparations. *Mikroskopie*, 40 (1–2), 9–34.

Urist, M.R., Mikolski, A.J. and Boyd, S.D. (1975) A chemosterilized antigen extracted bone morphogenetic allo implant. *Arch. Surg.*, 110, 416–420.

Ushatinskaya, R.S. (1993) Antifreezing importance of some polyhydric alcohols, sugars and carbohydrates in insect metabolism during winter diapause. *Izvestiya Akademii Nauk Seriya Biologicheskaya (Moscow)*, 1993, 447–459.

Van Harreveld, A. and Crowell, J. (1964) Electron microscopy after rapid freezing on a metal surface and substitution fixation. *Anat. Rec.*, 149, 381–386.

Van Harreveld, A., Crowell, J., Malhotra, S.K. (1965) A study of extracellular space in central nervous tissue by freeze-substitution. *J. Cell Biol.*, 25, 117–137.

Van Harreveld, A. and Trubatch J. (1979) Progression of fusion during rapid freezing for electron microscopy. *J. Microsc.*, 115, 243–256.

Vigneron, T., Arbault, S. and Kaas, R. (1997) Cryopreservation of gametophytes of *Laminaria digitata* (L) Lamouroux by encapsulation dehydration. *Cryo-Letters*, 18, 93–98 (1997).

Verkleij, A.J., Humbel, B.M., Studer, D. and Müller, M. (1985) 'Lipidic particle' systems as visualized by thin-section electron microscopy. *Biochim. Biophys. Acta*, 812, 591–495.

Verna, A. (1983) A simple quick-freezing device for ultrastructure preservation: evaluation by freeze-substitution. *Biol. Cell*, 49, 95098.

Villalba, R., Benitez, J., De No-Lowis, E., Rioja, L.F. and Gomez-Villagran, J.L. (1996) Cryopreservation of human skin with propane-1,2-diol. *Cryobiology*, 33, 525–529.

Walther, P. and Müller, M. (1997) Double-layer coating for field-emission cryo-scanning electron microscopy – present state and applications. *Scanning*, 19, 343–348.

Wares, C. and Bell, K.J. (1966) Leidenfrost, J.G. (1756), 'On the fixation of water in diverse fire'. *International Journal of Heat and Mass Transfer*, 9, 1153–1166.

Watanabe, K. (1988) Sub-ice microalgal strands in the Antarctic coastal fast ice area near Syowa Station. *Jpn J. Phyco. (Sôrui)*, 36, 221–229.

Watermeyer, J.M., Hale, V.L., Hackett, F., Clare, D.K., Cutts, E., Vakonakis, I., Fleck, R.A., Blackman, M.J. and Saibil H.R. (2016). Spiral scaffold underlies cytoadherent knobs in *Plasmodium falciparum*-infected erythrocytes. *Blood*, 127 (3), 343–351.

Webb, M.S., Gilmour, S.J., Thomashow, M.F. and Steponkus, P.L. (1996) Effects of COR6.6 and COR15am polypeptides encoded by COR (cold-regulated) genes of *Arabidopsis thaliana* on dehydration-induced phase transitions of phospholipid membranes. *Plant Physiol.*, 111, 301–312.

Whiteley, G.S.W., Fuller, B.J. and Hobbs, K.E.F. (1992a) Deterioration of cold-stored tissue specimens due to lipid peroxidation modulation by antioxidants at high subzero temperatures. *Cryobiology*, 29, 668–673.

Whiteley, G.S.W., Fuller, B.J. and Hobbs, K.E.F. (1992b) Lipid peroxidation in liver tissue specimens stored at subzero temperatures. *Cryo-Letters*, 13, 83–86.

Wildhaber, I., Gross, H. and Moor, H. (1982) The control of freeze-drying with deuterium oxide (D_2O). *J. Ultrastruct. Res.*, 80, 367–373.

Williams, R.C. (1953) A method of freeze-drying for electron microscopy. *Experimental Cell Research*, 4 (1), 188–201.

Williams, R.C. (1954) The application of freeze-drying top electron microscopy, in *Biological Applications of Freezing and Drying* (ed. R.J.C. Harris), Academic Press, New York, pp. 303–328.

Williams, R.J. and Hope, H.J. (1981) The relationship between cell injury and osmotic volume reduction. III. Freezing injury and frost resistance in winter wheat. *Cryobiology*, 18, 133–145.

Williams, R.J., Willemont, C. and Hope, H.J. (1981) The relationship between cell injury and osmotic volume reduction. IV. The behavior of hardy wheat membrane lipids in monolayer. *Cryobiology*, 18, 146–154.

Wyckoff, R.W.G. (1946) Frozen-dried preparations for the electron microscope. *Science*, 104, 36–37.

Wyckoff, R.W.G. (1949) *Electron Microscopy – Technique and Applications*, Interscience Publishers, New York, pp. 151–178.

Yang, G., Zhang, A. and Xu, L.X. (2009) Experimental study of intracellular ice growth in human umbilical vein endothelial cells. *Cryobiology*, 58, 96–102.

Yu, Z.-W. and Quinn, P.J. (1994) Dimethyl sulphoxide: A review of its applications in cell biology. *Biosci. Rep.*, 14, 259–281.

12

High-Resolution Cryo-Scanning Electron Microscopy of Macromolecular Complexes

Sebastian Tacke[1], Falk Lucas[1], Jeremy D. Woodward[3], Heinz Gross[1] and Roger Wepf[1,2]

[1]*ETH ScopeM, Swiss Federal Institute of Technology, ETH-Hönggerberg, Zürich, Switzerland*
[2]*UQ, CMM University of Queensland, Brisbane, Australia*
[3]*SBRU, University of Cape Town, Cape Town, South Africa*

12.1 SUMMARY

This chapter outlines the requirements to investigate macromolecular complexes at highest possible resolution by cryo-SEM. Preservation of the native state, exposure of the region-of-interest, the avoidance of artefacts as well as signal enhancement by metal coating are covered in this chapter, to guide through common obstacles typically associated with cryo-SEM experiments. The reader is offered a first 'tool kit' to set up their own cryo-SEM experiments, which have the potential to give a unique view on biological or soft-matter surface structures. Moreover, the reader will gain useful background information for the 3D SEM in Chapter 27 by Woodward and Wepf in Volume II.

12.2 INTRODUCTION

The beauty of scanning electron microscopy (SEM) is its capacity to describe and integrate structural details, mainly surface related details, within the context of a complex system. Its unique ability compared to other (transmission) electron microscopy techniques (STEM and TEM) is that handling and imaging of bulk samples is, in principal, possible and hence

Biological Field Emission Scanning Electron Microscopy, First Edition.
Edited by Roland A. Fleck and Bruno M. Humbel.

sectioning, thinning or replicating of the specimen is not essential when surface structures are to be investigated.

With the introduction of field emission SEM (FESEM), especially in combination with efficient signal detection (e.g. 'in-lens' detection), signal filtering (stage biasing or ExB filtering) and new types of detectors (Si-photon-multipliers (SPM), silicon drift detector (SDD), pn-semiconductor detectors, YAG garnet, pixelated CMOS, MCP – to name a few developments), high-resolution SEM (HRSEM) has become a powerful approach to resolve structural details, down to macromolecular dimensions (1–2 nm) in biology and soft-material science. Consequently, FESEM has a potential role as a bridging microscope between light microscopy and molecular or atomic imaging techniques for any surface interaction studies.

Natural surfaces of living systems are usually not directly accessible due to their glycocalyx or natural hydrated microenvironment. As a consequence, and due to the fact that HRSEM is only possible under high-vacuum conditions, the specimens have to be very carefully processed to preserve close-to-native surfaces at nanometre level for HRSEM investigations. The information density and the accuracy of structural information are strongly influenced by the preparation steps and imaging conditions. This problem becomes more pronounced when studying intra-cellular surfaces.

The preparation workflow for any SEM investigation can be briefly given as follows: (1) isolation of the specimen; (2) immobilization, which includes adsorption or embedding as well as chemical or physical fixation; (3) surface exposure, which includes chemical or physical dehydration and detail-of-interest exposure (e.g. fracturing) techniques; (4) signal enhancement, consisting of contrasting techniques (to introduce contrast between features in the final image); (5) transfer into the microscope without structural changes; (6) artefact free imaging at highest possible resolution (e.g. exposure time, beam current, specimen temperature, etc.).

In this chapter, we outline the current collected knowledge of the different steps employed for HRSEM of macromolecular assemblies and describe technical approaches to preserve and enhance structural details at the resolution limit of modern FESEM. To achieve this, we used a range of test specimens (CCV, TMV, Semliki Forest Virus, HPI-layer, F-actin filaments) specifically selected to investigate various parameters and techniques associated with cryo-SEM. The challenge for the future will be to use the acquired knowledge to move to a level of higher structural complexity whilst preserving the macromolecular resolution achievable by current generation HRSEM, for example macromolecular resolution in tissues or soft-matter samples as demonstrated at the end of this chapter.

12.3 PREREQUISITES FOR HIGH-RESOLUTION SEM (HRSEM)

12.3.1 Macromolecular Structure Preservation

The goal of structural studies is to image or reconstruct the specimen in its native shape and conformation in order to gain structural data, which is ultimately required for structure–function correlation. Therefore, any specimen manipulation step has to be carefully considered for its potential, to introduce structural artefacts to the specimen structure. Once approaching the native state, the structural details obtained increase in value, for example biological significance. The inherent problem is that the native state of living systems is highly dynamic and typically exist in an aqueous environment, which has to be transferred to a 'solid state' to withstand the vacuum conditions of the microscope and the bombardment with electrons.

In order to study the influence of different preparations steps on macromolecular structures such as fixation, washing and dehydration, we have established clathrin coated vesicles (CCVs) as a useful experimental model. CCVs are especially suitable for these experiments since their structure comprises a lipid part, the vesicle, surrounded by a protein part, the vesicle stabilizing scaffold (Crowther *et al.* 1976). With this model system we are able to show that any change of osmolarity, chemical cross-linking, pH changes and the removal and interference with the hydration states has a high potential to reorganize such a macromolecular complex system. The only way to preserve its structure is to keep the CCV in their native 'buffer' and image them in a frozen-hydrated state. To study single steps such as the influence of washing after adsorption, the dehydration procedure and gain in stability by chemical fixation, we kept the conditions for freezing, coating and imaging constant. Preparations of CCV were all coated at low temperature by planar magnetron sputtering (−120 °C; 2 nm tungsten (W)), high-vacuum cryo-transferred and imaged at 30 kV with a nominal spot size of 1.5 nm at low temperature (−120°C) in a cryo-SEM (Figure 12.1).

As a positive control, representing optimal preservation, CCVs were fast frozen (physically fixed) by plunging into liquid ethane in their natural stabilizing buffer (used during CCV isolation) and exposed only by freeze-fracturing and partial freeze-drying (Figure 12.1d). Such CCV fracturing exposed a spherical shape with hexagonal and pentagonal faces with very little distortion and shrinkage of the overall convex surface and concave dips, representing the turgescent 3D structure (see Crowther *et al.* (1976)). The most common way to preserve isolated specimens for electron microscopy is chemical fixation. Routinely, a chemical cross-linker such as glutaraldehydes and formaldehydes or OsO_4 are used to stabilize the structure for the subsequent preparation steps (for an overview and special cross-linkers, see Bell *et al.* (1988)). For comparison with the physical fixation (cryo) method, we have tested different concentrations of glutaraldehyde at different temperatures and different dehydration procedures (critical-point-drying (CPD) and freeze-drying (FD)). Some of these results are presented here (see Figure 12.1).

For an optimal freeze-drying result, the specimen should ideally be washed prior to freezing to avoid the formation of salt and protein aggregates, referred as eutecticum, which otherwise covers the sample after removing the water by sublimation (Miller *et al.* 1983). However, when CCVs were washed in double-distilled water after the adsorption to the supporting film to remove residual salt and protein prior to freezing, their spherical shape vanished completely (Figure 12.1b). The vesicle and shell with its protein framework collapsed onto the supporting film and very few polygonal faces were preserved, resulting in a severe loss of volume and structural detail. The residues of the protein framework are entirely distorted. Since freeze-drying prevents the sample from air-drying flattening (surface tension), this effect is caused only by the change in the osmotic solution to distilled water. This artefact can be partially avoided by chemical fixation in the buffer solution prior to the washing procedure (Figure 12.1a) or fully by freeze-fracturing the frozen sample in its close to 'natural' buffer media (Figure 12.1d).

Figure 12.1a shows CCVs that were fixed with 2% glutaraldehyde at 4 °C for 5 min after adsorption to the supporting film. Thereafter, fixed CCVs were washed in water prior to freezing and afterwards freeze-dried and metal coated. Such CCVs are obviously less collapsed and flattened onto the supporting film. Some distorted hexagonal and pentagonal faces of the CCV are still preserved but their 3D volume is clearly lost. The change in osmolarity and some surface tension effects of water in the sample (Figure 12.1a and b) are stronger than any chemical fixation can compensate for. We saw a slight improvement

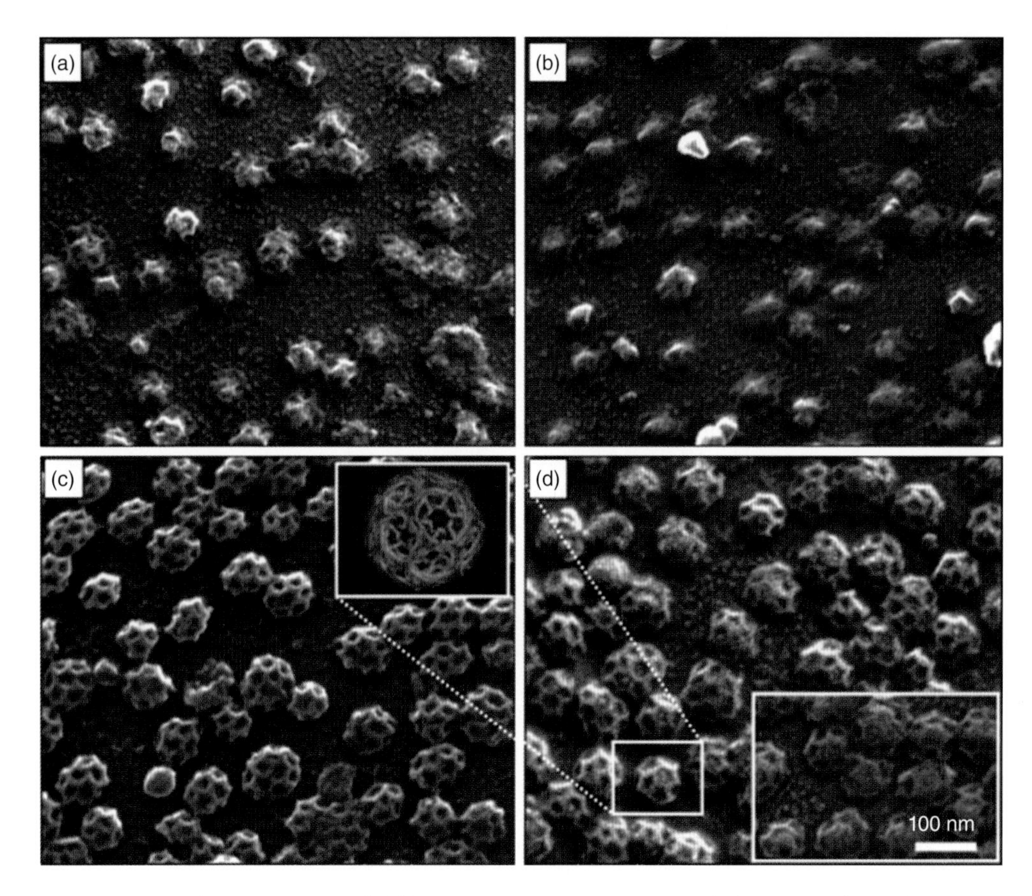

Figure 12.1 Quality assessment of different preparation strategies. Clathrin coated vesicles were utilized as test specimens consisting of a lipid and a protein part. Images were taken at 30 kV with a nominal spot size of 1.5 nm. (a) CCVs were adsorbed to a glow discharged carbon coated grid, fixed in 2% glutaraldehyde at 4 °C for 5 min, washed prior to freezing, freeze-dried and finally coated with 2 nm of tungsten. (b) CCVs prepared as in (a) but, additionally, the sample material was washed twice with double-distilled water after the adsorption to the grid, prior to rapid freezing. (c) Unfixed CCVs prepared by rapid freezing in liquid ethane in their natural buffer and processed by subsequent freeze-substitution. (d) CVV's were vitrified in their natural buffer and exposed by freeze-fracturing and partial freeze-drying. The inset shows the structural change due to total freeze-drying (dehydration at RT). The zoom-in shows a high-resolution model of a CCV.

at lower glutaraldehyde concentration down to 0.5% but still the 3D volume was clearly distorted and collapsed after such treatments, most probably by reduced crosslinking and osmotic effects at lower fixative concentration.

A better structural preservation was obtained by freeze-substitution of unfixed frozen CCVs, which were not washed in water but directly frozen in their isolation buffer (see Figure 12.1c; for a thin aqueous film on a carbon grid 5000–10 000 K/s is sufficient). Even after a gentle dehydration process and refreezing in ethanol, followed by freeze-drying from the frozen ethanol, CCVs still exhibited some shrinkage as compared to only freeze-fracture exposed samples (Figure 12.1d). If fully hydrated CCVs (Figure 12.1d) were completely dehydrated by increasing the temperature to room temperature (RT) during freeze-drying (Figure 12.1d inset) the vesicles lost volume and flattened onto the support; see also

Wildhaber *et al.* (1982) and Gross (1987). The polyhedral faces were, however, better preserved than after a washing step in either pure water (Figure 12.1b) or after chemical fixation (Figure 12.1a) or freeze-substitution (Figure 12.1c).

This implies that structural alterations already appear, by following the change of osmolarity (Figure 12.1b) and further by chemical cross-linking (Figure 12.1a) as well as through the dehydration process (Figure 12.1c) itself (CPD samples were usually totally disintegrated – data are not shown here). Therefore, one can conclude that the collapse of CCV onto the supporting film is due to the complete dehydration of the specimen and not to an adsorption artefact. The CCVs have proven to be very sensitive to any change in their environment such as changes in the pH, salt concentration, etc. (Figure 12.1a versus b and Figure 12.1c versus d). A certain level of hydration (Figure 12.1d inset) has to be maintained to preserve the *in vitro* (native) shape of these protein coated lipid vesicles. Chemical fixation itself causes some shrinkage of CCV similar to reports for tissues, cells, etc. (Boyde and Tamarin 1984; Boyde 1978; Murk et al. 2003) and dehydration in ethanol and/or CO_2 results in further shrinkage of biological specimens. Freeze-substitution with its gentle dehydration with or without simultaneously chemically cross-linking can to some extent overcome the collapse of CCV but not completely prevent shrinking.

For high-resolution studies, it is therefore essential to meticulously evaluate each preparation step concerning its influence on the structural integrity and to optimize the preparation protocol for every single specimen. For some rigid macromolecular complexes such as F-actin filaments (rabbit g-actin), an hexagonal packed protein layer, isolated from the cell wall of *Deinococcus radiodurans* (HPI-layer) and T-phages, a washing step in water does not change the protein complex arrangement, whereas in most cases distortion or disassembly may take place as shown in Figure 12.1. Essential for a good preservation of these fragile samples is the physical arresting in their natural media by vitrification and the subsequent exposure of their surface by freeze-fracturing and freeze-etching (partial freeze-drying). As an alternative method for removing excess salts and proteins, freeze-substitution can be utilized. Followed by rapid-freezing of the organic solvent and its subsequent sublimation, this approach has also proven to be of great value for larger samples such as complete cells, etc. (Walther *et al.* 1992). In both cases, if freeze-fractured or after freeze-substitution, ideally the sample needs to be kept under cryogenic conditions until and during imaging to avoid further collapse of the structures by freeze-drying.

12.3.2 Controlling the Freeze-Drying Process: Partial Freeze-Drying

As outlined in the previous section, the gentlest approach for a partial or total dehydration of the sample material is the freeze-drying process. In the following section, we want to have a closer look at this essential preparation step. The process of freeze-drying can be easily explained by looking at the phase diagram of water, as shown in Figure 12.2. This diagram is calculated for pure water and can be used as a reference graph for any cryo-EM work and for controlling freeze-drying processes (for details see Umrath 1983; Murphy and Koop 2005).

Sublimation of water (freeze-drying) is the process of removal of water molecules from a frozen surface and only takes place if the energy (temperature) of the samples matches a given local partial pressure of water in the surrounding environment (vacuum). The graph shows the saturation curve for water under vacuum conditions from $1–10^{-14}$ mbar and over a temperature range of 20 °C to –160 °C. The blue line follows the 'steady-state' conditions where the total number of sublimating and condensing molecules equals zero. Above this curve (white area), deposition dominates and these are the conditions that need to be avoided in cryo-EM applications and where special anti-contamination strategies come into play.

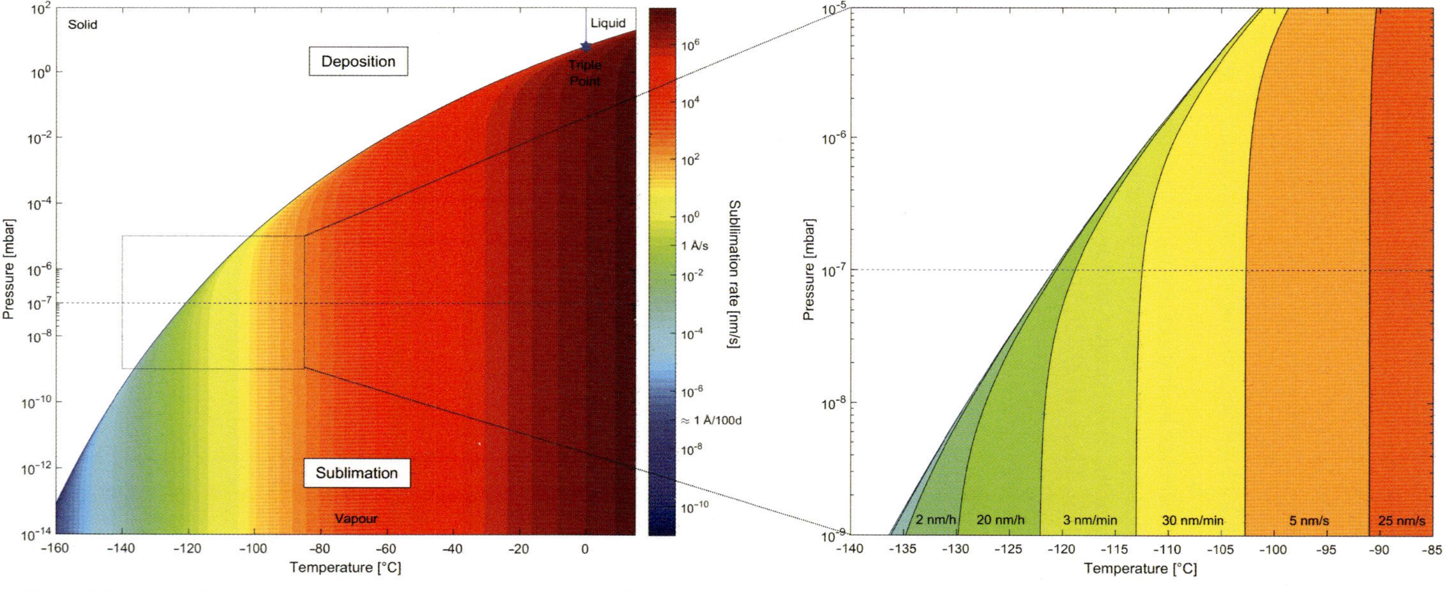

Figure 12.2 Theoretical phase diagram of pure water. On the left side, the phase diagram in the temperature regime of −160 °C to 20 °C and in the pressure regime of 10^{-14} to 100 mbar is shown. The blue line indicates the saturation vapour curve where the amounts of deposition and sublimation are equal. The decrease in thickness, in the case of sublimation, is shown in rainbow colour. Note that some rates are highlighted individually. The framed area is shown on the right side. Here, individual sublimation rates are highlighted. Note that the colour coding was simplified for a better visualization. All calculations are based on formulas taken from Umrath (1983).

The area below (rainbow colour) marks the conditions were water molecules can move directly from the solid phase into the gas phase (sublimation). The number of water molecules leaving the frozen bulk of material can be controlled by the amount of energy delivered to the sample by simply changing the temperature under a given vacuum condition.

As an example, at a pressure of 10^{-7}mbar at a temperature well below –120 °C no sublimation takes place. If the temperature is raised to –100 °C the sublimation amounts to approximately 5 nm/s, at –80 °C it increases to about 100 nm/s and at –60 °C the rate is approximately 150 nm/s. Keep in mind that only water will be removed from an exposed surface. This simple relationship allows us to carefully control the condition of drying and hence exposure of fine structures of our samples and stop it at any time by just lowering the local temperature of our samples again.

Knowing the physical background, one can now design gentle (partial) freeze-drying routines to preserve the 3D structure of a frozen macromolecular complex depending on the quality one would like to achieve. An often forgotten issue is the removal of structural water (see Wildhaber *et al.* 1982; Gross 1987; Wepf *et al.* 1991). For most protein complexes this strongly bound water (hydration shell and structural water) is typically removed above –40 °C, which leads to a further and final structure collapse (see Figure 12.1d). This collapse is still less than the surface tension collapse by air-drying (see Wildhaber *et al.* 1982; Wildhaber and Gross 1985).

12.4 A VERSATILE HIGH-VACUUM CRYO-TRANSFER SYSTEM

Partially freeze-dried samples, where the hydration shell is preserved, must be handled at cryogenic temperatures at all times to avoid any structural alteration, as discussed in the previous section. Also, totally freeze-dried samples need to be handled with care, since they turn hygroscopic due to the dehydration process. Therefore, these samples must be kept under anhydrous conditions during the entire handling.

To avoid delicate transfer steps between different devices, all sample preparation steps can be performed in a cryo-preparation chamber permanently mounted to an SEM (e.g. SCU Bal-Tec, LT7400 VG Microtech, Alto Gatan or PP3010T Quorum, just to name a few). Here the SEM is occupied for the whole preparation and, moreover, the preparation device potentially interferes with the SEM performance by adding extra weight and vibrating pumps permanently to the microscope. Alternatively, the cryo-preparation can be undertaken in an off-line cryo-preparation device. Consequently, samples need to be transferred between the different instruments, ideally avoiding any artefact related to the cryogenic or hygroscopic nature of the processed sample.

There are two different approaches for the transport of the specimen to the desired device. First, the transfer in liquid nitrogen and, second, the transfer in a high-vacuum (HV) environment at cryogenic temperatures. To quantify the structure preservation of these two different approaches at highest possible resolution, cryo-TEM data (instead of cryo-SEM data) of LN_2-transferred samples were compared with HV-cryo-transferred samples. After freeze-drying and coating, the samples were either removed under a counterflow of dry nitrogen from the cryo-preparation chamber and immediately immersed into LN_2 or directly transferred into a cryo-TEM under HV-cryo-conditions. The samples submersed in LN_2 were stored for at least 2 h in LN_2 to simulate an extended storage before they were transferred into the cryo-TEM and imaged at about –170 °C (100 °K) under low-dose conditions (<1000 e^-/nm^2).

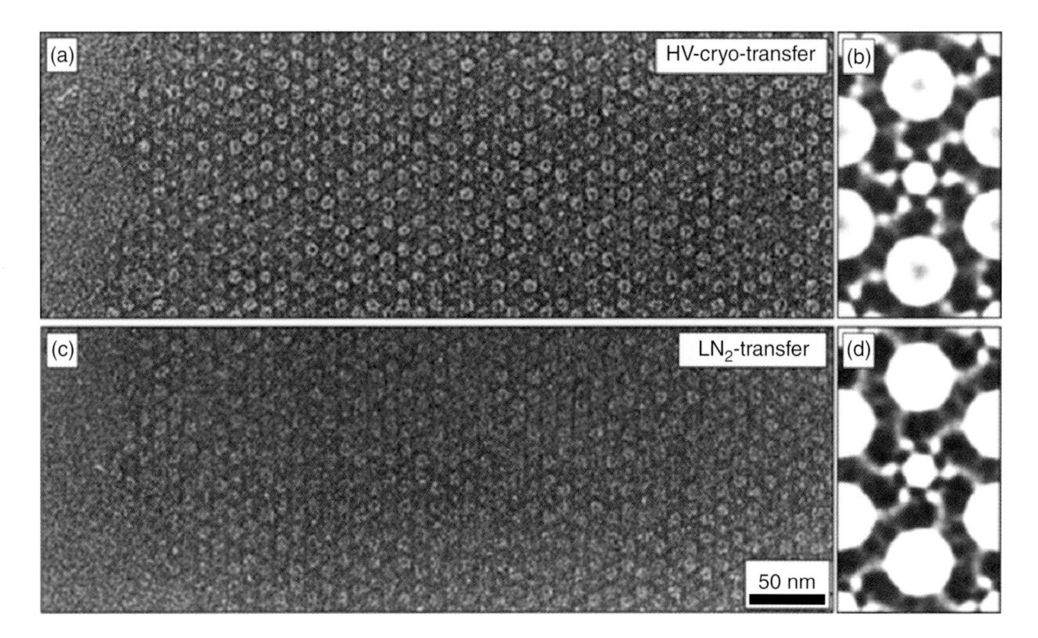

Figure 12.3 HPI layer transfer under (a) high-vacuum cryo conditions or (c) in a LN_2 environment; (b) and (d) show the averaged HPI layer taken from (a) and (c), respectively.

The HPI layer was selected as a test specimen because its outer surface is sensitive to dehydration (Wildhaber *et al.* 1982; Wepf *et al.* 1991) and it allows statistical analysis due to its 2D crystalline arrangement by cross-correlation averaging techniques. Figure 12.3 shows a comparison of the sample material transferred under HV-cryo-conditions (Figure 12.3a and b) or in an LN_2 environment (Figure 12.3c and d). The raw TEM image of the specimen stored in LN_2 appears a little blurred compared to the HV-cryo-transferred specimen but after averaging the resulting structural features are very similar (Figure 12.3b and d). No severe structural changes of the 'pseudo replica' sandwich due to the transfer in LN_2 can be observed. Different resolution assessments revealed an averaged resolution level of <1.8 nm. This demonstrates that LN_2 transfer is able to protect the freeze-dried and metal coated HPI layer from structural alterations down to a resolution level of better than 2 nm. The main problem during LN_2 transfer arises from an inefficient cryo-shielding from the HV-preparation chamber into the liquid nitrogen during submersion in liquid nitrogen (impurity, ice) and back into the HV chamber of a microscope. Because neither surface contamination nor oxidation, which both interfere with SE surface imaging, can be completely avoided during transfer under liquid nitrogen, a versatile high-vacuum cryo-transfer system (VCT) was developed (Wepf *et al.* 2004). This VCT system had from the beginning the potential to establish connectivity between different instruments and microscope types. Over the years it was adapted not only to cryo-SEMs but also to environmental SEMs for inert gas transfer, to cryo-focused ion beam/SEMs, X-ray photoelectron spectrometers, time-of-flight secondary ion mass spectrometers, atom-probe microscopes and finally to in-lens SEM systems (Tacke *et al.* 2016). Figure 12.4 shows historical versions of the VCT and the latest version for in-lens systems. Moreover, this figure illustrates the desired concept of connectivity, which finally allows a protected workflow to be established by transferring sensitive samples in an either inert-gas or high-vacuum environment at room or cryogenic temperatures.

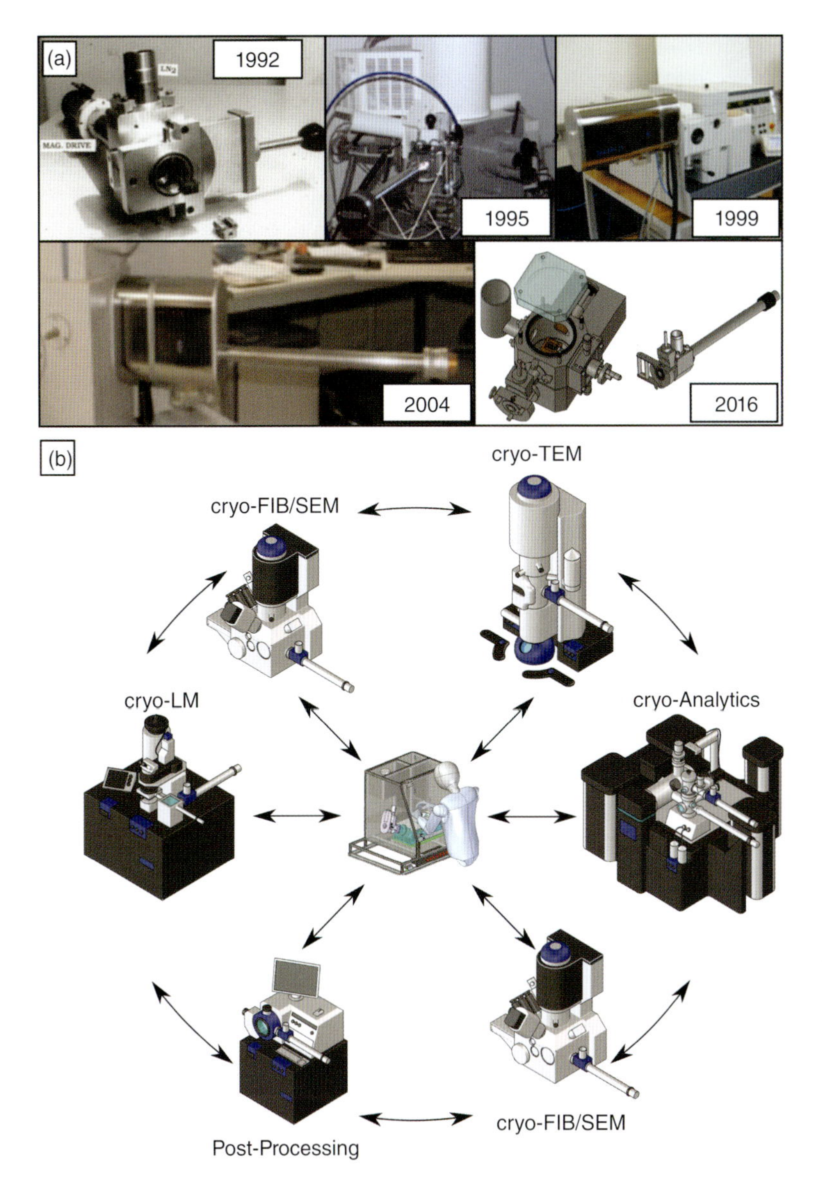

Figure 12.4 (a) History of high-vacuum cryo-transfer systems. (b) Extended cryogenic workflow, which combines cryo-preparation, cryo-imaging and cryo-analytics.

12.5 BLURRING, NOISE AND OTHER ARTEFACTS

Conservation of a close-to-native state during sample preparation does not necessarily mean that this state is also imaged. Direct imaging of nm-surface structures depends very much on avoiding 'blurring' of structures and taking images with a sufficiently high signal-to-noise ratio (SNR). Besides the artefacts related to sample preparation and handling, the electron impact itself may also induce some structural re-arrangements, which 'blur' the high-resolution features in the image. One artefact is related to the dose-dependent mass loss of the specimen caused by the electron bombardment and the other artefact is caused by the opposite, the contamination built up during observation and imaging in an SEM. In the following, we will discuss the best ways to avoid resolution loss due to blurring caused by beam damage or contamination build-up. Moreover, we present some strategies to increase the signal strength and to reduce obstructive background signals during HRSEM imaging.

12.5.1 Mass Loss During Imaging – Beam Damage

An exact determination of the dose-dependent mass loss can be performed by mass determination via dark-field STEM (Tacke *et al.* 2016; Krzyzánek *et al.* 2009; Sousa and Leapman 2007; Engel and Colliex 1993; Wall and Hainfeld 1986). Figure 12.5 shows two test series on freeze-dried tobacco mosaic virus (TMV) particles imaged under low-dose conditions (300 e^-/nm^2). For each series, the TMV particles were consecutively imaged to a cumulative dose of more than 10 000 e^-/nm^2. The left row shows the results from samples observed at RT and the right row shows another imaging series taken at −140 °C. Just looking at the images of the TMV structure it is hard to tell if there is a big difference, whereas in the mass determination plots it becomes clear that the major mass loss happens in the first few images at RT and this mass loss follows an exponential decay for any further exposure. This degradation of the structure can be prolonged by simply lowering the specimen temperature to −140 °C, allowing 1–2 images to be acquired at low dose (300 e^-/nm^2) conditions with a minimal amount of mass loss.

12.5.2 Contamination During Imaging

The interaction of electrons with the specimen surface produces various numbers of breakdown products (hydrocarbons, water, etc.), which themselves can be attracted by the charges deposited in the imaged area (Jiang 2016; Reimer 1998). Such effects become even more pronounced as the amount of electrons per area increases, which is the case when working at high magnification. To reduce lateral movement of surface contamination and breakdown products, it is absolutely essential for high-resolution imaging to work when performed at low temperatures. Working at low temperatures can avoid buildup of large contamination layers and hence blurring and obscuring SE imaging (Jiang 2016; Hermann and Müller 1991, 1992).

In Figure 12.6 freeze-dried and metal coated Semliki Forest virus were imaged at −90 °C (Figure 12.6a) or at room temperature (RT) (Figure 12.6b). The inset in Figure 12.6b shows the amount of contamination built up during the first slow scan on a specimen area at a primary magnification of 150 000 (Figure 12.6b) and with a beam current of 200 pA at RT in an XL-30FEG (FEI Company, The Netherlands) in a freshly pumped-down microscope chamber, whereas the first slow scan at a specimen temperature of −90 °C (Figure 12.6a) under identical imaging conditions and with the cold-trap at −187 °C (anti-contaminator) shows no blurring or covering of fine specimen details by contaminants. Therefore, we can conclude that the amount of contamination caused by the electron bombardment can be dramatically reduced in the presence of an anti-contaminator at liquid nitrogen temperature. In addition, the mobility of the surface contaminants is massively reduced under imaging conditions at cryogenic specimen temperatures (for more details see Hren (1986)). Our experience is that at specimen temperatures below −80 °C several slow scans can be acquired before a significant contamination layer obscures fine specimen details (data not shown). One must bear in mind, however, that beam damaging causes shrinkage of the investigated specimen (Figure 12.5) and hence more than one contamination-free slow scan

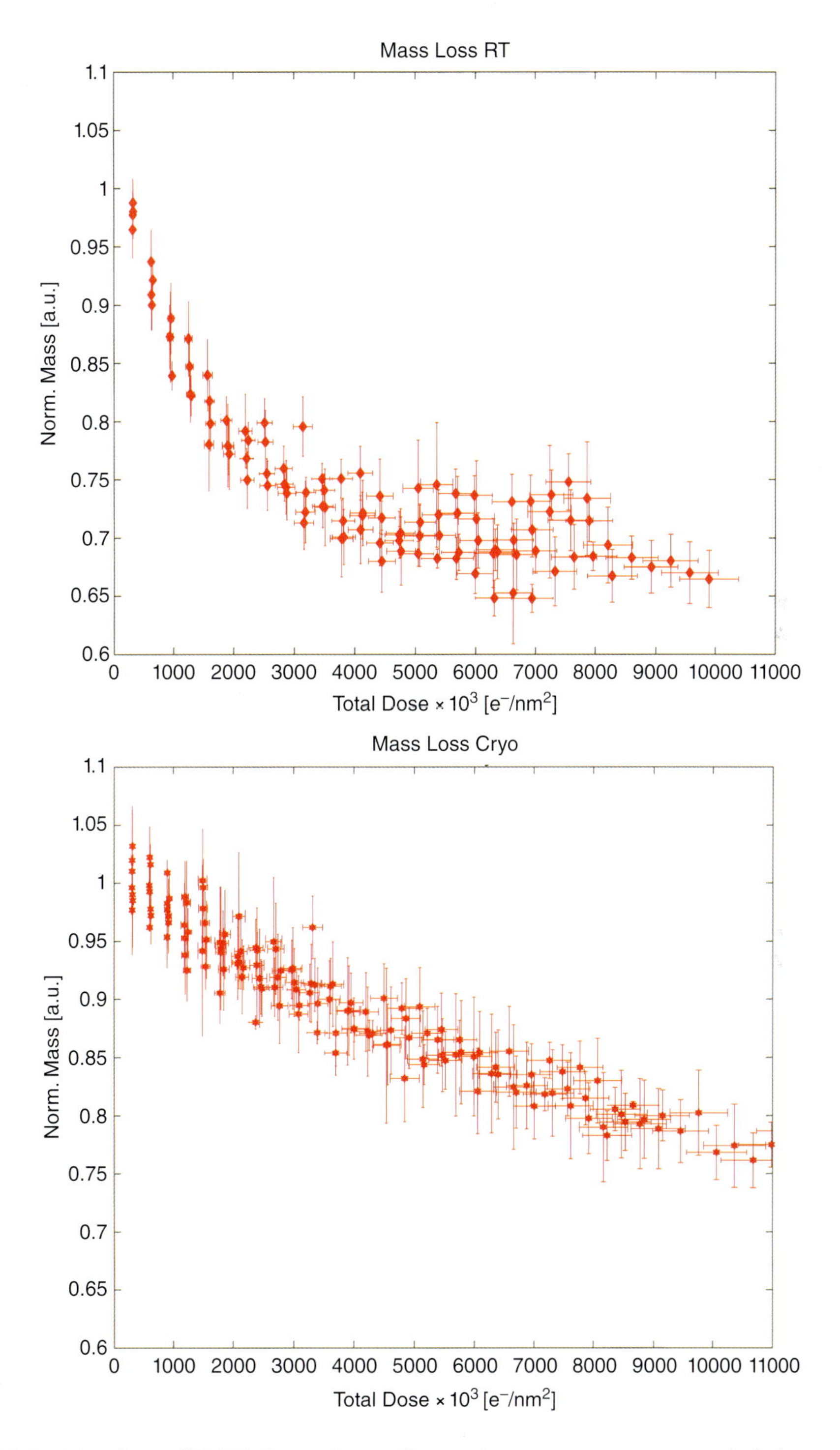

Figure 12.5 Mass loss of TMV. For each mass loss series, an area was consecutively imaged under low-dose conditions (300 e^-/nm^2). For each image the MPL of TMV was calculated, which is shown in the upper graphs. On the upper graph the mass loss series taken at room temperature is shown and on the lower graph the mass loss series taken at cryogenic temperatures (–140 °C) is shown. The images on the next page show a representative area where a mass loss series was determined. The inset shows some features at the beginning and at the end of the series.

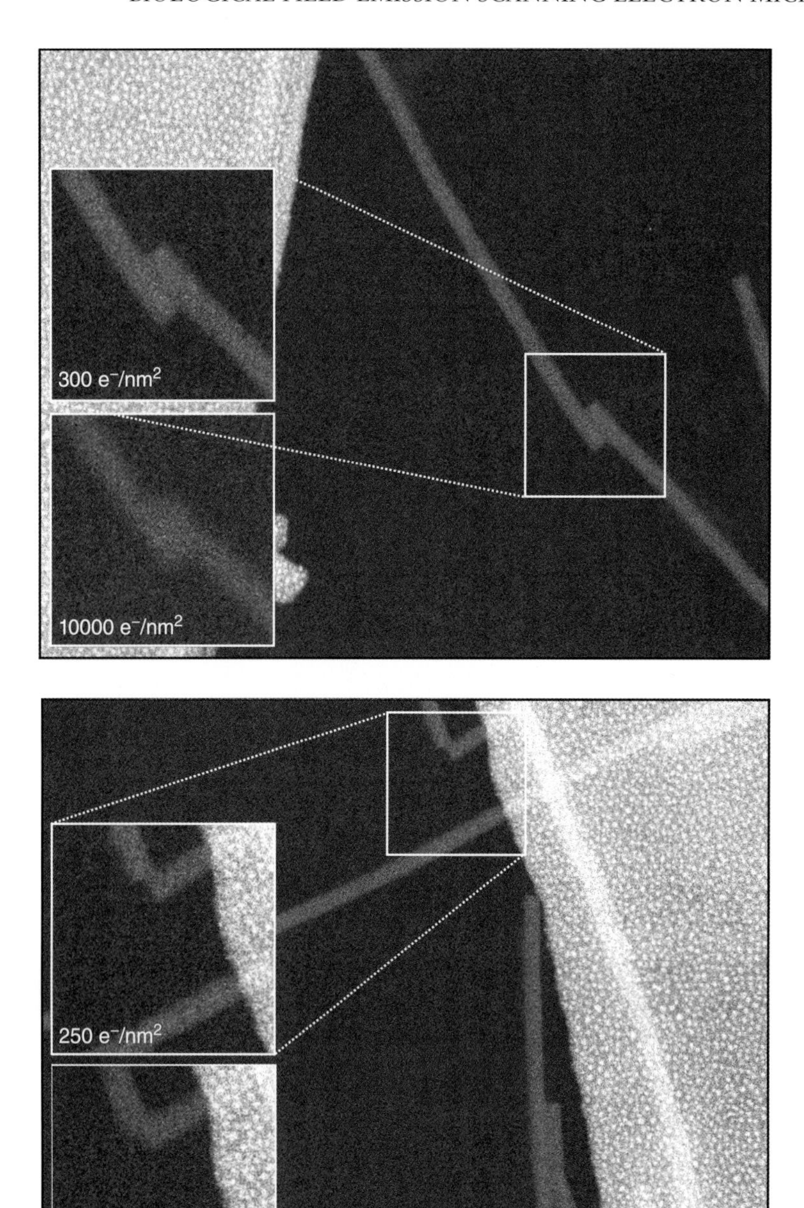

Figure 12.5 *(Continued)*

is ideally not performed. In addition, the amount of contamination is proportional to the total beam current (electron dose) and can be minimized by using the minimal possible beam current in a given SEM-system that still delivers a sufficient signal-to-noise (SNR) ratio for imaging (see Figure 12.8 later in Section 12.5.4).

Low specimen temperatures and adequate cold shields shrouding the sample area are essential not only to preserve a partially hydrated structure until and during imaging but also

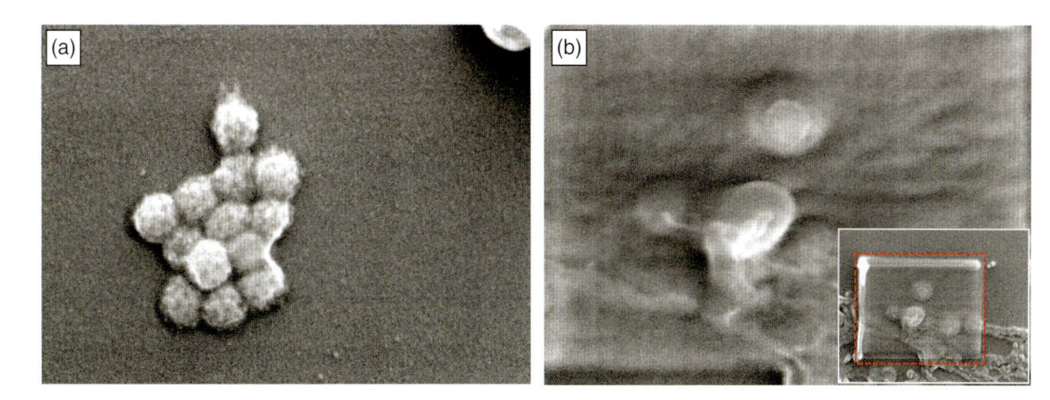

Figure 12.6 Freeze-dried and metal coated Semliki Forest Virus. (a) Samples imaged at −90 °C or (b) room temperature. The inset in (b) shows an overview of the imaged area whereas the red frame highlights the contamination build-up. Courtesy of Dr S. Fuller, EMBL.

to reduce the mass loss in the low-dose regime and the massive build-up of a contamination layer in the high-dose or RT regime. From Figure 12.2 we can deduce that any cryo-shield (cryo-trap, anti-contaminator, etc.) temperature has to fall into the condensation area for a given vacuum condition (white area).

12.5.3 The Sample Support: Strategies for HRSEM for 'Beam Transparent' Samples

Figure 12.7 shows four different kinds of support types one can use for HRSEM and a corresponding typical SE-signal along a line scan through a coated and uncoated F-actin filament cross-section. It is clear from the line scan that there is a much higher noise level in cases 1, 2, 4 and 5 than in case 3 and 6. This is especially the case if the background of the supporting film is made of an inhomogeneous material (case 1), such as the bar of a Cu-EM grid or a polished metal surface with its own grain structure, etc. The interaction with the background material can produce backscattered electrons, which may generate a delocalized SE signal (so-called SE2) and therefore adds an inhomogeneous background signal, which is comparable with an increased noise level. With such a contribution from the background, it is nearly impossible to observe fine signal variations and hence achieve high-resolution imaging of fine structural details. One way to circumvent this artefact is to reduce or match the total interaction volume by operating the microscope at very high magnification. If the imaged area matches the corresponding total excitation volume the background signal remains almost constant over the entire scanned area (Joy 1991). This solution is accompanied with the risk of a too-high beam dose, and hence high damage on sensitive samples, and is therefore only suitable for samples that can withstand higher electron doses. Another alternative is to use a very thin 5–10 nm carbon support film (cases 2 and 3), as used in TEM studies. Within such a thin and low atomic number film, the broadening of the primary beam at high acceleration voltages and the BSE yield are minimized. Therefore, the signal from the contrasted specimen is not as strongly buried in the reflected signal (case 3) caused by the transmitted electrons hitting matter behind the grid (case 2). Similar results can be obtained with a homogeneous solid support such as a glass or sapphire disc coated with 20 nm of carbon or the conductive yttrium tin oxide (ITO), mica sheets (see Furuno *et al.* (1992)

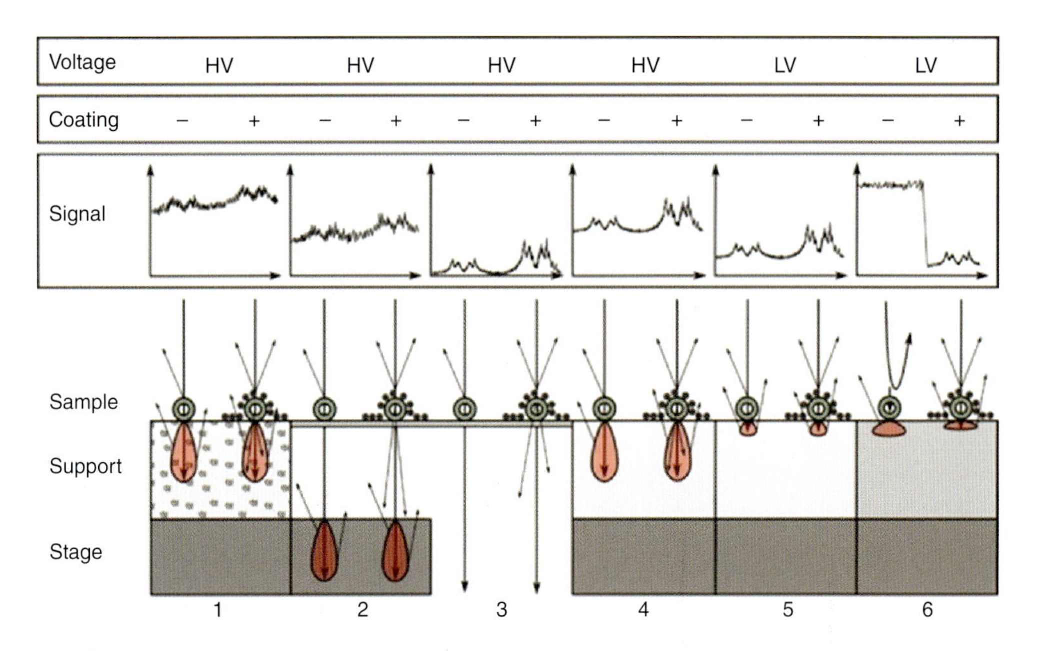

Figure 12.7 Assessment of different kinds of support materials and voltage magnitude with regard to the quality of the SE signal. For each case, the coated and uncoated sample material is compared. From top to bottom: type of voltage, high voltage (HV) (do not mistake this abbreviation of HV with high vacuum, also abbreviated to HV) or low voltage (LV), and coating (+) or no coating (–), approximated line trace of the SE signal, sample material, support material and the stage. The interaction volume of the impinging electrons is shown in red. The following exemplary cases are discussed: (1) inhomogeneous support film, (2) thin support film with and (3) without stage, (4) homogeneous support film, (5) homogeneous support film and low-voltage and (6) insulating support film and low voltage.

and Figure 12.7) or silicon wafers (Furuno and Sasabe 1993) and highly oriented pyrolytic graphite (HOPG). Such a support adds a constant offset to the signal but theoretically does not influence the SNR ratio. In practice, the resolution on such supports is limited to about 2–4 nm (Furuno *et al.* 1992; Furuno and Sasabe 1993).

With the modern dedicated low-voltage SEMs (LVSEM - equipped with cold-emission FEG, monochromator, decelerations modes and/or Cc correction), 1 nm resolution with landing energies below 2 kV became available. Imaging of uncoated or coated samples on conducting surfaces is possible by tuning the imaging conditions (high voltage and exposure time) to reduce background contributions (case 5). If the support is a non-conducting material (case 6, e.g. mica, glass, sapphire), even LV will not help to avoid local charging and especially SE-imaging of beam-transparent thin samples will become impossible.

12.5.4 Low Dose Imaging versus Signal-to-Noise Ratio

An important factor to consider in high-resolution imaging of irregular aperiodic structures is the signal-to-noise ratio (SNR). Especially when operating at low-dose conditions, direct observation of features can easily be obscured by the noise in an image (see Figure 12.8). There are two options to increase the SNR: either the noise is reduced or the signal strength

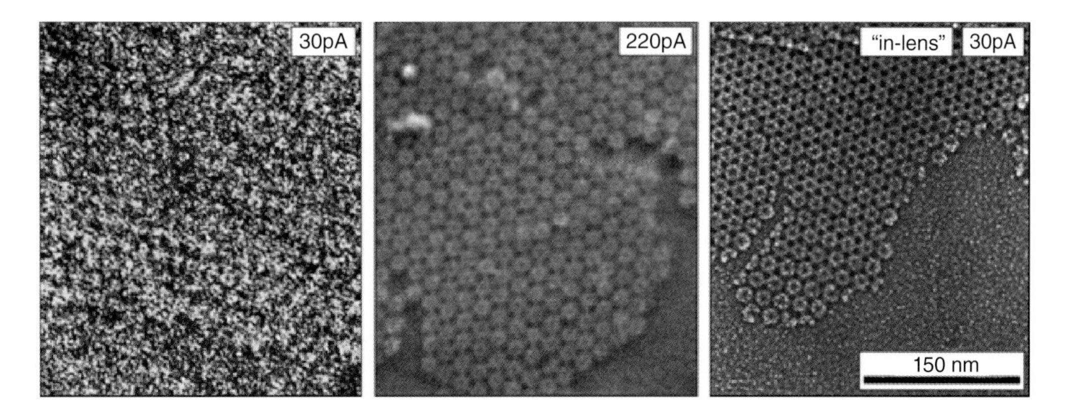

Figure 12.8 HPI layer imaged with different microscopes under different imaging and detection conditions. From left to right: Hitachi S-800 with a beam current of 30 pA (approx. 4400 e^-/nm^2); FEI XL-30FEG with a beam current of 220 pA (approx. 30 000 e^-/nm^2); Hitachi S-900 operating with a beam current of 30 pA (approx. 100 e^-/nm^2) with an 'in-lens' detector system and specimen mounted inside the immersion lens (in-lens stage similar to an STEM).

is increased. Noise reduction is typically achieved by appropriate read-out electronics and sufficiently stable electron guns. In order to enhance the signal strength, either the overall signal yield is increased or collection and counting efficiency of the detector are optimized. Whereas signal enhancement is discussed in Section 12.6, we would like to discuss a few technical parameters of existing detector systems, which should be considered when aiming for the highest possible resolution.

As mentioned above, one way to reduce the destructive beam damage and obscuring contamination build-up is to operate the microscope under low-dose conditions, which leads inherently to a low signal yield and hence a potential loss of image quality. As Figure 12.1 shows, working at low-dose conditions is not possible for all SEM setups. The three images have all been taken in different FE-SEMs equipped with a cryo-stage on identical samples – 1 nm W coated, freeze-dried 2D HPI-protein-crystal (HPI layer). The picture on the left was taken in a conventional post-lens system (Hitachi S-800) with 30 pA beam current (which, at a spot size of about 1 nm and an exposure time of 0.5 microseconds per pixel, corresponded to about 100 $e^-/nm^2 = 1\ e^-/Å^2$). The second image was taken in another post-lens FE-SEM (Philips XL-30), which required the electron dose to increase up to 220 pA (~30 000 e^-/nm^2) to see the best possible protein crystal lattice. Both systems are so-called post-lens detection systems and show that the lattice is resolved but very noisy, which can only be compensated for by increasing the amount of electron dose. Whereas an in-lens detection system (Hitachi S-900) operates at a low-dose condition (30 pA for ~1000 e^-/nm^2), the SNR is sufficient to directly visualize structural details of the HPI layer down to 1.5 nm. This image series shows clearly that at such a low dose and hence low signal yield a high signal collection efficiency of the detector system is a key issue for high-resolution macromolecular structure SE imaging and likewise for BSE imaging.

12.6 ALL ABOUT COATING

Best structural preservation is worthless if the structure of interest remains veiled in a noisy image. To improve the SNR, appropriate detectors must be utilized. Besides the optimization of hardware components, the SNR can be also increased by suitable coating of the sample material. In the following paragraph different coating techniques and important coating parameters are discussed.

12.6.1 Signal Enhancement – Contrasting Techniques for HRSEM

Several contrasting methods have been used in SEM to enhance the SE or BSE yield of biological specimens, such as negative staining (Furuno *et al.* 1992), positive staining (Seligman *et al.* 1966) and metal coating by different coating techniques (see Echlin 1979; Peters 1986; Walther *et al.* 1990; Wepf and Gross 1990; Hermann and Müller 1991). These contrasting techniques are summarized and sketched on a cross-section of a tubular structure in Figure 12.9, where they are divided into direct and indirect contrasting methods. Corresponding experimental results are shown on F-actin filaments in Figure 12.9.

Direct imaging of macromolecules reveals a 'mass thickness'-dependent signal, which means a contrast that is related to the atomic density (atoms per volume) and composition (atomic number) of the specimen. This signal is localized, so contains high-resolution information if the penetration depth of the primary electrons is much larger than the specimen thickness and the mean-free-path length of the SE and BSE is the range of the sample thickness (see also Section 12.4.1. This direct 'mass thickness' contrast can be slightly enhanced if the macromolecules are positively stained with uranyl acetate (UA) or uranyl formate (UF) salts prior to freezing and freeze-drying (Figure 12.10b) compared to the uncoated freeze-dried F-actin filaments in the shadow area in Figure 12.10a (upper left corner).

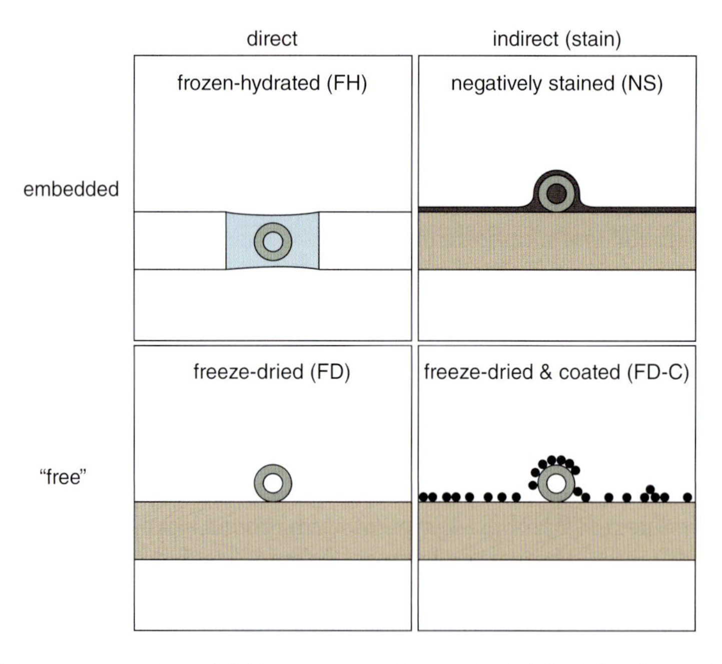

Figure 12.9 Schematic overview of different contrasting methods. The left column shows the direct imaging modes, whereas the right column includes the indirect imaging routines. In both categories, you can differentiate between preparation routines offering embedded or free sample material.

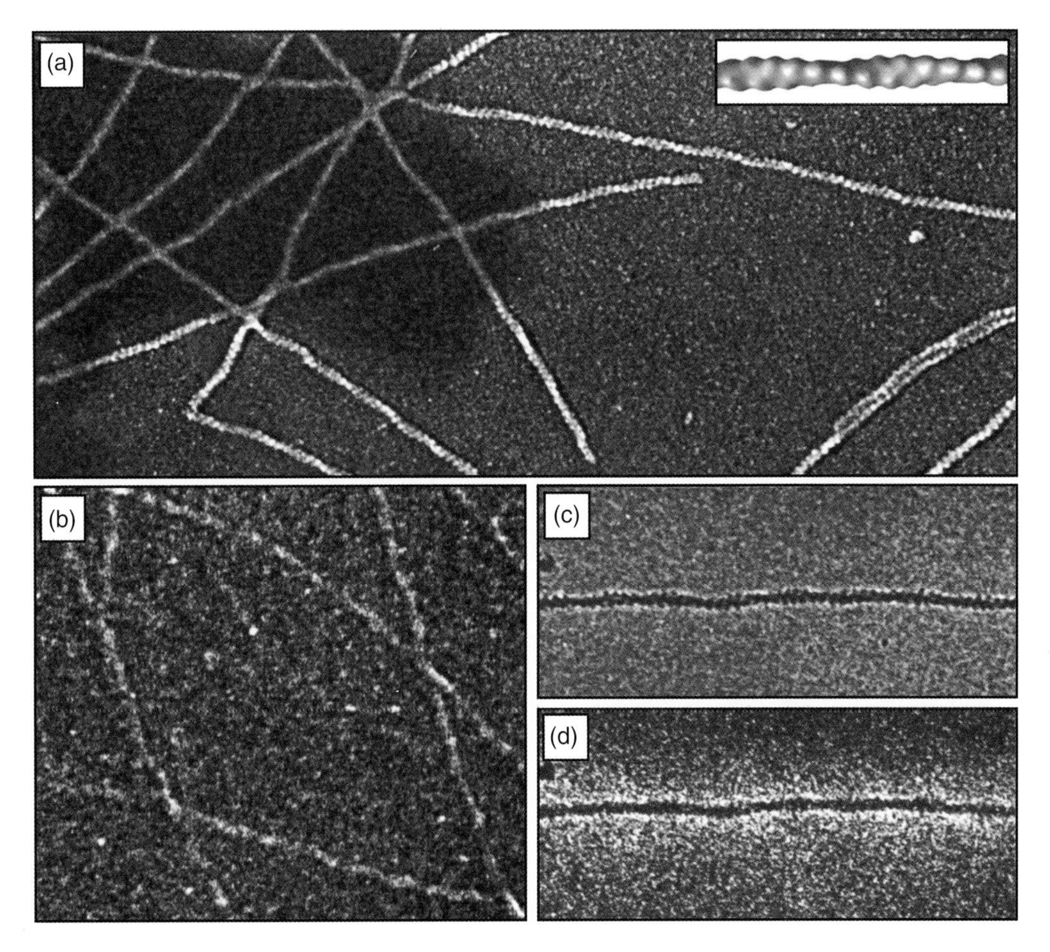

Figure 12.10 F-actin filaments prepared with different contrasting methods. (a) SE image of freeze-dried and metal coated (1 nm W from a 45° shadowing angle) F-actin. Note that there are also uncoated F-actin filaments in the top left corner. Inset shows a 3D high-resolution reconstruction of an F-actin filament from the SE image in Figure 12.11. (b) SE image of positively stained F-actin filaments. (c) SE and (d) BSE images of negatively stained F-actin filaments.

Note that with both 'mass thickness' methods, only a rod-like filament with some density fluctuations became visible in SE images. Neither structural details from the single actin monomers nor from the topographic features of the F-actin filament are visible. Periodically every 37 nm along an uncoated filament (Figure 12.10a) the SE signal reaches a maximum, which matches exactly with the location of the largest mass thickness at the cross-overs of the two monomer filament stands (Figure 12.10a). No structural details can be found in the noisy SE image of positively stained filaments where the filament signal is weakly enhanced compared to the background due to some metal incorporation into the F-actin filament stain pattern (Figure 12.10b), whereas in both indirect staining methods structural details down to 2 nm are revealed (Figure 12.10a, coated area, and c). The contrast of the 'heavy metal stain' in metal coated specimens (Figure 12.10a) reveals even a topographic contrast of the specimen surface by enhancing and localizing the SE signal within the replicating metal film grains. Surprisingly, it is also here that the strongest SE signal is found at the cross-overs of the two F-actin strands. In comparison with the uncoated naked structure in the shadow area of a unidirectionally coated F-actin sample (Figure 12.10a, top left), from which no BSE-signal could be detected (not shown here), the single subunits and their orientation within the filament are clearly resolved. Negative staining, on the other hand, reveals a stain excluding pattern as shown in Figure 12.10c and d. Here the main BSE signal emerges from the metal, which is surrounding the protein mass (Figure 12.10d), revealing a similar topographic view in the SE mode in reverse contrast (Figure 12.10c), as on coated F-actin filaments (Figure 12.10a). In the BSE mode (Figure 12.10d) a projection view of the surrounding metal is obtained, similar to the contrast in TEM image of negatively stained F-actin filaments (Bremer *et al.* 1991). At a closer look, it becomes obvious that the stain penetrates further into the filament structure, resulting in a more teetered projection view of the F-actin filament in the SE image compared to the topographic appearance of coated F-actin filaments (Figure 12.10a and c). Therefore, the mini 'arrowhead (Bremer *et al.* 1991) of the F-actin filament is better resolved by negative staining than by metal coating. Finest structural details on macromolecular assemblies, such as the F-actin filament, could only be obtained by inducing heavy metal 'replication' either by coating techniques or negative staining. Where coating enhances surface details of the macromolecule and negative staining penetrates further into the structure, a more teetered projection appearance of the F-actin filament results. Hence, the so-called mini 'arrowhead' of the F-actin filament becomes visible, which is difficult to resolve by unidirectional coating (Figure 12.10a) and becomes only weakly resolved by rotary shadowing (Figure 12.11, zoom-in).

12.6.2 Coating Techniques

As already stated in the previous sections, a soft-matter specimen, for example macromolecular complexes, consist of light elements, mainly carbon. Consequently, the secondary electron (SE) yield from such a specimen is typically low and hence the SNR is inappropriate for HR imaging. In order to increase the signal and to enhance the localization of the SE signal, the sample material is typically coated with metal, consisting of nanometre small metal grains (see Figures 12.10, 12.11 and 12.12).

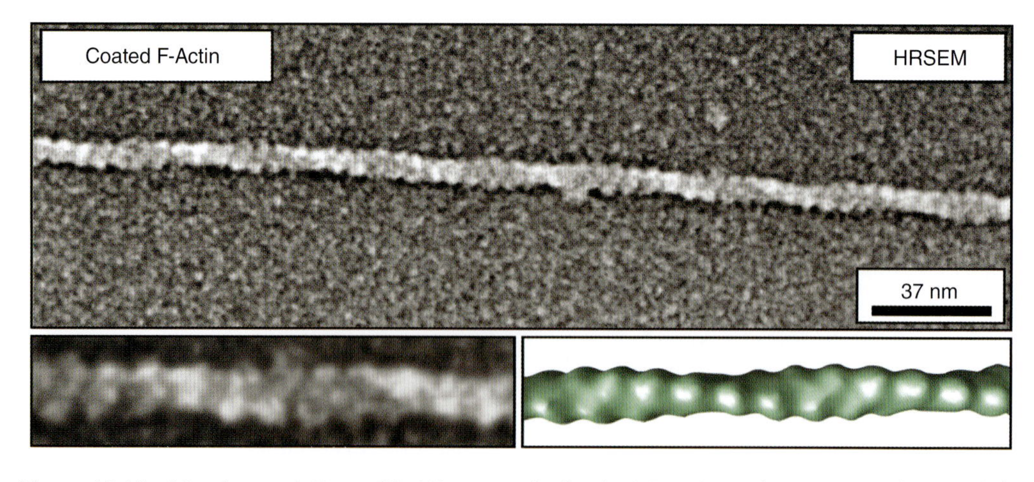

Figure 12.11 Metal coated (1 nm W, 65° rotary shadowing) F-actin with a zoom-in (bottom left) and 3D reconstruction (bottom right).

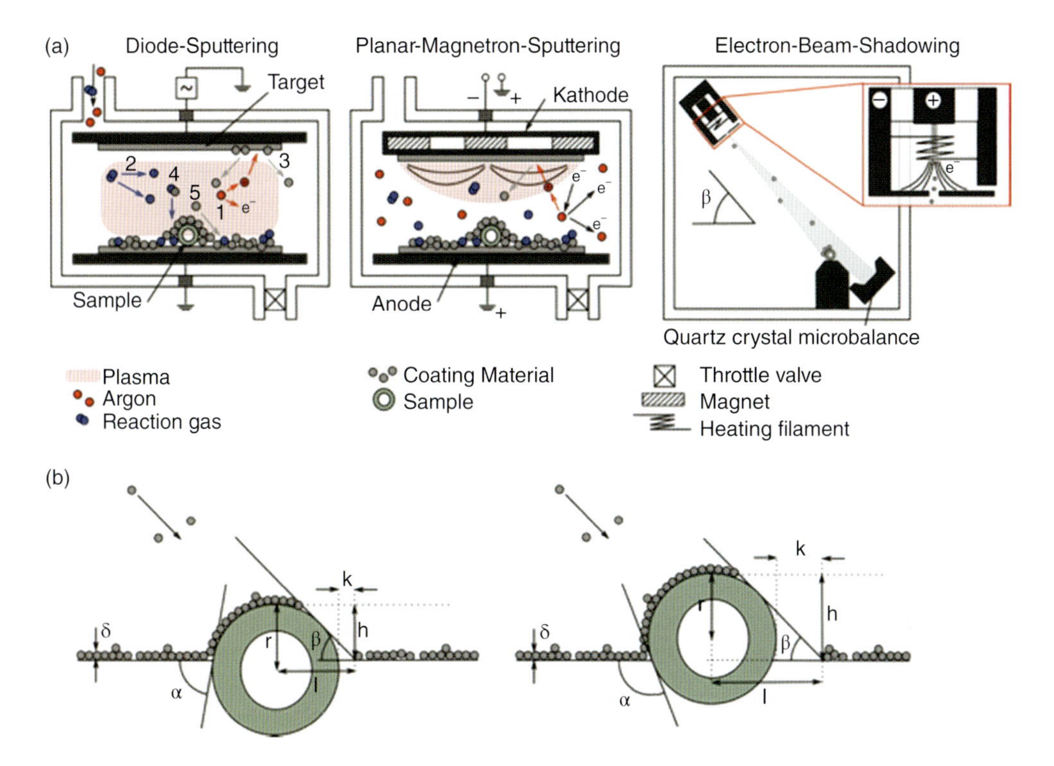

Figure 12.12 (a) Diode sputtering (DS), planar magnetron sputtering (PMS) and electron beam evaporation (EBE). (b) Model of electron beam shadow casting of metal on to two interface particles. Knowing the shadowing angle β and the metal film thickness allows one to determine also that the contact angle α of a particle on a (liquid) interface may have the height of a particle – see the formula and calculation in Lucio *et al.* (2011).

Since there are different methods available for lab usage, we introduce here only the basic principles and discuss the best way to perform HRSEM imaging of macromolecular complexes.

The two most widely employed metal coating techniques in electron microscopy belong to the so-called physical vapour deposition (PVD) technologies. Here, each device can be classified into one of two major categories: either the device utilizes 'sputtering' or 'evaporation' as the technical concept for releasing metal from the solid phase into a vapour phase. Sputtering is based on the bombardment of a metal target by ions and evaporation by inducing heating and herewith melting the metal. Both techniques are broadly used in materials science and the semiconductor industry and have many more variations, as discussed below.

Sputtering of metals from a solid target is achieved by generating a plasma and accelerating its ions towards the target such that they have enough energy to release (sputter-off) metal atoms or clusters from it. The ions may be produced by exposing a gas load (argon, nitrogen, xenon, air, etc., up to 10^{-2} mbar) between two diode plates (diode-sputtering (DS), Figure 12.12a). Here, charged particles (electrons or ions) are accelerated between the plates and may collide with other gas molecules, generating gas ions – the so-called plasma (pink). Finally, the positive gas ions of the plasma are accelerated towards the target to generate neutral metal clusters, which diffuse through the plasma until they find a surface upon which to be adsorbed (Figure 12.12a, left). During diode sputtering, the sample is exposed to a highly energetic plasma and sensitive samples will be destroyed, but not, for example, insects, pollen and other samples with hard exocuticula containing biological structures. To avoid samples being exposed to energized particles, the plasma can be localized with the help of strong magnetic fields, submersing the target in a local plasma (e.g. planar magnetron sputtering (PMS), Figure 12.12a, middle). Other ways to sputter metal use a focused ion beam, ion-sputtering, radio frequency or a triode system; these methods have proven not to be of any use for high-resolution SEM or TEM and are only used in industrial coating applications or for general conductivity or fixation (see FIB/SEM GIS coating techniques).

Alternatively, metal atoms can be released from a target by resistance heating, by 'arc heating', by inductive heating or heating with an electron beam (e-beam) (Figure 12.12a, right). For carbon coating, resistance and arc heating are still commonly used, but no longer for metal evaporation. Nowadays, for TEM and SEM applications, where high atomic number material depositions are favoured, evaporation with an e-beam (Moor 1971) has proven to be the most reproducible way for sample coating. E-beam evaporation (EBE) (Figure 12.12a, right) can be achieved with a focused e-beam impacting a metal target or with electrons being accelerated towards a small metal target cathode (Figure 12.12a, right) releasing metal atoms and ions from a very small area. Depending on the distance from this source to the specimen, one can treat it as a point source evaporation, which 'shadow casts' the specimen under an elevation angle beta β. Compared to sputtering, where the metal atoms are randomly scattered and deposited from any angle and direction on to the sample surface, e-beam evaporation (EBE) allows shadowing to be controlled and adapted by the operator. To successfully control the thickness, 1–2 nm, of the deposited films, a thin film microbalance (quartz crystal monitor) is used for all high-resolution coating experiments.

12.6.3 Tungsten Planar Magnetron Sputtering versus e-Beam Evaporation

A comparison of PMS and EBE coating on a freeze-dried HPI layer is shown in Figure 12.13. The left column shows the SE, BSE and TEM images from samples coated by PMS at –80 °C with 1 nm W and the right column shows the corresponding images from samples that were rotary shadowed (EBE) at an angle of 45° at –80 °C with 1 nm W average film thickness (calculated film thickness from a quartz crystal monitoring system). The grains after planar magnetron sputtering (PMS) are significantly larger in all imaging modalities, giving at a first glance a crisper image. Carefully comparing the PMS with the EBE images shows less fine structural details in the raw images. The cross-correlated averaged images of about 250 unit cells are shown in the right upper corner for each imaging mode and will be called the 'average structure'. By comparing the fine structures from EBE coating it becomes clear that the PMS coated samples reveal a more 'blurred' structure with less details and only in the TEM averaged image does the handiness of the HPI outer layer become evident. In contrast, the EBE averaged structures show the handiness of the HPI unit cell as well as finer details on the arm-like connections. Resolution assessments (FRC, S-Image and SSNR) of PMS and EBE coated HPI layers revealed 2–2.2 nm for W-PMS and 1.5–1.8 nm for W-EBE coating. The measured metal grain sizes are for PMS values between 1.5 and 3 nm and for EBE between 1.2 and 2.2 nm.

12.6.4 Coating Film Thickness

For most biological applications, it is interesting to know the maximal metal film thickness for an 'optimal visibility' of fine structural details without the necessity to use averaging techniques. We have therefore coated freeze-dried T4 bacteriophage capsid polyhead

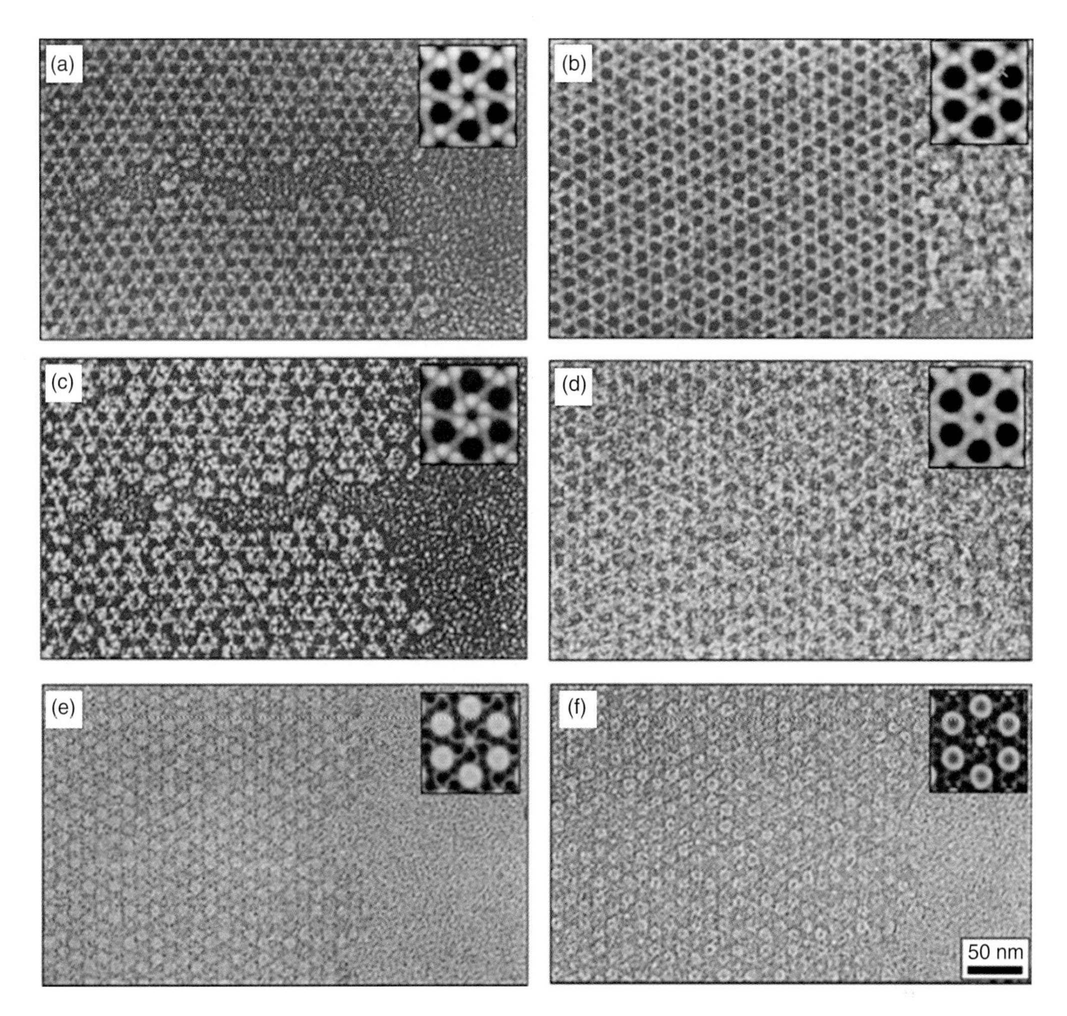

Figure 12.13 Comparison of PMS and EBE coating on a freeze-dried HPI layer. The left column shows the (a) SE, (c) BSE and (e) TEM images from samples coated with PMS at –80 °C with 1 nm W. The right column shows the corresponding (b) SE, (d) BSE and (f) TEM images from samples rotary shadowed from 45° at –80 °C with 1 nm W.

(T4 polyhead) with an increasing amount of metal; here we show 0.4, 0.7 and 2 nm average thickness of the metal films (measured in-situ with a quartz crystal balance). At a film thickness of 0.4 nm W the TEM image (Figure 12.14a) is very noisy. Even though single metal grains can be resolved, the hexagonal arrangement just becomes visible in the raw image, whereas at film thicknesses around 0.7 nm (Figure 12.14c) the rosette-like single cores and the six subunits within the core become resolved. At film thicknesses of 2 nm, the six subunits become 'directly' visible in the raw data and become the dominant structures in the polymer assembly. In addition, the cores show a lateral enlargement. The averaged structures of the coated T4 polyhead capsids show very fine details down to 1.3 nm with little difference in structural features from 0.4 nm and 0.7 nm coatings. The 0.7 nm coating shows a little more contrast in the averaged image and the same number of subunits, due to a better SNR. The averaged image from the 2 nm W coated T4 polyhead reveals clearly a broader and less detailed structure due to too-large metal gains 'covering' and masking

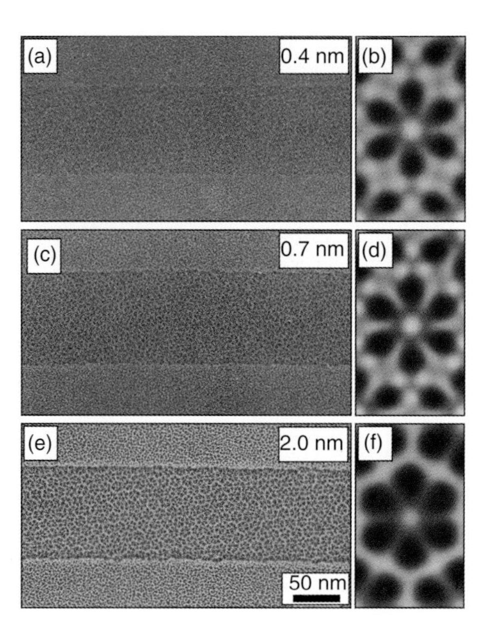

Figure 12.14 Resolution assessment in dependence of different coating thicknesses for the example of a freeze-dried T4 polyhead specimen. (a) Coating thickness: 0.4 nm. (b) Averaged structure taken from (a). (c) Coating thickness: 0.7 nm. (d) Averaged structure taken from (c). (e) Coating thickness: 2 nm. (f) Averaged structure taken from (e).

underlying fine structural details. This artefact is known as the 'low-pass' filtering effect, which one can see in the raw image. For direct visibility, one would preferentially use a film thickness between 0.7 and 2 nm, and not more, in order to avoid further masking of thin surface features.

12.6.5 Elevation Angle

Besides the material and the average thickness of the coating film, the elevation angle – the angle between the specimen plane and the direction of the incoming impinging metal atoms – also influences massively the image contrast and hence the structural details that can be extracted or directly resolved. At low elevation angles, metal grains and small surface features become themselves shadowing objects and tend to grow laterally faster towards the metal source. By rotating the specimen this effect is enhanced in all azimuthal directions and helps in an understanding of what actually happens. Figure 12.15 shows the T4 polyhead structure image in TEM and the corresponding averaged unit cell after rotary shadowing under different elevation angles (15°, 30° and 45°). In Figure 12.15a, the sixfold capsid becomes directly visible, whereas the images from a 30° or 45° deposition reveal less structural definition. However, the averaged images of these areas show the finest structure and the handiness of the capsids and their interconnection best from elevation angles above 45° (65° is similar to 45° data, not shown here). At 15° elevation angle an overall enlarged and broader capsid symmetry is revealed with no structural details between the capsid elements. It seems that these connective structures of the capsid assembly are either not contrasted or simply burned out by the lateral deposition of the metal until a homogeneous contrast (grey level) is formed. At higher angles, areas with no metal (white)

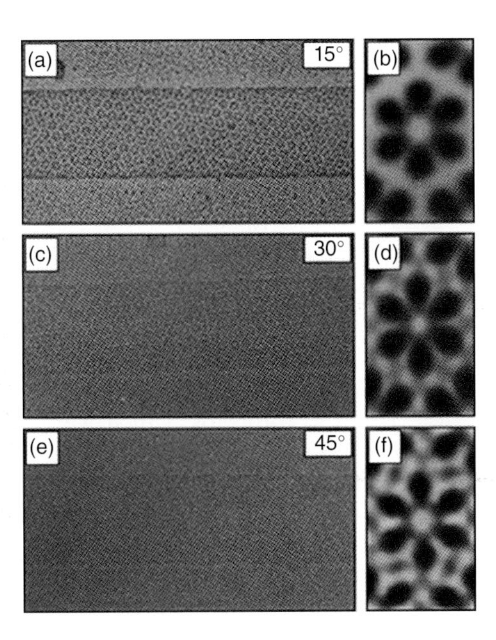

Figure 12.15 Dependence of resolution assessment on different coating angles for the example of a freeze-dried T4 polyhead specimen. (a) Coating angle: 15°. (b) Averaged structure taken from (a). (c) Coating angle: 30°. (d) Averaged structure taken from (c). (e) Coating angle: 45°. (f) Averaged structure taken from (e).

and fine arm-like structures are enhanced with the help of the metal depositions, allowing fine features to be extracted down to 1.5 nm and at an elevation angle of 45° or above the handiness in these connecting arm-like structures is also revealed.

These high angle deposited metal films with film thicknesses of less than 2 nm (see Figure 12.14) are so-called 'optimal granularity films', meaning that the amount of metal grains and size are optimized for a given sample to extract, with the help of image averaging techniques (e.g. cross-correlation averaging), the highest possible structural resolutions; for SEM this is in the range of <1 nm and for TEM demonstrated on Aquaporin (Walz *et al.* 1996) it is around 0.5 nm. A prerequisite for the use of 'optimal granularity films' is the use of periodic structures, which allow for averaging to obtain a 'virtual homogeneous metal coating film', not achievable with physical deposition techniques at cryogenic temperature or at room temperature. This is 'in-kind' a work around the physical natural barriers for metal coating used to enhance and stabilize structural features. Any metal film deposited under cryogenic conditions in HV or higher process gas pressure will always condense in small metal grains and would only form a homogeneous continuous film under UHV and at temperatures above 600 °C, which is incompatible for biological EM applications (see molecular beam epitaxy (MBE) or other physical vapour deposition (PVD) techniques technology for more details on homogeneous metal film depositions).

In summary, electron beam evaporation performed at the highest possible elevation angle (between 45 and 65°) with an average film thickness below 1 nm (W, Re) reveals the highest structural 'replication' and hence minimizes replication artefacts and maximizes resolution. The film thickness needs to be carefully adjusted for a given specimen corrugation. If image averaging techniques cannot be applied and direct visibility is requested, a lower elevation angle, a thicker metal film or, for sputter coating, a higher process gas pressure, will enhance low-angle metal deposition and are the solutions to adopt. There is no general rule and depends on the object size, corrugation and dimensions. You may start from a film thickness above 2 nm. A few iterations may be necessary to get the right balance between extraction of the highest possible resolution and avoiding obscuring the fine specimen details, while keeping in mind that the charge balance must be maintained during imaging.

12.6.6 Which Metal?

There remains one big question after investigating the different parameters in metal film deposition technology – which metal is the best to be used for HRSEM? This depends very much on the final application, since some metals tend to oxidize or produce a high backscattered signal or are not inert for replica production. There are hundreds of publications on coating films, mainly from the 1960s to the1990s, which were chiefly done to show that a certain product is the best. Most of the examples are unfortunately not comparable nor run under best possible HRSEM conditions that are available today. However, let us try to summarize a few experiences from more than 20 years of testing metals and metal oxides used for coating from room temperature down to 15 K sample temperature (Gross 1987; Gross *et al.* 1985). After comparison with HPI, T4 polyhead and other structures, a general empirical rule emerged – 'the higher the melting temperature', the finer the metal grains at specimen temperatures between 15 and193 K. Mainly W, Ta, Ta/W and rhenium (Re) revealed the finest metal grains when not exposed to oxygen or atmospheric conditions. An alternative way to reduce the grain size is to deposit simultaneously C or SiO_2, as, for example, Pt/C (Moor 1971) or Pt/Ir/C (Wepf *et al.* 1991).

In addition, for HRSEM imaging of insulating samples or samples on insulating supports the 'conductivity' of thin films becomes one of the major selection criteria. It has been reported (e.g. Frethem *et al.* (1993)) that the electrical resistance of metal films depends on the metal type, the deposition conditions, the sample temperature and the average film thickness. To summarize these findings, we focus just on the electrical resistance in Ω/mm^2 (sheet resistance), where 1 nm W was found to be in the range of 10^4 Ω/mm^2 for Ir about 10^5 Ω/mm^2, for Cr about 10^6 Ω/mm^2 and for Pt about 10^6 Ω/mm^2, making W, Cr and Ir the top candidates to prevent charge instability. This has also been reported empirically, for example, in Peters (1986) and Hermann and Müller (1991) and in many more HRSEM publications.

Figure 12.16 tries to 'highlight the benefit of very thin W films in a simple experiment. For testing the electrical conductivity (mainly charge redistribution) of sub-nm films, 1 μm latex beads were adsorbed to an nm C-film or to a freshly cleaved mica surface (insulator). These latex beads were coated from different orientations at an elevation angle of 45° with various amount of tungsten, leading to areas that were not coated to films of 0.2 nm, 0.5 nm, 0.7 nm, 0.8 nm and 1 nm total average film thickness, as shown in Figure 12.16a. Figure 12.16b shows the experiment performed on an 8 nm carbon supporting film and imaged with 1000 e^-/nm^2 in an in-lens FESEM. A 0.2 nm W film already enhances the SE yield compared to the pure carbon film and allows stable imaging of such a coated area on mica. Charge build-up and electron reflection (generally known as charging) on the insulator is prevented from expanding towards these sub-nm W metal film areas several nm before the edges of these coated areas. An average film thickness of 0.2 nm would be less or about 2 atoms thick! A closer look at the film shows that even in these sub-nm thin films one can still find distinct W grains of 0.8–1.5 nm in size. These grains act as distinct SE signal emitting entities – 'lantern like' emitters – and hence amplify the signal yield and prevent SE from moving as far as their mean free pathway (about 3–4 nm) would allow in a continuous film. In other words, these small W grains improve not only the signal enhancement by

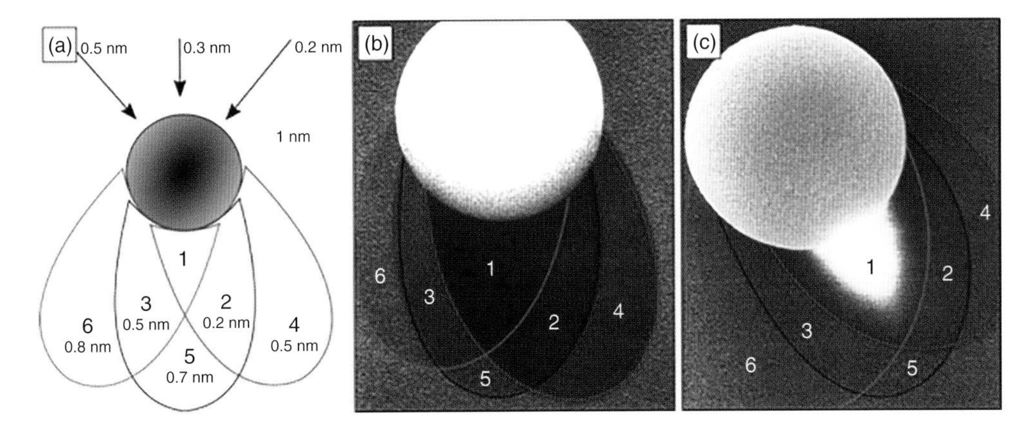

Figure 12.16 The benefit of very thin tungsten films on electrical conductivity. (a) Schematic of the experimental setup: 1 μm latex beads were coated from different directions at an elevation angle of 45° with different amounts of tungsten (in total 1 nm) on either (b) an 8 nm carbon support or (c) freshly cleaved mica (insulator). The numbers mark the final film thickness: (1) 0 nm; (2) 0.2 nm; (3) 0.5 nm, (4) 0.5 nm; (5) 0.7 nm; (6) 0.8 nm; total: 1 nm.

their higher atomic number but also the localization accuracy of the emitted signal by preventing the signal from travelling within a continuous metal film. The single grains have fractal shape and are laterally more expanded than in the z-direction (height) for W, as determined by STM and STEM investigations (results are not shown here). The individual grains are separated by vacuum on the support film and once an SE signal is emitted at a metal grain surface, the electron is attracted immediately by the collecting field or potential gradient towards the SE detector. The distance between the grains is sufficiently small that the charge can still flow laterally, avoiding charging artefacts. This most probably happens by some kind of charge tunnelling effect due to an average distance of 1.5–2 nm between neighbouring grains.

This means that metal films from 1 nm thickness and greater are absolutely sufficient to locally prevent charging for HRSEM imaging (see the film on the mica support area in Figure 12.16). The only remaining risks are that such films are 'discontinuous' by shadowed areas, cracks, etc. Such film interruptions across the whole sample surface or badly grounded support film and/or sample stage would allow charge build-up during imaging and thus reduce resolution and image quality.

12.6.7 Example: Comparison of W, Cr and Pt/Ir/C Coating of Protein 2S Crystal Layer

Comparing the most potential coating metals on HPI protein layers reveals at a first glance very little difference in the 2D protein crystal appearances, all metal coating films revealing the sixfold symmetry of the protein unit with the broadest appearance after a Cr coating (see Figure 12.17). Cr coated unit cells at the edge of the layers revealed an enhanced secondary signal (edge effect) compared to Pt/Ir/C and W coated layer edges. The connected arms between the sixfold protein core are best visible on Pt/Ir/C and W coated HPI layers, where W coating reveals the finest lattice details and a handiness without averaging due to the fine replication and high SNR ratio found in such SE images. Fine structures down to 1 nm (individual fibres or connecting arms) are present after Pt/Ir/C and W coating but not after Cr coating.

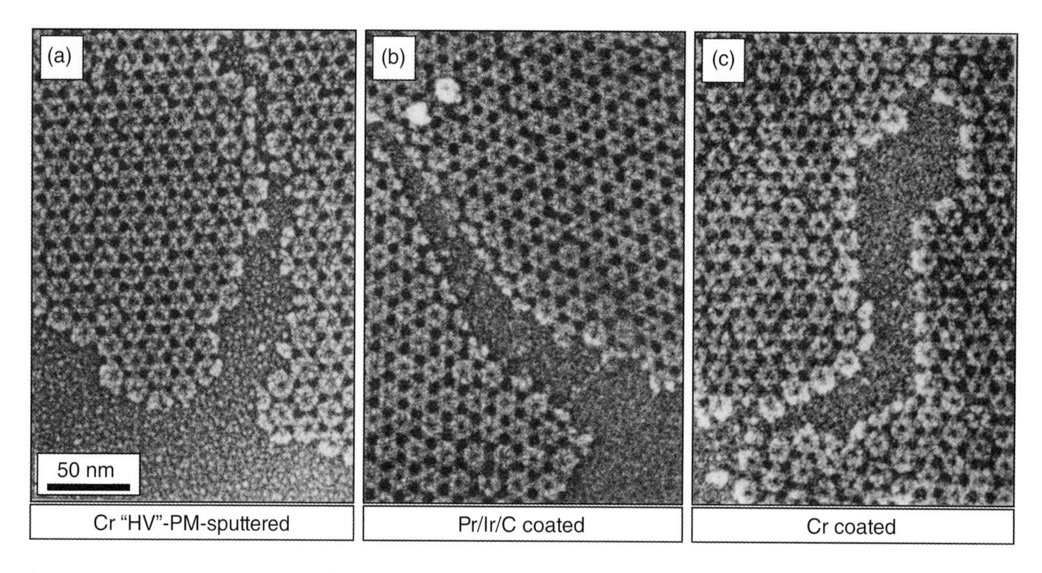

Figure 12.17 Comparison of freeze-dried, (a) Cr 'HV'-PM-sputtered, (b) Pt/Ir/C coated and (c) Cr coated HPI layers.

12.7 FINAL RESOLUTION OBTAINABLE DURING HRSEM WORK FROM METAL COATING

A resolution of 3–5 nm on freeze-dried thin biological specimens, coated with Cr by planar-magnetron sputtering (PMS) or double-axis rotary shadowing (DARS) at low temperature (193 K) has been reported in Hermann and Müller (1991) and 2 nm resolution by high angle rotary shadowing in Wepf (1992). High-resolution SE images were also shown on heavy metal coated samples, for example on Pt, Cr or W sputter coated, frozen-hydrated yeast cells (Walther *et al.* 1990, 1992; Hermann and Müller 1993) and on coated freeze-dried macromolecular structures (Wepf *et al.* 1991, 1992; Wepf and Gross 1990; Chen *et al.* 1994).

To finally determine the resolution limit of metal coating of biological samples for HRSEM, we compared coated and non-coated areas of freeze-dried HPI layers and determined the resolution power by averaging techniques, using standard methods used in TEM 2D protein-crystal structure research.

Figure 12.18 shows a latex bead adsorbed on top of an HPI layer after freeze-drying and metal shadowed from one direction (unidirectional coating) with 1 nm W from an elevation angle of 45°, resulting in a coated and an uncoated area on the same HPI layer. The area in the shadow shows no BSE signal (Figure 12.18a) and a strong SE signal (Figure 12.18b), which allows the protein crystal lattice and some fine structure to be extracted without a surface contrast enhancement. Averaging of coated and uncoated HPI areas from SE images reveals the unit cells of coated (Figure 12.18c) and uncoated (Figure 12.18d) HPI layer. The uncoated area reveals an averaged core structure with a strong sixfold core structure and very weakly resolved connecting arm-like structures; even they can be from time to time directly seen in the raw images. Using standard TEM frequency-based resolution criteria (SSNR, FRC, S-image) a resolution of 3–5 nm was found on such areas, whereas the coated area reveals a much more detailed unit cell with clear arm-like connections to the core proteins and resolves a right-handed core structure. The resolution criteria show structural details down to 1.5 nm. The mean-free-path length of secondary electrons is known to be in carbon around 3–5 nm and in heavy metal around 3 nm until secondary electrons lost

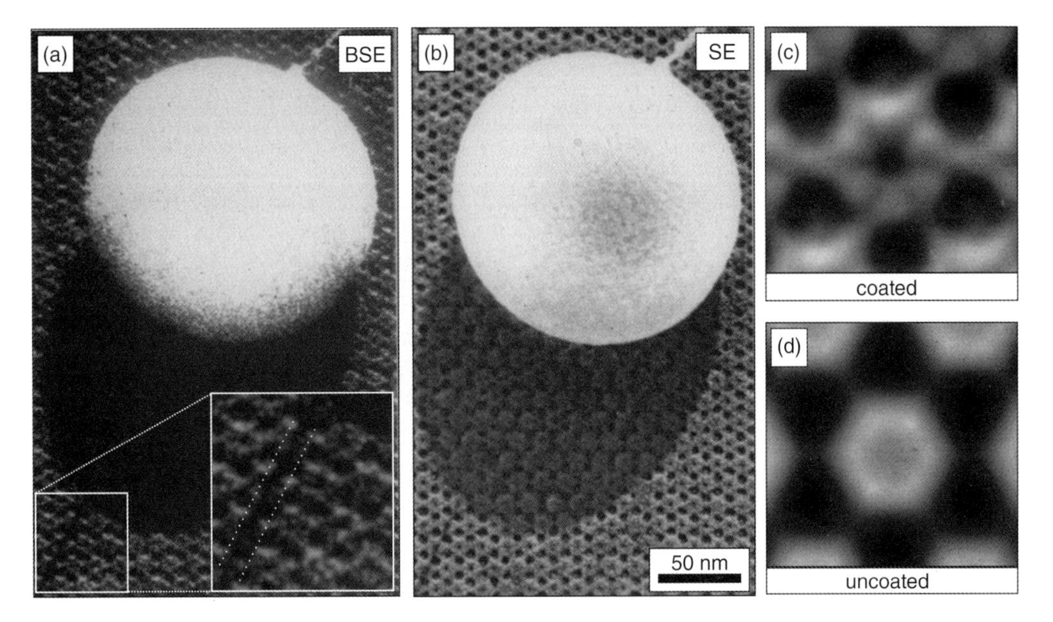

Figure 12.18 Latex beads were adsorbed to a freeze-dried HPI layer and metal coated from one direction with 1 nm W from an elevation angle of 45°. (a) BSE image. (b) SE image. (c) Averaged HPI structure revealed from the coated area. (d) Averaged HPI structure revealed from the uncoated area in the shadow of the latex bead.

their energy or to be released at the surface before they get 'stuck' in the sample. Hence, one has to conclude that the single W-metal grain also helps here to contain the signal laterally and more importantly localized to the surface area where the metal was deposited. In other words, uncoated areas with SE images more readily resemble a 'mass-density' representation as long as the sample is not much thicker than the mean free pathway of electrons. Metal coating, on the other hand, allows with its defined and discrete grains the replication of a soft surface area more accurately and hence the signal is generated preferentially at the surface of a specimen. After averaging of 30–100 unit cells from SE images and 100–200 unit cells from BSE images, these statistical grain locations are lost, revealing a continuous signal, which virtually assembles as a continuous 'film' on a given surface topography with a replication potential of 1.5 nm. The image pair also reveals small changes after coating – for example one of the fine fibre's (eutectica) in the top right area remained after coating, transferring and imaging, whereas in the bottom left area only the shadow of a 1 nm structure reminds us of a fine structure forming on the latex bead. Since it was maintained until the deposition of W, the change must have occurred during transfer or imaging and shows that a 1 nm W film cannot stabilize such a fine structure from being modified during sample transfer or imaging.

Finally, to summarize the findings, best structural preservation can only be achieved by cryo-fixation and partial freeze-drying and the best SNR ratio is obtained by coating the surface with 1–3 nm thin metal films, which are not continuous but with single grains enhancing SE and BSE signal generation at the site of deposition. Such a grainy film replicates fine structural details more accurately than a continuous metal film. Which coating techniques may be used is summarized in Figure 12.19, where diode sputter coating (DS) can help to visualize sample details in the μm range, planar magnetron sputtering down to the 2–3 nm range and only electron beam evaporation is capable of resolving surface details down to 1 nm in SE or BSE images.

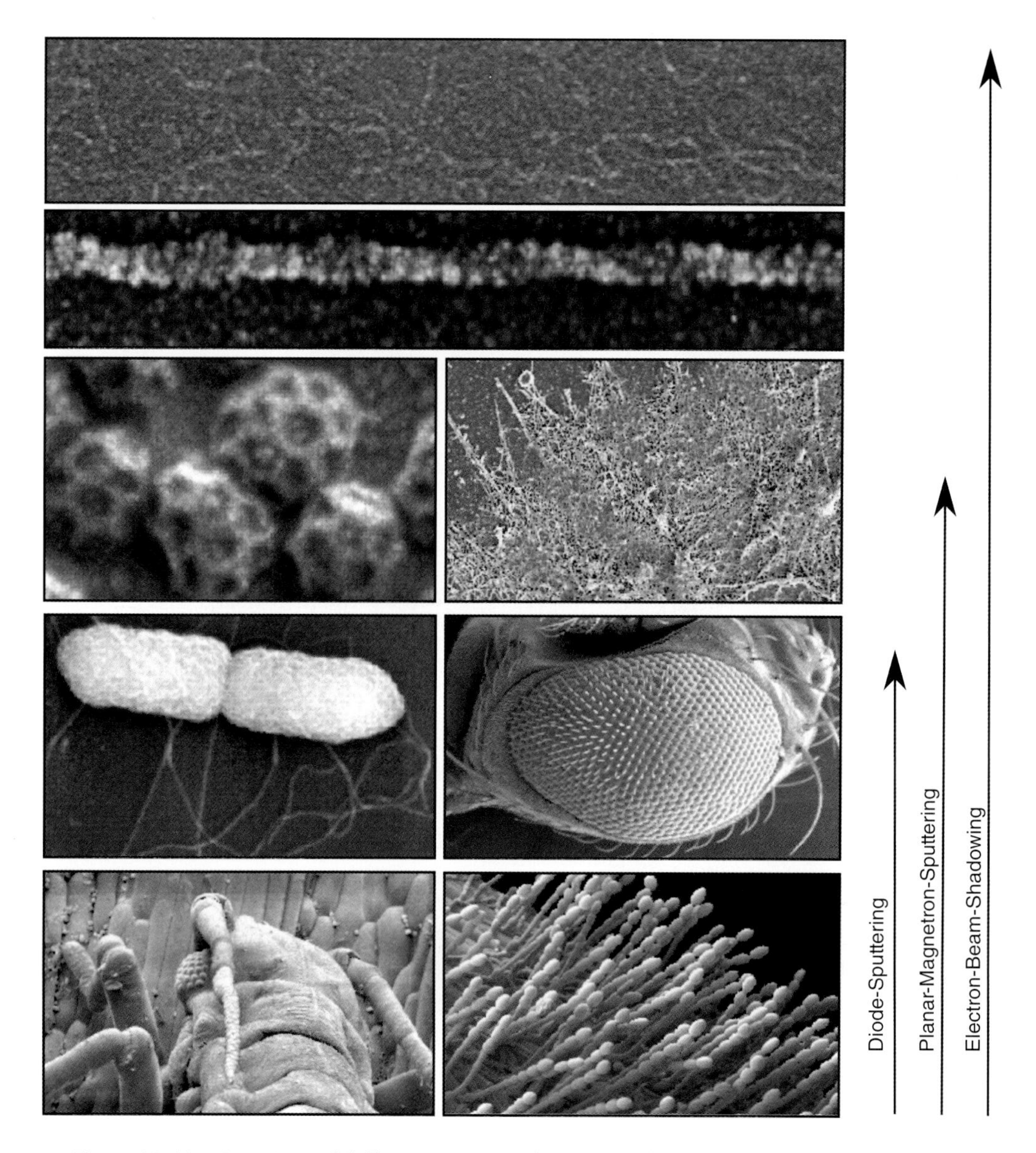

Figure 12.19 Overview of different coating techniques and their possible application fields.

12.8 MOLECULAR HRSEM IMAGING ON FROZEN HYDRATED BULK TISSUE

Once the final resolution enhancement with metal coating has been characterized, the final goal is to achieve this resolution also in bulk samples (see Figure 12.20) and not only on isolated macromolecular complexes. As an example of what can be achieved, Figure 12.20 shows a freeze-fracture plain through a high-pressure frozen human skin biopsy sample (in Figure 12.20a the image length is about 100 μm) with images taken from different areas (Figure 12.20b to h) throughout the epidermis. These samples show no ice crystal damage and macromolecular resolution could be achieved throughout. These samples have been

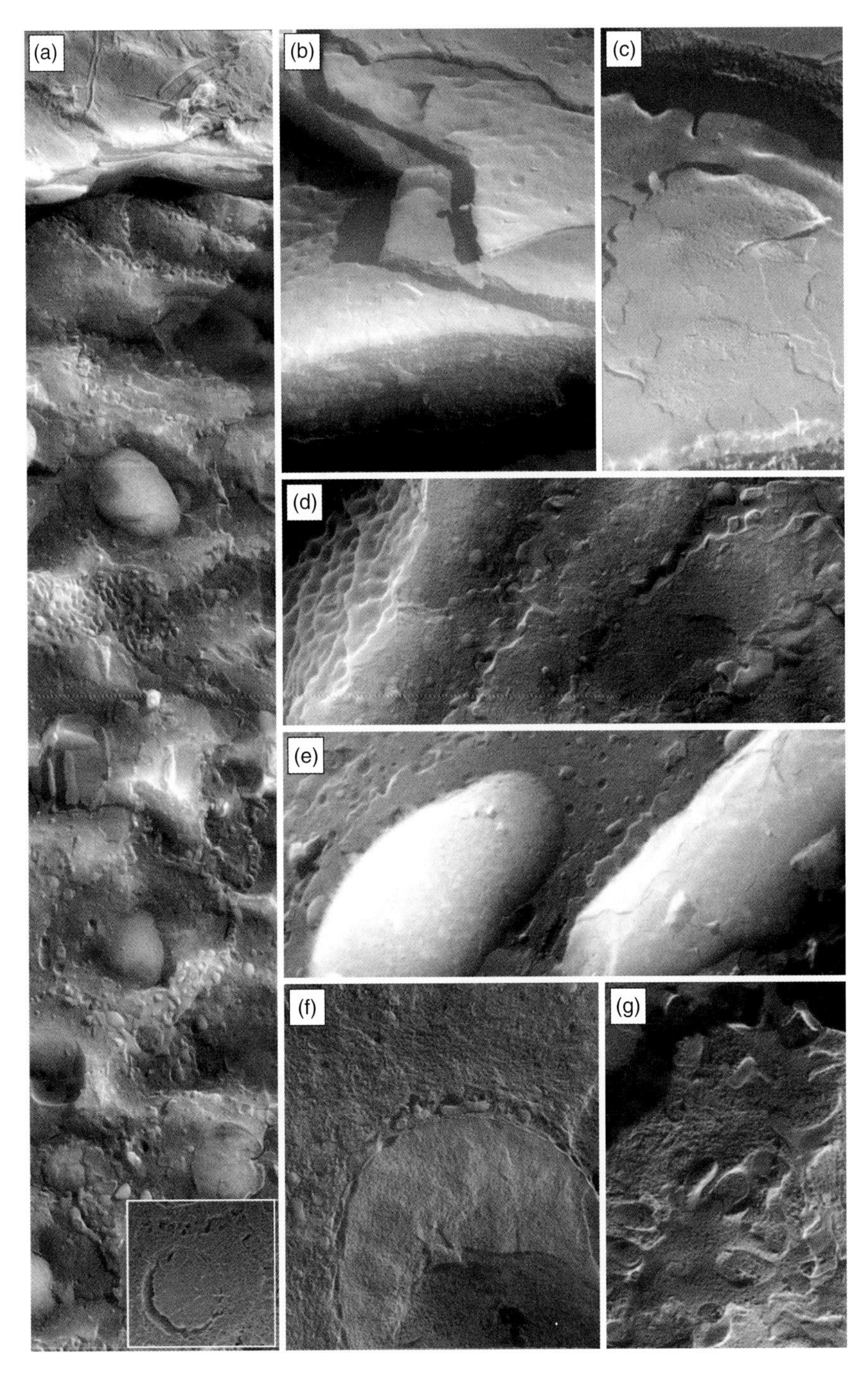

Figure 12.20 Bulk imaging of high-pressure frozen human skin biopsy samples imaged in an LEO1530 Gemini column (in-lens type detection). (a) Freeze-fractured plane, freeze-fractured and coated with 2 nm W from an elevation angle of 45° at −120 °C and imaged with a cryo-stage at −120 °C without any extra carbon layer and under low-dose conditions. (a) Overview of a freeze-fracture plane through the sample (image length about 100 μm). The inset shows a cross-sectioned Stratum Spinosum cell and nucleus after freeze-drying with clear shrinkage artefacts and empty space where water has been removed. (b), (c) and (d) Freeze-fracture plane through a stratum corneum cell. (e) Freeze-fractured nucleus in the live–death transition zone of the Stratum Granulosum. (f) Cross-fractured nucleus imaged in the frozen hydrated state. (g) Cross-fractured interface between two Stratum Spinosum cells.

freeze-fractured and coated with 2 nm W at an elevation angle of 45° at −120 °C and imaged in a cryo-SEM (in-lens type) at −120 °C without any extra carbon layer under low-dose conditions. Image Figure 12.20a inset shows a cross-sectioned (diamond knife) Stratum Spinosum cell and nucleus after freeze-drying showed clear shrinkage artefacts and empty spaces where water has been removed. Compared to Figure 12.20g, where a nucleus was cross-fractured and kept frozen-hydrated for imaging, the double membrane layers of the nuclear membrane are only clearly resolved as two independent membranes with nuclear pore contact sites as long as the structure is in its native conformation. Hence to prevent artefacts, it is essential to maintain the sample in the frozen-hydrated state for imaging (Figure 12.20b versus g) and avoid freeze-drying. In addition, many intermediate filaments (mainly keratin fibres) are resolved in the cytoplasm surrounding the nucleus, which is also decorated with vesicular or extended membrane structures attaching to the nuclear envelope, whereas a freeze-fractured nucleus in the live–death transition zone of the Stratum Granulosum in Figure 12.20f exposes nuclear core complexes (vivid nucleus) or multiple lipid fracture plains (death nucleus) as the facture plain follows the organelle surface. These cells also expose a very dense cytoplasm area filled with macromolecular structures and many membranes and vesicles (known as cell crowdedness), with no gaps or holes when preserved in their native state. Cross-fracture plains in the dead layers of the human skin barrier – the Stratum Corneum (Figure 12.20c, d and e) - show an even denser packaging of intracellular structures (filaments and vesicles) and molecular lipid facture plains between the densely packed corneocytes layers exposed in intercellular fracture plains (Figure 12.20c and d). Looking at a cross-fracture biopsy sample under frozen hydrated preservation, it becomes obvious that the native state of macromolecular assembly can be studied by cryo-HRSEM within the cellular context if one considers all essential sample preparation steps and uses the adequate coating technique as demonstrated in this chapter.

12.9 CONCLUSION

Finally, we can conclude that HRSEM, combined with cryogenic sample preparation, interconnecting tools and appropriate cryo-adaption of the SEM, is capable of preserving biological macromolecular structures on the nanometre scale. With the appropriate coating technique – electron beam evaporation – cryo-HRSEM provides SE and BSE images on isolated and freeze-fractured tissue samples down to 2–3 nm resolution.

Which coating techniques should be used is summarized in Figure 12.19, where diode sputter coating (DS) can help to visualize sample details in the sub-μm range and planar magnetron sputtering down to the 3–5 nm range. Only electron beam evaporation coating is capable of resolving details down to 1 nm in SE or BSE images at low or high acceleration voltages.

The main factor limiting the resolution at these HRSEM images is no longer the electron probe size but a physical parameter of the signal electrons. Resolution is limited by the mean-free-path length of SE electrons, which is larger (3–5 nm) than the smallest metal grains (1 nm) and hence direct visualization with SE in organic material will be blurred compared to the localization of small metal grains 'lanterns' (preparative resolution enhancement with small 'signal yield' enhanced metal 'labels').

Only if averaging techniques over at least 20–30 identical features are applied can the final resolution from a 1 nm W, Cr or Pt/C coated structure be enhanced down to 1–1.5 nm for EBE and 2–3 nm for PMS coating. Averaging techniques allow one to extract structural features as the individual statistics distributing highly localized metal grain 'lanterns' are averaged out to a virtual continuous signal film replicating the underlying structures that depend on the metal–sample interaction properties. Finally, one can conclude that Cryo-SEM of thin and bulk specimens, combined with appropriate freezing and cryo-preparation workflow tools, is one of the fastest structure research investigation workflows; for example, within 2–3 hours after high-pressure freezing, frozen-hydrated samples can be investigated at the nanometre level down to 1–3 nm by electron beam evaporation metal coating signal enhancement.

In primary structure investigations, this allows a high-throughput, fast-feedback approach that facilitates fast understanding of even complex hierarchical structures (see Figure 12.20). In addition, for high SNR gained on macromolecular structures HRSEM allows simple extraction of configuration and handedness of structures directly (see Miller *et al.* (1970) on viral particles) without the need for averaging techniques. If averaging is possible due to multiple identical unit cells one can even extract 3D models of macromolecular complexes, as shown in Chapter 27 by Woodward and Wepf in Volume II.

ACKNOWLEDGEMENTS

The authors thank Martin Müller, René Hermann, Brigitte Joggerst, Derick Mills, Joel Lanoix, Joachim Zach and Max Haider for their constant support and help in collecting or processing data over a time period of more than 15 years to establish a better understanding of high resolution coating techniques for LV- and HVSEM.

REFERENCES

Bell, P.B., Lindroth, M. and Fredriksson, B.A. (1988) Preparation of cytoskeletons of cells in culture for high resolution scanning and scanning transmission electron microscopy, *Scanning Microscopy*, 2, 1647–1661.

Boyde, A. (1978) Pros and cons of critical point drying and freeze-drying for SEM, *Scanning Electron Microscopy*, II, 303–313.

Boyde, A. and Tamarin, A. (1984) Improvement to critical point drying technique for SEM, *Scanning*, 6, 30–35.

Bremer, A., Milloniq, R.C., Sütterlin, R., Engel, A., Pollard, T.D. and Aebi, U. (1991) The structural basis for the intrinsic disorder of the actin filament: The 'lateral slipping' model. *J. Cell Biol.*, 115, 689–703.

Chen, S., Centonze *et al.* (1994) Ultra-high resolution cryo-SEM and specimen preparation for cytoskeleton, *Acta Histochem. Cytochem.*, 27 (5), 507–509.

Crowther, R.A., Finch, J.T. and Pearse, B.M.F. (1976) On the structure of coated vesicles, *Journal of Molecular Biology*, 103, 785–798.

Echlin, P. (1979) Thin films for high resolution conventional scanning electron microscopy, *Scanning Electron Microsc.*, 2, 21–30.

Engel, A. and Colliex, C. (1993) Application of scanning transmission electron microscopy to the study of biological structure, *Current Opinion in Biotechnology*, 4, 403–411.

Frethem, Ch., Wells, C., Carlino, V. and Erlandsen, S.L. (1993) High-resolution field emission SEM (FESEM) of bulk biological samples: Relationship between different metal coatings, their thicknesses, and quality of backscatter electron (BSE) versus secondary-electron (SE) imaging, in *Proceedings of the 51st MSA Meeting*, pp. 460–461.

Furuno, T. and Sasabe, H. (1993) Two-dimensional crystallization of Streptavidin by nonspecidic binding to a surface film: Study with a scanning electron microscope. *Biophysical Journal*, 65, 1714–1717.

Furuno, T., Ulmer, K.M. and Sasabe, H. (1992) Scanning electron microscopy of negatively stained catalase on silicon wafer, *Microscopy Research and Techniques*, 21 (1), 32–38.

Gross, H. (1987) High resolution metal replication of freeze-dried specimens, in *Cryotechniques in Biological Electron Microscopy* (eds R.A. Steinbrecht and K. Zierold), Springer Verlag, pp. 205–215.

Gross, H. *et al.* (1985) High resolution metal replication, quantified by image processing of periodic test specimens, *Ultramicroscopy*, 16, 207–304.

Hermann, R. and Müller, M. (1991a) High resolution biological scanning electron microscopy: A comparative study of low temperature metal coating techniques, *J. Electron Micro. Technol.*, 18, 440–449.

Hermann, R. and Müller, M. (1991b) Prerequisites of high resolution scanning electron microscopy, *Scanning Microsc.*, 5, 653–664.

Hermann, R. and Müller, M. (1992) Towards high resolution SEM of biological objects, *Arch. Histol. Cytol.*, 55, Suppl., 17–25.

Hermann, R. and Müller, M. (1993) Progress in scanning electron microscopy of frozen-hydrated biological specimens, *Scanning Microscopy*, 7, 343–350.

Hren, J.J. (1986) Chapter 10, in *Principles of Analytical Electron Microscopy* (eds D.C. Joy, A.D. Romig and J.I. Goldstein), Plenum Press.

Jiang, N. (2016) Electron beam damage in oxides: A review, *Reports on Progress in Physics*, 79, 016501.

Joy, D. (1991) Contrast in high resolution scanning electron microscope images, *Journal of Microscopy*, 161, 343–355.

Krzyzánek, V., Müller, S.A. and Reichelt, R. (2009) MASDET – a fast and user-friendly multiplatform software for mass determination by dark-field electron microscopy, *Journal of Structural Biology*, 165, 78–87.

Lucio, I., Falk, L., Wepf, R. and Reimhult, E. (2011) Measuring single-nanoparticle wetting properties by freeze-fracture shadow-casting cryo-scanning electron microscopy, *Nature Communication*, 2, 438.

Miller, K.R., Prescott, C.S., Jacobs, T. L. and Lassignal, N.L. (1983) Artifacts associated with quick-freezing and freeze-drying, *Journal of Ultrastructural Research*, 82, 123–133.

Miller, J.L., Woodward, J., Chen, S., Jaffer, M., Weber, B., Hagasaki, K., Tomaru, Y., Wepf, R., Rosenman, A., Varsani, A. and Sewell, T. (1970) Three-dimensional reconstruction of *Heterocapsa circularisquama* RNA virus by electron cryo-microscopy, *Journal of General Virology*, 92, 1960–1970.

Moor, H. (1971) Recent progress in the freeze-etching technique, *Phil. Trans. Roy. Soc. Lond., B*, 261, 121–131.

Murk, J.L.A.N., Posthuma, G., Koster, A.J., Guze, H.J., Verkleij, A.J., Kleijmeer, M.J. and Humbel, B.M. (2003) Influence of aldehyde fixation on the morphology of endosomes and lysosomes: quantitative analysis and electron tomography, *Journal of Microscopy*, 212, Pt, 1, 81–90.

Murphy, D.M. and Koop, T. (2005) Review of the vapour pressures of ice and supercooled water for atmospheric applications, *Quarterly Journal of the Royal Meteorological Society*, 131, 1539–1565.

Peters, K.R. (1986) Metal deposition by high-energy sputtering for high magnification electron microscopy, in *Advanced Technique in Biological Electron Microscopy*, vol. 277, Springer-Verlag, Berlin, Heidelberg, New York, Tokyo.

Reimer, L. (1998) Scanning Electron Microscopy – Physics of Image Formation andMicroanalysis, *vol.* 45, Springer-Verlag, Berlin, Heidelberg, New York, Tokyo.

Seligman, A.M., Wasserkrug, H.L. and Hanker, J.S. (1966) A new staining method (OTO) for enhancing contrast of lipid-containing membranes and droplets in osmium tetroxide-fixed tissue with osmiophilic thiocarbohydrazide (TCH), *J. Cell Biol.*, 30, 424–432.

Sousa, A.A. and Leapman, R.D. (2007) Quantitative STEM mass measurement of biological macromolecules in a 300 kV TEM, *Journal of Microscopy*, 228, 25–33.

Tacke, S., Krzyzanek, V., Nüsse, H., Wepf, R., Klingauf, J. and Reichel, R. (2016) A versatile high-vacuum cryo-transfer system for cryo-microscopy and analytics, *Biophysical Journal*, 110, 758–765.

Umrath, W.W. (1983) Berechnung von Gefriertrocknungszeiten für die Elektronenmikroskopische Praeparation, *Mikroskopie*, 40, 9–34.

Wall, J. S. and Hainfeld, J. F. (1986) Mass mapping with the scanning tranmission electron microscope, *Annual Review of Biophysics and Biophysical Chemistry*, 15, 355–376.

Walther, P., Chen, Y., Pech, L.L. and Pawley, J.B. (1992) High-resolution scanning electron microscopy of frozen-hydrated cells, *Journal of Microscopy*, 168, 169–180.

Walther, P., Hentschel, J., Herter, P., Müller, T. and Zierold, K. (1990) Imaging of intramembranous particles in frozen-hydrated cells (*Saccharomyces cerevisiae*) by resolution cryo SEM, *Scanning*, 12, 300–307.

Walz, T., Tittmann, P., Fuchs, K.H., Müller, D.J., Smith, B.L., Agre, P., Gross, H. and Engel, A. (1996) Surface topographies at subnanometer resolution reveal asymmetry and sidedness of Aquaporin-1, *J. Mol. Biol.*, 264, 907–918.

Wepf, R. (1992) Surface imaging of protein complexes with high resolution coating, TEM, SEM and STM techniques, ETH Thesis 9920, Swiss Federal Institute of Technology, Zurich.

Wepf, R. and Gross, H. (1990) Pt/Ir/C, a new powerful coating material for high-resolution SEM, in *XIIth International Congress for Electron Microscopy*, Seattle, San Francisco, San Francisco Press, Inc.

Wepf, R., Amrein, M., Bürkli, U. and Gross, H. (1991) Platinum/iridium/carbon: A high-resolution shadowing material for TEM, STM and SEM of biological macromolecular structures, *Journal of Microscopy*, 163, 51–64.

Wepf, R., Bremer, A. *et al.* (1992) Surface imaging of F-actin filaments: A comparative study by SEM, TEM and STM, in *Proceedings of the 19th EUREM 1992*, vol. III, Granada, Spain, *European Microscopy* (eds L. Megias-Mgias, M.I. Rodriguez-Garcia, A. Rios and J.M. Arias), Universidad de Granada; ISBN: 84-338-1593-8.

Wepf, R., Richter, T., Sattler, M. and Kaech, A. (2004) Improvements for HR- and cryo-SEM by the VCT 100 high-vacuum cryo transfer system and SEM cooling stage, *Microscopy and Microanalysis*, 10, 970–971.

Wildhaber, I. and Gross, H. (1985) The effects of air-drying and freeze-drying on the structure of a regular protein layer. *Ultramicroscopy*, 16, 411–422.

Wildhaber, I., Gross, H. and Moor, H. (1982) The control of freeze-drying with deuterium oxide (D2O), *J. Ultrastruct. Res.*, 80, 367373.